THE TURBINE PILOT'S FLIGHT MANUAL

THE TURBINE PILOT'S FLIGHT MANUAL

SECOND EDITION

GREGORY N. BROWN AND MARK J. HOLT

GREGORY N. BROWN's love for flying is obvious to anyone who knows his column, "Flying Carpet," in *Flight Training* magazine or who has read his other books, *The Savvy Flight Instructor* and *Job Hunting for Pilots.* A flight instructor since 1979, Greg was 2000 Industry/FAA Flight Instructor of the Year, as well as winner of the 1999 NATA Excellence in Pilot Training Award. In addition, he has served as a professional pilot in both scheduled and corporate aviation. Mr. Brown holds an ATP pilot certificate with Boeing 737 type rating and Flight Instructor certificate with all fixed-wing aircraft ratings including glider. An active pilot since 1971, Mr. Brown was the first "Master CFI" designated by the National Association of Flight Instructors.

MARK J. HOLT, a pilot for a major airline based in Atlanta, soloed at age sixteen. He holds an ATP pilot certificate with Boeing 757/767 and BAE Jetstream 41 type ratings, a flight engineer (Turbojet) certificate, along with numerous flight and ground instructor ratings. His professional flying career includes extensive flight and ground instructing experience and service as a check airman for a large regional airline. Mark also served nine years as an electronics/radar technician for the U.S. Air Force Air National Guard.

The purpose of this book is to provide information on turbine operations. The user of this information assumes all risk and liability arising from such use. Neither Blackwell Publishing nor the authors can take responsibility for the actual operation of a turbine aircraft or the safety of its occupants.

© 2001, 1995 Gregory N. Brown and Mark J. Holt
All rights reserved

Blackwell Publishing Professional
2121 State Avenue, Ames, IA 50014

Orders: 1-800-862-6657
Office: 1-515-292-0140
Fax: 1-515-292-3348
Web site: www.blackwellprofessional.com

Source: Illustrations for the Airline, Regional, and Corporate Aircraft Spotter's Guide and several other figures by AvShop's Civil Aviation Clip Art CD-ROM. All other illustrations by Gregory N. Brown, except where noted.

Cover credits: Airbus A-340 and Boeing 777 courtesy of Lawrence Feir, Lawrence Feir Photography and Imaging; Cessna Citation, Embraer RJ, and Raytheon Beech King Air courtesy of Mike Fizer, Fizer Photography; Fairchild Dornier 328 turboprop courtesy of Fairchild Dornier; and Pilatus PC-XII by Mike Fizer courtesy of the Aircraft Owners and Pilots Association. All photos are copyrighted by the providers and reprinted by permission.

Authorization to photocopy items for internal or personal use, or the internal or personal use of specific clients, is granted by Blackwell Publishing, provided that the base fee of $.10 per copy is paid directly to the Copyright Clearance Center, 222 Rosewood Drive, Danvers, MA 01970. For those organizations that have been granted a photocopy license by CCC, a separate system of payments has been arranged. The fee code for users of the Transactional Reporting Service is 0-8138-0023-4/2001 $.10.

Printed on acid-free paper in the United States of America

First edition, 1995
Second edition, 2001

Library of Congress Cataloging-in-Publication Data

Brown, Gregory N. (Gregory Neal)
The turbine pilot's flight manual / Gregory N. Brown, Mark J. Holt.
—2nd ed.
p. cm.
Includes bibliographical references and index.
ISBN 0-8138-0023-4
1. Aircraft gas turbines—Handbooks, manuals, etc. 2. Jet planes—Piloting—Handbooks, manuals, etc. I. Holt, Mark J. II. Title.

TL 709.5.T87 B76 2001
629.132'5249—dc21

2001016829

Last digit is the print number: 9 8 7 6 5 4

CONTENTS

PREFACE, xi

ACKNOWLEDGMENTS, xii

1 Introduction, 3

How to Use This Manual, **3**
Transitioning Piston Pilots, **3**
Transitioning Military Aviators, **4**
- Crew Resource Management, **4**
- Training by Civilian Employers, **4**
- Civilian Aircraft and Civilian Aviation Terminology, **4**
- Aircraft Systems, **5**
- Contemporary Issues in the Aviation Industry, **5**
- About Your Civilian Counterparts, **5**

2 General Preparations, 6

Training, **6**
- Limitations, Systems, and Procedures, **6**
- Indoctrination Training, **6**
- Simulator and Flight Training, **7**

Preparing for New-Hire Training, **7**
- Preparing for Ground School, **7**
- Preparing for Simulator and Flight Training, **8**
- Preparing for the Flight Line, **8**
- Computers, **9**

Contemporary Issues in the Aviation Industry, **9**
- Crew Resource Management and Teamwork, **9**
- Unions, **9**
- Driving under the Influence, **9**
- Discrimination and Harassment, **9**

3 Turbine Engine and Propeller Systems, 10

Introduction to Gas Turbine Engines, **10**
- Centrifugal-Flow and Axial-Flow Compressors, **12**
- Multistage Compressors, **16**
- Multispool Engines, **16**
- Core Turbine Engine (Gas Generator) , **16**

Turbojets, Turbofans, and Turboprops, **18**

Turbojet Engine, **18**
Turbofan Engine, **20**
Turboprop Engine, **20**
Thrust versus Power, **22**
Engine Operating Parameters, **22**
Turbine Engine Controls in the Cockpit, **22**
Turbine Engine Starting, **24**
Turbine Engine Characteristics in Flight, **27**
Thrust Reversers, **29**
Thrust Reversers on Jets, **29**
Reverse Thrust on Turboprops, **29**
Use of Reversers, **31**
Turboprop Propeller Systems, **31**
Propeller Governors, **31**
Beta Range, **31**
Propeller Auto-Feather Systems, **33**
Propeller Synchronizers and Synchrophasers, **36**
Propeller Supplement for Transitioning Military Jet Pilots, **37**
Propeller Terminology, **37**

4 Turbine Aircraft Power Systems, 42

Basics of Aircraft Power Systems, **42**
Depiction of Aircraft Systems in Pilot Training, **42**
Understanding Aircraft Power Systems: The Reference Waterwheel, **43**
Comparing Aircraft Power Systems to the Reference Waterwheel System, **44**
Electrical Power Systems, **46**
Electrical Power Sources, **46**
Control Devices, **50**
Circuit Protection, **51**
Reading an Airplane Electrical Diagram, **54**
Troubleshooting, **60**
Hydraulic Power Systems, **60**
Benefits of Hydraulic Power in Large Airplanes, **60**
Hydraulic Systems and Components, **62**
Hydraulic System Characteristics, **69**
Pneumatic Power Systems, **69**
High-Pressure Bleed Air, **69**
Low-Pressure Air, **69**
Bleed Hazards and Protections, **70**
Auxiliary Power Units, **72**

5 Major Aircraft Systems, 75

Flight Controls, **75**
Control Surfaces, **75**
Flight Control System Redundancy, **81**
Flight Control Surface Position Indicating Systems, **81**
Fly-by-Wire Control Systems, **81**
Pressurization, **82**

Pressurization Indicators and Controls, **85**
Pressurization System Safety Features, **87**
Loss of Cabin Pressure in Flight, **87**
Environmental Systems, **91**
Heat Exchangers, **92**
Air and Vapor Cycle Machines, **92**
Aircraft Environmental System, **96**
Fuel Systems, **96**
Fuel Tanks, **96**
Fuel Pumps, **99**
Fuel Control Unit, **99**
Fuel Valves, **99**
Fuel Heaters, **101**
Fuel Quantity Measurement Systems, **101**
Fuel Quantity Measuring Sticks, **101**
Fuel Vents, **101**
Fuel Management, **102**

6 Dedicated Aircraft Systems, 103

Ice and Rain Protection, **103**
In-Flight Structural Icing, **103**
Engine Icing, **106**
Fuel System Icing, **109**
The Role of the Pilot, **109**
Ground Icing, **109**
Rain Protection, **110**
Landing Gear Systems, **110**
Landing Gear Squat Switch, **110**
Brakes, **111**
Nosewheel Steering, **113**
Annunciator and Warning Systems, **113**
Annunciator or Advisory Panels, **113**
Audio Advisory and Warning Annunciation, **116**
Fire Protection Systems, **116**
Fire Detection and Extinguishing Systems, **116**
Pilot Actions and Cockpit Controls, **117**
Electrical Considerations, **117**
Cabin and Cockpit Protection, **118**
Auxiliary Power Unit Fire Protection, **118**

7 Limitations, 120

Airspeeds, **120**
Engine Limits, **120**
Other System Limitations, **123**

8 Normal Procedures, 124

Crew Coordination, **124**
Captain and First Officer/Copilot, **124**
Flying Pilot and Nonflying Pilot, **124**

Crew Resource Management, **124**
- Optimizing Crew Communication, **126**
- Improving Overall Flight Management, **126**
- Development of a Team Performance Concept, **126**
- Crew Resource Management Training, **127**

Checklists and Callouts, **127**
- Checklist Procedures, **127**
- Types of Checklists, **128**
- Normal Checklists, **130**
- Standard Callouts, **131**

9 Emergency and Abnormal Procedures, 133

Emergency versus Abnormal Situations, **133**
Emergency Procedures, **133**
Abnormal Procedures, **134**

10 Performance, 137

Takeoff, Climb, Landing, and Engine-Out Performances, **137**
- Takeoff and Climb Performances, **138**
- Enroute Engine-Out Performance Planning, **140**
- Landing Performance, **140**
- Braking Performance, **140**

Routine Performance Planning, **140**
- TOLD Cards, **142**
- Airport Analysis Tables, **142**
- Cruise Performance: Fuel Planning, **142**

11 Weight and Balance, 145

The Weight in "Weight and Balance," **145**
- Aircraft Weight Categories, **145**

Balance Considerations, **147**
- CG as Percentage of MAC, **147**
- Performance Benefit of an Aft CG, **148**
- In-Flight CG Movement, **148**

How It's Done in the Real World, **150**
- Average Passenger Weights, **150**
- Random Loading Programs, **150**

12 Airplane Handling, Service, and Maintenance, 151

Flight Dispatch, **151**
Fueling Procedures, **151**
Standard Preflight, **152**
- Aircraft Documents Review, **152**
- Cockpit and Emergency Equipment Checks, **152**

Exterior Preflight Check, **152**
Final Preflight Preparations, **155**
Minimum Equipment List (MEL), **155**
Configuration Deviation List (CDL), **155**

13 Navigation, Communication, and Electronic Flight Control Systems, 156

Horizontal Situation Indicator, **156**
Autopilots, **158**
Flight Director, **158**
Electronic Flight Instrumentation Systems (EFIS), **158**
ACARS, **163**
Head-Up Displays, **163**
Area Navigation (RNAV), **163**
VOR/DME-Based RNAV, **167**
LORAN, **167**
Global Positioning System (GPS), **168**
Inertial Navigation System (INS), **169**
Using RNAV, **169**
Latitude and Longitude, **169**
Flight Management System (FMS), **171**
Basic FMS Components and Operating Principles, **172**
Basic Operation of a Generic FMS, **172**
Pilot Operations in the Glass Cockpit, **184**

14 Hazard Avoidance Systems, 185

Weather Avoidance Systems, **185**
Airborne Weather Radar, **185**
Doppler Radar, **194**
Combined Weather Radar and Navigation Displays, **196**
Electrical Discharge or Lightning Detectors, **196**
Traffic Alert and Collision Avoidance System (TCAS), **198**
Ground Proximity Warning Systems and Enhanced Ground Proximity Warning Systems, **200**

15 Operational Information, 201

Aerodynamics of High-Speed/High-Altitude Aircraft, **201**
High-Speed Flight and the Sound Barrier, **201**
Swept Wing Aerodynamics, **203**
Fixed Aerodynamic Surfaces, **205**
Stalls, **209**
IFR Operations in Turbine Aircraft, **209**
Profile Descents, **209**
Jet Routes, **209**
Altimetry and IFR Cruising Altitudes at Flight Levels, **209**
Category I/II/III Approaches, **210**
Holding, **210**

Extended Range Twin-Engine Operations (ETOPS), **211**
International Flight Operations, **211**
Wake Turbulence, **211**
  Wing Tip Vortices, **211**
  Identifying Likely Areas of Wake Turbulence, **213**
Air Rage, **216**

16 Weather Considerations for Turbine Pilots, 217

Low-Altitude Weather: Wind Shear and Microbursts, **217**
  Wind Shear, **217**
  Microbursts, **217**
  Effects of Microbursts on Aircraft, **218**
  Avoidance Procedures, **218**
  Low-Level Wind Shear Alerting Systems, **218**
  Recognizing and Responding to Wind Shear, **221**
  Training for Wind Shear Encounters, **221**
High-Altitude Weather, **221**
  Icing, **221**
  Wind, **221**
  The Jetstream, **222**
  Clear Air Turbulence (CAT), **224**
  Avoiding CAT, **224**

APPENDIX 1: Handy Rules of Thumb for Turbine Pilots, 227

APPENDIX 2: Airline, Regional, and Corporate Aircraft Spotter's Guide, 233

GLOSSARY: Airline and Corporate Aviation Terminology, 249

BIBLIOGRAPHY, 255

INDEX, 257

PREFACE

THIS MANUAL grew from the need to summarize in one place the information a pilot is expected to know when moving up to high-performance turbine aircraft. Flying professionally in today's competitive environment demands good basic knowledge of aircraft systems and procedures. This knowledge pays off at job interviews, at ground school, and of course, in flight operations.

Most initial pilot training programs cover only those areas of "aeronautical knowledge" required to pass FAA knowledge and practical tests up through commercial and CFI certificates. By nature, such training emphasizes low-altitude piston airplane systems and operations. This book is designed to be a ready, readable source for pilots to learn and prepare for that first step up into turbine equipment and operations and for the subsequent transition into more advanced types. Such preparation is important for several reasons.

Interviewers for turbine flight positions expect a certain basic level of knowledge among applicants. This book is designed to capsulize that knowledge in one place for purposes of interview preparation.

Initial training ground schools for turbine operators are relatively similar. They anticipate basic knowledge from participants and dive right into the detailed specifics of their own aircraft. We wanted to provide the basics in a form that would allow new-hire pilots to prepare for first-time turbine ground schools.

Review for recurrent training is equally important. When annual checkride rolls around the first few times in a pilot's career and at upgrade time to the captain's seat, many of the basics have been forgotten. This is the place to refresh understanding of the basic principles of aircraft systems.

Each step of a pilot's flying career takes him or her into more-advanced aircraft—from piston aircraft to turboprops, then to corporate and regional jets, and for many pilots, on to transport category aircraft. This book is designed to make every one of those transitions easier.

Finally, this manual is also for those pilots who simply enjoy the opportunity to learn about more-advanced aircraft, even if they don't anticipate flying them anytime soon.

New and returning readers alike will appreciate the many enhancements made in this second edition of *The Turbine Pilot's Flight Manual.* Of interest to most readers will be extensive new material detailing concepts and operational principles of latest-generation cockpit instrumentation, including flight management systems (FMS), head-up guidance systems (HGS or HUD), and global positioning systems (GPS).

Other updates throughout the book's text and illustrations include significant additions in such areas as ice protection, cockpit and cabin fire and smoke protection, cabin depressurization and emergency descent procedures, supplemental oxygen, automatic braking, takeoff warning systems, flaps and leading-edge devices, hydraulic backup systems, and many more. Additional useful rules of thumb have been included, and the Turbine Aircraft Spotters Guide has been revamped to include recently-introduced models.

Also newly incorporated in this edition are symbols keying book illustrations to the companion *Turbine Pilot's Flight Manual Aircraft Systems* CD-ROM. Now, while reading the book and examining relevant illustrations, you can at the same time access narrated color animations and illustrations on the CD-ROM, making understanding of complex devices and systems easier than ever.

It has been our goal since first writing *The Turbine Pilot's Flight Manual* to ease and accelerate pilot transition into each level of turbine aircraft. We have pulled out all the stops in this latest edition to make that next step in your flying career smoother and more exciting than ever.

ACKNOWLEDGMENTS

AMONG the great difficulties of assembling a book like this is gathering and checking all of the information. Aircraft systems vary by manufacturer, type, and model. Procedures for flying any one aircraft type vary tremendously from one operator to the next. Our objective is to cover turbine flying in a general way, but with enough detail to provide all of the basics. No one person can be knowledgeable of all these things.

Many individuals and companies helped us greatly by sharing their expertise. We'd like to thank, first and foremost, the individuals who spent their valuable time reviewing our manuscript and sharing suggestions and information. Captain Pat O'Donnell is a production test pilot for a major manufacturer of large transport aircraft. Captains Don Cronk and Dick Ionata are senior captains with one well-known major airline. Captain Ray Holt retired from another. All four are ex-military pilots. Shane LoSasso is president of Jet Tech, a well-known type rating training company. Vick Viquesney is an engines consultant and former Senior Principal Engineer for a major turbine engine manufacturer, while aeronautical engineers Bruce Haeffele, Richard W. Thomas, and Paul S. Sellers work for a major aircraft manufacturer. Bill Niederer, Dan Moshiri, and John Trimbach fly for major airlines. Tom Carney is Professor and Associate Department Head of the aviation technology department at Purdue University. Dr. Carney holds a Ph.D. in meteorology and also captains executive jet and turboprop aircraft for the university. The input from each of these individuals was invaluable. We are grateful to them all.

We also wish to thank all the many companies and their people who answered our questions and provided supporting information. While it is impossible to name them all here, among them are AlliedSignal Garrett Engine and AiResearch Los Angeles Divisions, United Technologies Pratt & Whitney, Jet Tech, Honeywell Business and Commuter Aviation Systems, and Aerospace Systems and Technologies.

Finally, our appreciation goes out to Joe Statt, Nick Apostolopoulos, and Michelle Statt, for their fine work on "The Turbine Pilot's Flight Manual Aircraft Systems CD-ROM," and to Lawrence Feir, Mike Fizer, the Aircraft Owners and Pilots Association, and Fairchild Dornier for generously providing cover photos.

THE TURBINE PILOT'S FLIGHT MANUAL

CHAPTER 1

Introduction

How to Use This Manual

This manual is designed for both comprehensive reading and quick reference. We recognize that some readers will want to familiarize themselves with turbine operations and will read it from beginning to end. At the same time, we feel that the manual should be in a form that allows easy access to information in order to answer a question or review a system.

Pilots will immediately notice the familiar format of an aircraft *Pilot's Operating Handbook* or *Pilot's Information Manual,* so they should find it easy to access reference information. While most sections of the book correspond to those found in aircraft manuals, we have rearranged the order of the sections in order to make logical reading for straight-through readers.

We have tried our best to minimize specific references to federal aviation regulations (FARs) due to their constant state of change. (Holding speeds changed at least twice during the writing of this manual.)

In general, procedures discussed in this book are based on commercial operations conducted under FAR Parts 135 and 121. This is because the largest percentage of turbine aircraft and pilots operate commercially. Many corporate flight departments also elect to operate under more conservative commercial rules, and in any case most of our readers are already familiar with less-stringent Part 91 operations.

One of the most exciting aspects of a flying career is the continuing personal growth that comes from mastering new knowledge. We've tried our best to convey turbine aircraft information in a manner that's as enjoyable and interesting for you as it is for us. We hope you'll agree.

Transitioning Piston Pilots

You've earned your commercial pilot certificate, perhaps your CFI, and even an ATP. You're over the hump and off to a good start on your professional aviation career.

Now is a good time to reflect upon the knowledge and experience you've gained to date. You're an expert on piston aircraft operations and systems, sectional charts, basic aerodynamics, pilot certification requirements, low-altitude weather, and the basics of instrument flying. But there's a lot more to learn as you transition to ever more sophisticated aircraft. The hours you've invested in learning the system probably have been spent flying at slow airspeeds at altitudes below 12,000 feet. Chances are that you'll soon be operating faster equipment at higher altitudes. Looking up the career ladder you probably won't be flying those piston aircraft you've mastered for long. Turbine engines power the airplanes at the next levels. Their mechanical and electrical systems vary considerably from the aircraft you've trained in. You'll be dealing with big-time hydraulic systems, additional flight controls, and computerized flight management systems. Some interesting new aerodynamic issues also arise with the move into turbine aircraft—when did you last worry about the sound barrier? Even the terminology is different. There is no V_{NE} in turbine-powered aircraft. Maximum operating speeds are instead defined in terms of V_{MO} and M_{MO}. Approaches are conducted relative to V_{REF}.

Even your psyche as a pilot must change. As part of a two- or three-pilot crew, you'll have to master careful and precise team coordination.

Why should you care? Employers will expect at least rudimentary knowledge of turbine systems and operations at your interviews. Your first turbine ground schools will be a heck of a challenge, and plenty of pilots are waiting to fill the shoes of any washouts. You'll want to prepare as much as possible ahead of time. Finally, knowledge will help get you hired. How can you network with a friendly corporate or airline captain if you don't know what kind of jet he or she is flying?

Where does a pilot learn about these things? Most entry-level turbine jobs require a commercial pilot's certificate with multiengine and instrument ratings. The FAA doesn't require any turbine aircraft knowledge or experience to earn any of those ratings. And many private flight schools don't even cover turbine topics since graduates are still 1000 hours away from their ATPs. Graduates of university flight programs are sometimes introduced to turbine operations and systems, but years may pass before the knowledge is applied.

"I want a turbine job, but I don't know anything about it. Somewhere there must be a straightforward book that explains this stuff in general terms!" This book is designed to do just that: introduce you to the basic concepts and terminology of multipilot turbine aircraft.

We recommend that you read the book from front to back, with an eye toward picking up the principles and the terminology. If you have a ground school scheduled, follow up by seriously reviewing topics that directly relate to your upcoming job. Then go back to the training section in Chapter 2. You'll be able to prepare a specific study program for yourself by following the suggestions there.

Obviously, in one book we can't cover every aspect of every turbine aircraft out there. Our goal, rather, is to familiarize the upgrading pilot with turbine aircraft in a broad introductory manner. You won't be ready to fly a turboprop or jet after reading this book, but you *will* know what they're talking about in ground school.

Transitioning Military Aviators

Among our target readers are current or ex-military pilots interested in pursuing civilian flying careers. Military fliers have always been rated among the best candidates for civilian jobs. However, some aspects of civilian flying are markedly different from what military pilots may be used to. Military pilots, while having received excellent training, often lack exposure to certain topics that civilian employers expect them to know. Throughout the book, we have made a thorough effort to point out issues of special interest to you, the military pilot. Based on the comments of military pilots who've already transitioned to civilian aviation, we especially encourage you to concentrate on certain topics as you proceed through this book. Even minimal understanding of the following areas should pay off at interview time, in ground school, and on the flight line.

Crew Resource Management

If there is a single, most important topic impacting your success in today's civilian aviation market, it is "CRM": "crew resource management" (also known as "cockpit resource management"). This term refers to the latest procedures for interaction and coordination of multipilot crews. At interviews, on your simulator checkride, during flight training, and on line you can expect constant evaluation of your performance in this context.

Crew resource management skills become virtually a state of mind among multipilot crews. Your career may be heavily impacted based on whether employers perceive you as part of the CRM process or as a lone-wolf pilot. (See "Crew Resource Management" in Chapter 8.)

If you've been flying single-pilot operations, pay particular attention to CRM issues in your reading. It would also be well worth your time to take one of the excellent CRM workshops offered around the country.

A related topic is checklist procedures. These vary significantly between civilian and some military operations, especially if you've been flying single-pilot aircraft. We recommend covering that section (Chapter 8, also) in some depth and perhaps practicing checklist procedures with friends who've already made the civilian transition.

Training by Civilian Employers

Once hired by a civilian employer, you'll probably enjoy your training. While often challenging and intensive, civilian ground and flight training is generally based on the concept of "train to proficiency." The intent is to train all hired pilots until they're sharp, rather than wash people out. Standard training procedures and preparation suggestions are covered in Chapter 2.

If there's anything to be careful of in civilian training, it's to avoid coasting, due to the relaxed and supportive nature of classes. Pilots do flunk out of training. Participants need to be self-motivated because there are plenty of other applicants waiting if anyone drops out. (It's not uncommon for outside pilots to show up uninvited at smaller operations on the first day of ground school, with hopes of filling any unexpected slots.)

Civilian Aircraft and Civilian Aviation Terminology

As a military pilot, you probably haven't spent much time around regional or corporate aircraft or the latest airline equipment. You may be familiar with the Boeing 747s and the McDonnell Douglas DC-10s operated by the majors. But can you tell the difference between an Airbus A-300 and a Boeing 767? How about the forty or so most common commuter and corporate aircraft?

You may lump civilian aircraft identification skills into the "nice to know but not really important" file. However, it takes only one detailed taxi clearance at a major airport to prove that civilian aircraft identification skills are a necessity (for example, "American 71, wait for the A-300, then taxi via the inner, hold short of K, wait for the Embraer RJ145, then transition to the outer behind, and follow the Falcon Jet to 25R").

Review the Airline, Regional, and Corporate Aircraft Spotter's Guide (see Appendix 2) to improve your civilian aircraft identification skills. Spend some time at the airport checking out these aircraft. The knowledge will pay off for you on the line. Aircraft familiarity is also important when deciding where to apply for a flight position. Your life on the job will vary tremendously depending on whether or not the aircraft you fly are pressurized, carry flight attendants and refreshments, and are equipped with lavatories or autopilots.

A civilian airline and corporate terminology section has also been included in this book, with you in mind. (See the Glossary.) Like the military, civilian aviation has its own lingo. The more familiar you are with the system, the more comfortable your transition to civilian aviation will be.

Once you've gotten through the book, we strongly encourage you to subscribe immediately to some of the excellent civilian aviation magazines currently available. You'll learn more about the airplanes, lingo, and issues facing the industry. Go hang around the airport, too. Pilots everywhere love to show off their airplanes. Besides, some may turn into job contacts.

Aircraft Systems

Depending upon what types of aircraft you've been flying, there are probably significant differences in aircraft systems that you'll need to learn. We recommend skimming all of the systems chapters (Chapters 3–6) for minor differences and then returning in more depth to those systems unfamiliar to you.

One system, in particular, is brand new and challenging for many transitioning military pilots: propellers. We have included a special supplement on propeller basics for those who may not be going straight into jets. (Review "Propeller Supplement for Transitioning Military Jet Pilots" and then "Turboprop Propeller Systems" in Chapter 3.)

Contemporary Issues in the Aviation Industry

While brief, "Contemporary Issues in the Aviation Industry" in Chapter 2 is very important to you. Social and union issues have become extremely sensitive in today's civilian aviation industry. It's important to maintain the proper mind-set if you want to get hired and to keep and enjoy your job.

About Your Civilian Counterparts

Finally, a few words are in order regarding your civilian counterparts. Many military pilots wonder about the credentials of the civilian pilots they'll be flying with. Most civilian pilots earn their ratings through university flight programs, at private flight schools, or with private flight instructors. Civilian flight training varies tremendously in quality. While some pilots graduate from top-notch programs comparable in quality with military training, others collect their training from many different sources, a la carte. (The predictable nature of military training is one of your competitive points as a job candidate.)

To gain flight hours and professional experience, the typical newly graduated civilian pilot works first as a CFI for a year or so, then moves on to cargo, air-tour, or air-taxi operations. ("Air-taxi" refers to commercial, on-demand charter operators.) From there, he or she moves to a regional (commuter) airline or corporate flight department. (The luckier ones may skip a level or two on their way up the ladder.) To be sure, by the time civilian pilots make it to the higher professional levels, they're pretty sharp. In particular, many have excellent all-weather flying experience in commercial operations. Like your military peers, most civilian pilots aspire to corporate or airline jet captains' positions.

In the course of flying you'll probably hear about some interesting "time-building" jobs held by your civilian counterparts during their careers. Bush flying, island cargo hopping, emergency medical flying, water bombing, mercenary and missionary flying make some of their stories almost as good as yours! Most pilots would agree that a combination of military and civilian backgrounds makes for a great flight department.

CHAPTER 2

General Preparations

IN MOST RESPECTS the actual flying of large and sophisticated turbine aircraft is not much different than what you're used to. Basic flight controls are similar, and so are system operations like those for flaps and landing gear.

There are some fundamental differences, however, between these larger aircraft and the smaller piston or military planes that you've likely been flying. Effective coordination of a multiperson crew requires training and development of interpersonal skills. Many complex systems must be operated and monitored. Finally, there are huge differences in performance and response of turbine powerplants in heavy aircraft, especially when compared with fighters and lighter piston models.

Training

Fortunately, the days are largely over for operations where someone inexperienced and untrained hops into a sophisticated aircraft and blasts off. Statistics show that safety is greatly enhanced by good training. Besides, professional pilots find that the inconveniences of training are well justified by the pride and confidence that come from being sharp on the aircraft.

When the time comes for you to move up to sophisticated turbine aircraft, you'll likely get some serious training: ground, flight, and often simulator. Extensive initial training is required by the FARs for Parts 135 and 121 operations. Insurance requirements (and common sense) usually dictate similar training for Part 91 operators. Of course, if you're getting an aircraft type rating for any of these operations, you'll again be getting the training.

Recurrent training is also required for most turbine operations. Pilots under Parts 135 and 121 must receive regular, recurrent ground instruction and flight checks to remain qualified in their positions. Quality Part 91 operators generally have similar recurrent training policies, even though not required by the FARs.

Once you've attended a few training programs, you'll find that most are organized in a similar fashion. Let's talk briefly about the types of information you'll need to learn in ground school. Most ground schools are broken into two sections—aircraft and indoctrination. For the aircraft and engines there are limitations, systems, and procedures to learn. For indoctrination training (or "indoc"), pilots review applicable federal regulations and learn company policies.

Limitations, Systems, and Procedures

Limitations are generally memorization items including aircraft and engine limit weights, speeds, temperatures, capacities, and pressures. This can be a lot of information to learn in a short time, so experienced pilots often check with the instructor ahead of time and start the memorization before beginning class.

Aircraft systems are normally taught at the schematic level, meaning what each system does, how it works, key components, how the parts are connected, and what happens under various operational and failure scenarios. The idea is to understand each system well enough to visualize it in a coherent manner and to thereby be able to operate and troubleshoot it effectively.

Procedures are the specific operating methods and sequences that pilots are to use when operating the aircraft. Topics include everything from when and how checklists are to be performed to when ground deicing must be ordered.

Indoctrination Training

It should be no surprise that pilots must be familiar with the federal aviation regulations governing their operation.

Indoctrination training also covers related regulations not directly associated with flying of the plane, such as handling of hazardous materials, security measures, and firearm requirements. Finally, there are important company policies, such as required record keeping, handling of passengers and cargo, and aircraft dispatch procedures. These are all covered under each company's operations manual. Under Parts 135 and 121 operations, the FAA must approve the manual. This is important to you because via FAA approval each company's operations manual becomes part of the federal aviation regulations for that operation. Pilots who don't conform to the company's operations manual are likely to be violating federal aviation regulations.

Simulator and Flight Training

Once ground school has been successfully completed, most new hires are sent on to simulator training. While some corporate and commuter operators still do all their training in the airplane, this is becoming rarer, due to higher cost and lower effectiveness. These days, most regional and corporate outfits supplement thorough "sim" training with some flight time in the airplane. In the major airlines, it's common for all flight training to be done in the simulator. (Major airline simulators are certified even for landing qualifications.) Successful completion of flight training is marked by passing oral and flight checks, not unlike the checkrides you've taken to earn ratings in the past. For Part 121 and some 135 operations, there is additional *IOE* (*initial operating experience*) training. In this case, the newly qualified pilot goes to work for a specified period with one or more "IOE captains." These experienced training captains help new employees learn the ropes of operations on line.

It should be obvious that the quality of the extensive training we've just described is important to your safety, your knowledge, and your career. Unfortunately, while each training department may, on paper, cover the same topics, there are big variations on how well the job is done in different companies. A good training program offers a special opportunity for you to grow professionally as a pilot. Make quality training one of your top priorities in choosing a company to work for.

Preparing for New-Hire Training

If you've already been hired and will be training for a turbine flight position, you may be interested in some tips on how to prepare. Among your first acts after getting your acceptance letter, should be to call and learn the details of training, including the aircraft type to which you're assigned. Ask which training materials may be checked out early for ground school preparation. Some operators are understanding about lending these materials out early, but a surprising number are not.

If your new employer falls into the latter category, check with other recent hires who are already on line. They'll have collected extensive notes and probably copies of many required manuals and training materials. They'll also know the training priorities and where you might best invest your pre-ground school study time. Find out if your new employer uses computer-based training (CBT) to supplement or replace traditional ground school training. If so, it might be possible to obtain a copy of the CD-ROM used in class to study ahead of time. Some pilots also recommend various study tapes or CD-ROMs offered in aviation magazine classifieds.

On the other hand, do *not* try to prepare using materials collected from companies other than the place where you were hired. Even though they may be flying the same types of aircraft, many operators prepare their own training materials, with a surprising variety in interpretation and emphasis. Besides, the difference of a dash number on a given aircraft model (say a 737-200 versus a 737-500) can mean a tremendous difference in the details of the aircraft. All Jetstreams are not alike! (Nor are all Beech 1900s, nor all Boeing 747s.) Consider also that many larger operators order custom-tailored aircraft and work with the manufacturer to develop their own operating procedures. It's a drag to study three weeks for a ground school and then find that all the specs and procedures you've memorized are wrong.

Preparing for Ground School

Ground school preparation is best divided into three areas. First, try to gain a general understanding of each aircraft system. You can't possibly learn every detail of every system before ground school, but if you know the basics, you'll have little trouble absorbing the instructor's favorite details in class.

Secondly, get a list of the "must know" aircraft limitations from the training department or a recent hire. You can bet that there's a long and specific list to memorize. Even if you don't know what half the limitations mean before ground school, you'll avoid trauma by memorizing them ahead of time. Most people do this by recording the limitations on index cards and then using them as flash cards. (You may be able to find flash cards ready-made for your aircraft. If so, for the reasons already discussed, be sure that they're current and came from someone in your own company. For many people making your own cards is best because it helps in the memorization process.)

Finally, see if you can get your hands on your company's operations manual. This will include the many policies and procedures applying to day-to-day operations, as well as samples of forms you'll use on the line. A quick review of federal regulations applying to your operation would also be in order. In ground school itself, each information section will be covered separately and then followed with quizzes. You can then anticipate a written final exam that must be passed to complete the course. These days, most operators work under the concept of "train to proficiency." That means that the instructors are on your side, with the objective of properly training everyone. The material is challenging, however, and some people do wash out of ground school. It's definitely important to prepare ahead of time.

Preparing for Simulator and Flight Training

Once you've completed ground school, it's on to simulator and flight training. (Depending on the operator, your training might include any combination of simulator training and actual flying in the airplane. The trend is toward simulators for much of or all flight training.) It's possible to prepare pretty effectively for sim training ahead of time. Obviously, if you're not sharp on instruments, you should brush up before attending training. This can be done with any number of private training companies. Your employer expects to train you on the airplane, so any hands-on sim practice beforehand should simply emphasize basic instrument competence. Even if you can access only a rudimentary simulator, practice as closely as possible to the approach airspeeds for the plane you'll be flying.

The next challenge, and for most people the biggest one, is mastering crew coordination aspects of flying in a multipilot crew. To prepare, get copies of the checklists you'll be using on line. Find out when each checklist is to be called for. (This is usually in the company operations manual, but any line pilot can tell you.) Remember how you used to practice engine-out procedures in your car while waiting for a red light? Well now you can spend that time thinking through each flight sequence and calling for checklists at the appropriate times.

Another trick is to pick up some large-scale cockpit posters or schematics for your aircraft, and mount them on the wall of your room. Then place a couple of chairs in front as pilot seats. (Some pilots go so far as to mock up consoles, side panels, and overhead panels.) This arrangement is known as a *CPT* (*cockpit procedures trainer*) and is something you may experience in a grander form at class. Gather some other new-hire pilots and practice normal, abnormal, and emergency checklists together using your mock-up. It's best to include an actual line pilot in your first few sessions to teach you the ropes. This way you'll quickly learn about the cockpit, locations of switches and controls, checklists, and basic crew teamwork. (Expect some interesting comments from your nonflying roommates!)

As for the flying itself, turbine aircraft procedures and maneuvers are designed to be performed in the same way each time. This standardization comes in the form of "flight profiles," which are standard sequences for performing everything from steep turns to engine-out go-arounds.

Memorizing your company's flight profiles ahead of time is very important. It'll reduce your stress during simulator or flight training and will free you up to concentrate on the flying itself. Simulator and/or flight training programs conclude with challenging oral exams and practical tests, much like what you've come to expect on other checkrides.

Preparing for the Flight Line

Once you've completed ground school, and simulator and flight training, preparing for life on the flight line is simple. Talk to others in the same crew position who've been on line for awhile; make a thorough list of your anticipated duties, and learn how they're to be accomplished. If time permits, it also pays to ride along on a few routine trips so you can see how the operation works.

While each of us wants to be a great stick on the day we go on line, other skills are just as important. For example, if a new copilot can get the chores done, preflight the airplane, and properly load the passengers, the captain will be greatly assisted. Sharp pilots learn how to do this routine stuff before going on line. It keeps the operation on time and allows more energy to go into the flying, where it's most important.

From a flying standpoint, perhaps the toughest challenge for "step-up" pilots is staying ahead of the airplane through the course of a flight. Piston pilots moving up to turboprops or jets for the first time face this challenge in a big way. Along with their civilian counterparts, military pilots transitioning from King Airs, C-130s, and rotary-wing aircraft learn that things happen *much* faster in jets. Know the procedures before going on line, and study your assigned trips ahead of time. To quote one seasoned airline captain, "By far, the biggest problem I've seen with newer first officers is [lack of] preplanning. Things just happen so damned fast in a modern jet. Reviewing and setting up early is essential, especially at high-density airports."

For IOE in multi-pilot operations, the company will assign you to an experienced captain who'll help you learn the ropes. Those pilots working single-pilot turbine operations for the first time should arrange for a type-experienced instructor to ride along for the first several flights; there's a lot to learn, even after all the training.

Expect a few challenges and a few chuckles from others during your first few days on line as you learn the ropes. But you won't mind—you're there!

Computers

There's one other more general training topic for you to consider. Aircraft systems, particularly those in the cockpit, are rapidly computerizing. If you ask long-time Boeing 767 captains about their toughest professional challenges over a career, a good many will name the transition to a computerized cockpit. Sophisticated, computerized flight management systems now appear in most corporate jets, regional jet airliners, and even in many turboprops. In fact, the odds are good that you'll face them in your first turbine aircraft. So if you're not particularly knowledgeable about computers, start learning immediately. Even basic operating skills on a desktop or notebook computer will help you greatly in the transition from analog instruments to a computerized cockpit. Don't wait until ground school and your first digital cockpit. It may be too late. (For more information on computerized cockpits see Chapter 13, Navigation, Communication, and Electronic Flight Control Systems.)

Contemporary Issues in the Aviation Industry

As you advance in the aviation industry, it's valuable to recognize some of the issues that you're likely to face. Getting hired in this industry is challenging enough without getting on the wrong side of sensitive issues. In addition, most companies have probation periods for newly hired pilots. (One year of probation is common for the airlines.) It's now easier to get released during probation for inappropriate behavior unrelated to flying skills. Some of the sensitive issues are related to teamwork, union affiliations, driving under the influence (DUI), and harassment.

Crew Resource Management and Teamwork

Due to the ever-increasing emphasis on crew resource management (CRM), teamwork comes up time and time again as one of the most addressable areas of flight safety. What does it mean to you? From the interview onward, pilots must convey teamwork to their employers and peers. While this may sound easy enough, it's not. Pilots accustomed to single-pilot operations, for example, must bend over backward to learn crew concept procedures and attitudes. If you come from such a background, it'd be well worth your while to read up on the topic, attend a CRM workshop, and discuss related issues with friends in the industry. This CRM sensitivity must be conveyed throughout the interview process and continue on through employment. Those coming from a single-pilot background should expect to be asked about teamwork issues at job interviews and should be prepared to make the right impression.

Unions

Union issues are among those that theoretically are never supposed to come up at interviews but often do. There are widely divergent opinions about the roles of pilot unions in aviation. Some companies are unionized, while others are not, and among the employees and management of every company are differing perspectives on the roles of unions. As an applicant, the most important thing is to be sensitive to union issues at companies to which you apply. If you're lucky enough to have a choice, you may find that such issues may impact your decision of who to work for. In any case, you'll want to handle any union issues as sensitively as possible. Along with politics and religion, union issues are best not discussed with people you don't know!

Driving under the Influence

It cannot be overemphasized how seriously DUI (driving under the influence of alcohol or drugs) offenses can impact a pilot's career these days. You'll have real difficulties getting hired with a history of such offenses. If charged with DUI after you're hired, you won't make it through your probationary period. It's just not worth the risk.

Discrimination and Harassment

As you know from reading the daily papers, discrimination and harassment have become major public issues. The aviation business, as well as others where minorities and women were largely excluded until recently, is highly visible and highly sensitive to charges of discrimination and harassment.

As a pilot, it's imperative to present yourself as a team player and as an open-minded professional. If you are suspected of discrimination, you can bet it won't help your career. As you might imagine, flight departments want to deal with flight operations rather than harassment and personnel issues. Employers want to present only the best possible image to the public. Charges of harassment are particularly devastating to business these days and present huge legal liabilities. Accordingly, it should be no surprise that interviewers and management are trained to be alert for such problems all the time.

Besides, as a professional pilot you'll likely be flying with the same folks for years. This is a wonderful experience, given that you have the right attitude. If you harass people or otherwise make enemies, your job will no longer be any fun, even if you're lucky enough to keep it. As a pilot applicant, and as an employee, you'll want to present and sustain only the best possible image of sensitivity, tolerance, and getting along with others.

CHAPTER 3

Turbine Engine and Propeller Systems

AMONG A PILOT'S responsibilities is to understand the basics of each system on the aircraft flown. Sometimes this seems excessive to pilots in training (especially those attending ground school for a new aircraft). It's important, though, that a pilot be able to identify, understand, and rectify failures and problems that may occur in emergency situations. Therefore, most systems training emphasizes understanding of each system's operating sequence, with special focus on what happens if various components fail.

The systems of most modern turbine aircraft are similar in operating principles and components. However, they differ greatly in the details of implementation. The result is that, once you've learned the systems on one turbine aircraft, others are much easier to comprehend. Let's take a look at the basics of the major turbine aircraft systems. Then when you attend ground school, you can concentrate not on "What is it?" but on "How is it implemented here?"

Introduction to Gas Turbine Engines

The basic operating principle of a gas turbine engine is simple. Jet propulsion is described by Isaac Newton's third law, which states that for every action there's an equal and opposite reaction. In the broadest sense, gas turbine engines share jet propulsion with party balloons that have been inflated and released. In each case pressurized gases escaping from a nozzle at one end create an equal and opposite force to drive each of these "reaction engines" in the opposite direction (Fig. 3.1).

The main difference between the balloon and the engine is in the source of the propulsive gases. The balloon is an energy storage device. Someone blows air into it, and once the stored air has escaped, the propulsion is over.

Gas turbine engines, on the other hand, are heat engines. Through combustion of fuel with intake air, they continuously create expanding gases, the energy of which is converted into propulsive force to move an aircraft. For this reason, the core of a turbine engine is known as the *gas generator;* it generates the expanding gases required to produce thrust. A *nozzle* is then used to accelerate the velocity of the high-energy gases escaping from the gas generator. By increasing gas pressure through its tapered chamber, a nozzle forces gases to exit through its smaller exhaust opening. The effect of this process is an increase in the momentum of air passing through the engine, producing thrust (Fig. 3.1).

The gas turbine engine has the same basic stages of operation as its distant heat engine relative, the internal combustion engine (also known as the "piston" or "reciprocating engine"). As you remember, most aircraft reciprocating engines operate in four-stroke sequence: intake, compression, power (or combustion), and exhaust. Power developed by a piston engine is intermittent since only one stroke in four (the power stroke) actually creates power. The same stages occur in gas turbine engines but with a fundamental difference: turbine engines operate under more efficient continuous-flow conditions (see Fig. 3.2).

Instead of compressing intake air with a piston, turbine engines use one or more rotating "wheels" in the compressor section known as *compressors.* Another set of wheels, known as *turbines,* is driven by the exhaust gases in the

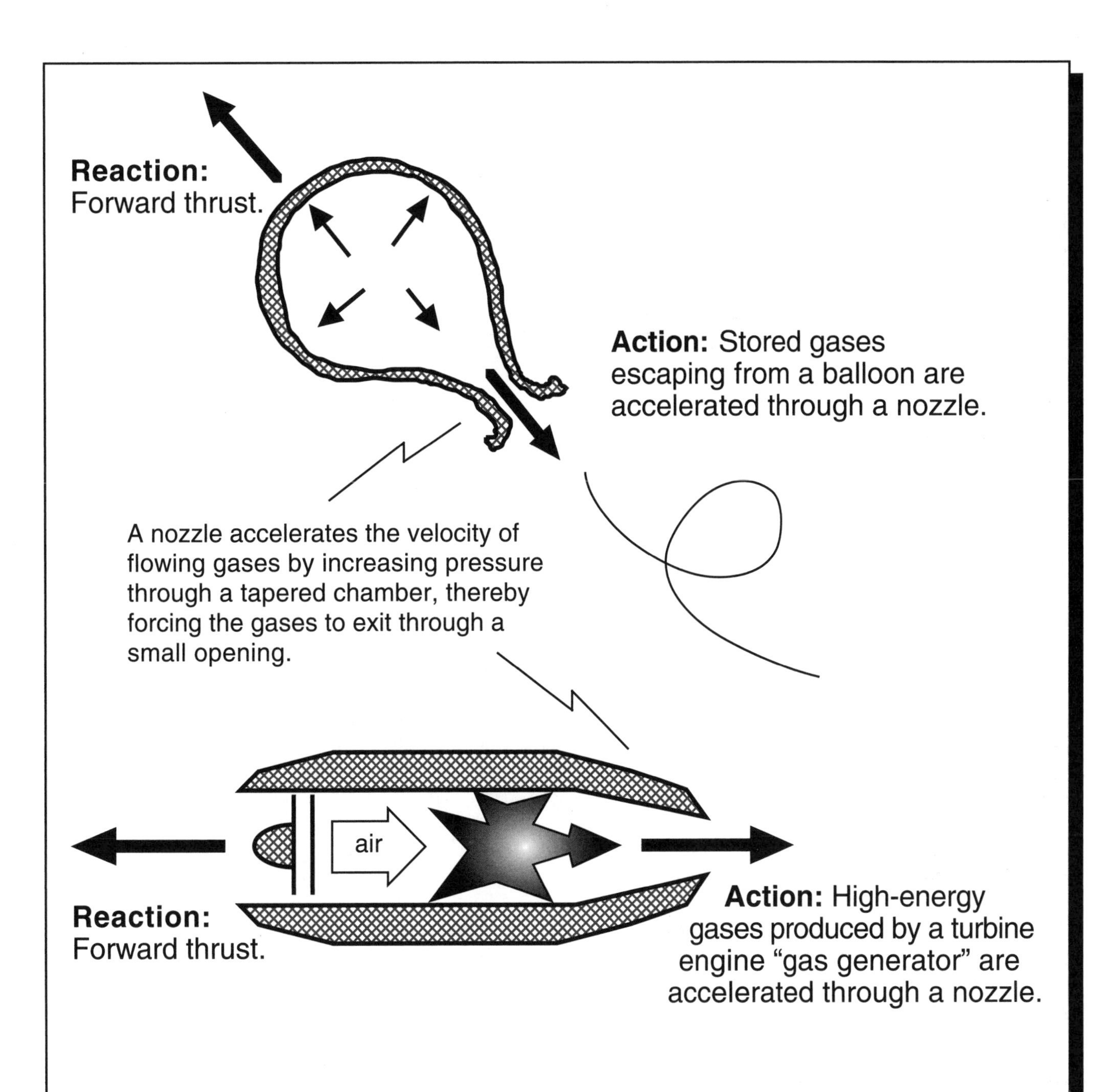

Newton's third law states that for every action there is an equal and opposite reaction. The "action" of accelerated gases escaping rearward from the nozzle of a turbine engine produces an "equal and opposite reaction," in the form of forward thrust to drive an airplane through the air.

3.1. Reaction engines: a balloon and a gas turbine engine.

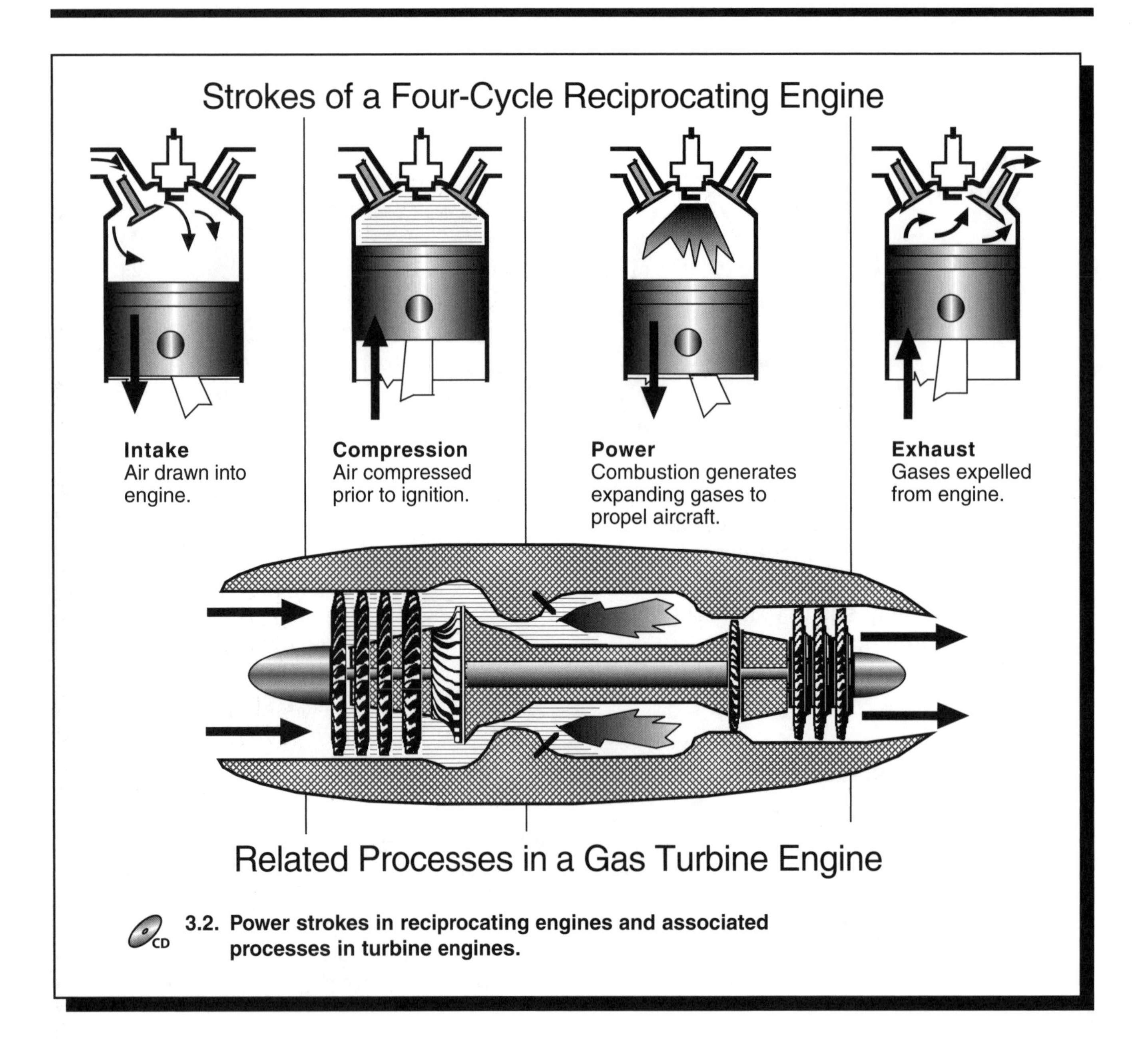

CD **3.2. Power strokes in reciprocating engines and associated processes in turbine engines.**

turbine section. Compressors and turbines are similar in that each is basically an extremely sophisticated fan, composed of a large number of high-tolerance blades turning at very high speeds inside a tightly ducted cowl. Compressors, however, are used to compress air at the "front" of the engine, while turbines serve to tap the engine's energy aft of the combustion chamber (Fig. 3.2).

The basic operating principle of all gas turbine engines is the same. Much like a turbocharger, each compressor and turbine of a simple gas turbine engine is mounted on a common shaft. Intake air is compressed by the compressors and forced into the combustion chamber (*combustor*). Fuel is continuously sprayed into the combustion chamber and ignited, creating exhaust gases that drive the turbines. The turbines, through shafts, drive the compressors, starting the whole process over again. Turbines also harness the engine's energy to drive all of the engine's accessories, such as generators and hydraulic pumps (Fig. 3.3).

Among the main differences between various types of gas turbine engines are the types of compressors they utilize. There are centrifugal-flow compressors and axial-flow compressors.

Centrifugal-Flow and Axial-Flow Compressors

A *centrifugal-flow compressor* utilizes an impeller much like that of a turbocharger. The engine intake directs

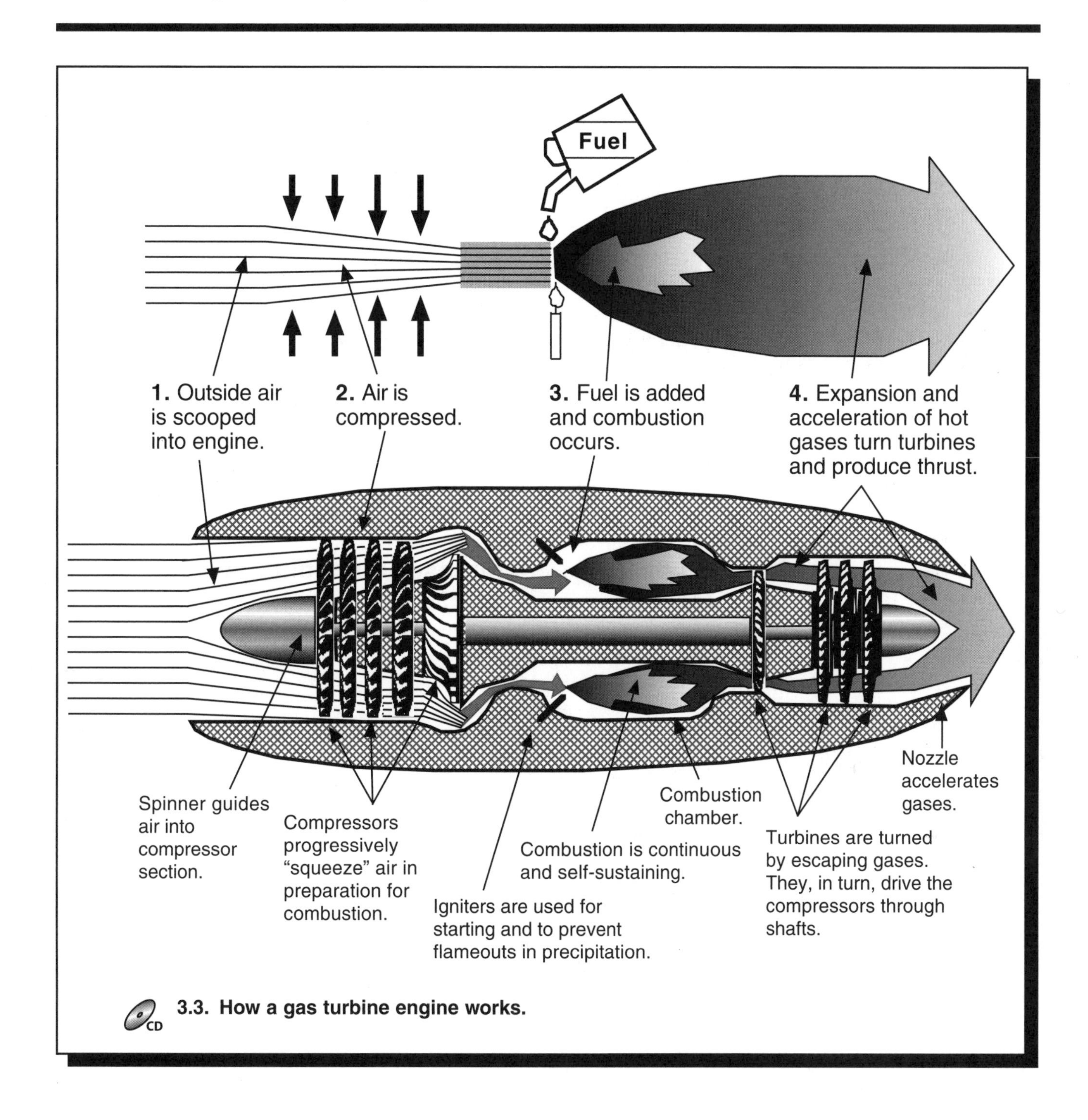

3.3. How a gas turbine engine works.

the intake air into the center of the impeller, where it is centrifugally slung outward into a carefully designed chamber known as the "diffuser." A diffuser is simply a divergent duct that slows the velocity of the impeller's output air, thereby increasing the air pressure before it enters the combustion chamber (see Fig. 3.4).

Air flowing through an *axial-flow compressor,* on the other hand, remains essentially parallel to the longitudinal axis of the engine; the air is not slung outward into the diffuser as with the centrifugal-flow compressor. An axial-flow compressor is made up of an alternating series of rotating *rotor blades* and stationary *stator vanes.* Inlet air enters the first set of rotor blades, where the air is deflected in the direction of rotation. Stationary stator vanes between each set of rotor blades help to direct and compress the flow of air. The objective is to keep the airflow essentially parallel to the longitudinal axis of the engine between each set of rotor blades. In this sense, an axial-flow compressor works more like a window fan, whereas a centrifugal-flow compressor throws air to the outside like a slingshot (see Fig. 3.5).

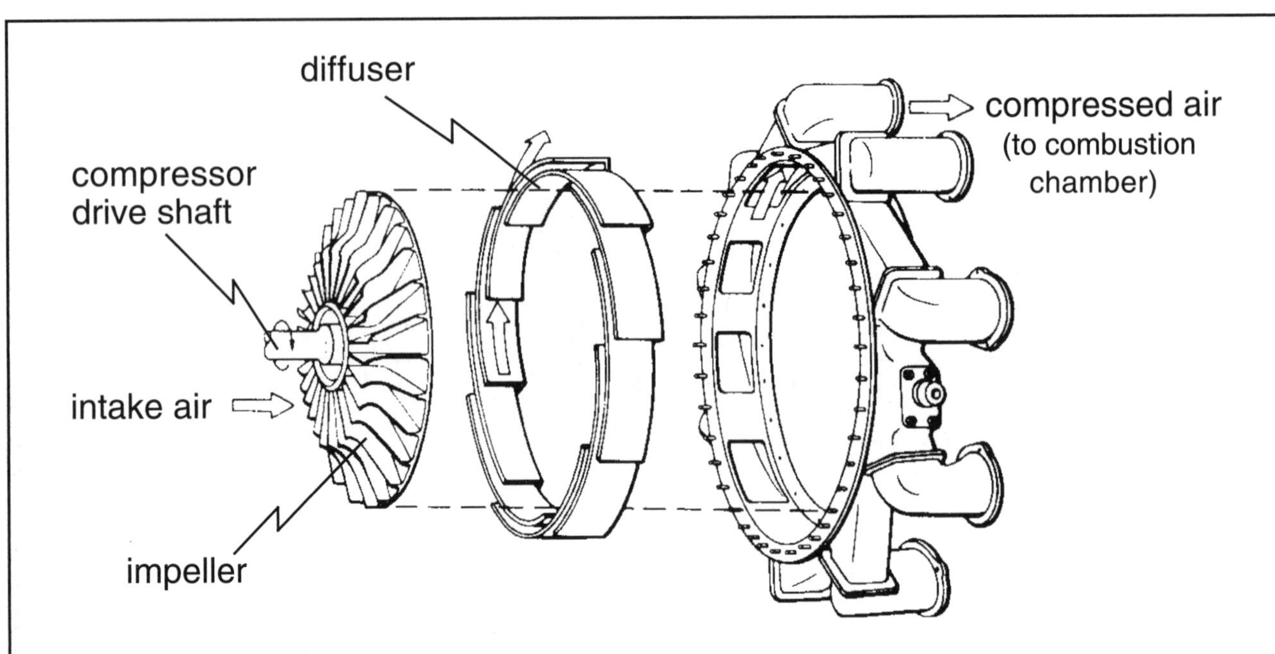

Exploded view of centrifugal compressor components.

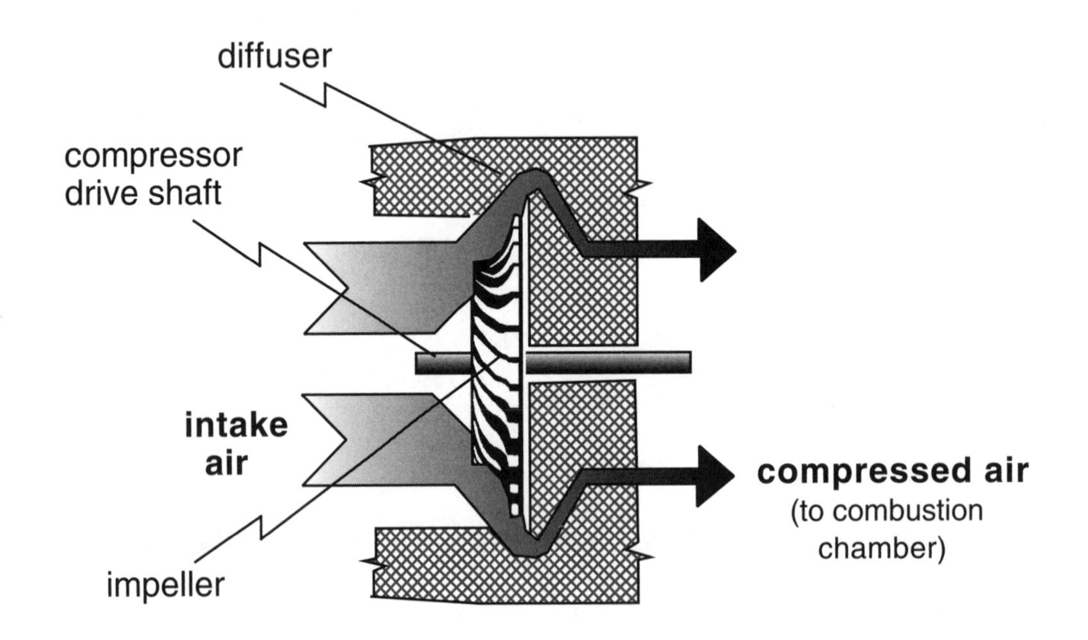

A centrifugal compressor slings air to the outside, compressing it into a diverging chamber known as a diffuser.

3.4. Centrifugal-flow compressor. *Exploded view courtesy of United Technologies Pratt & Whitney.*

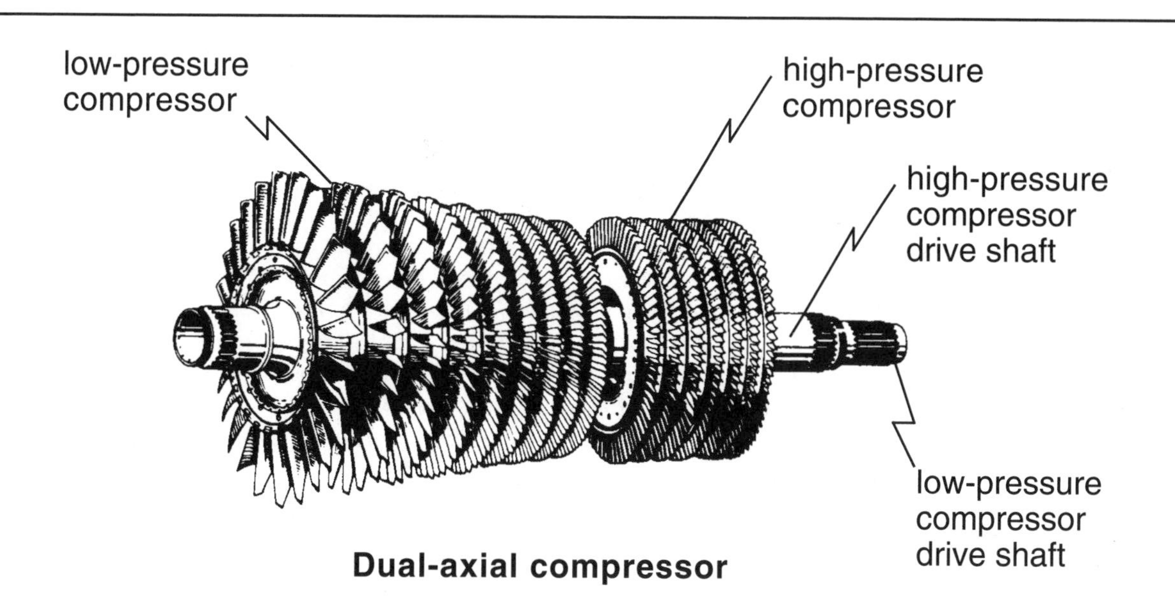

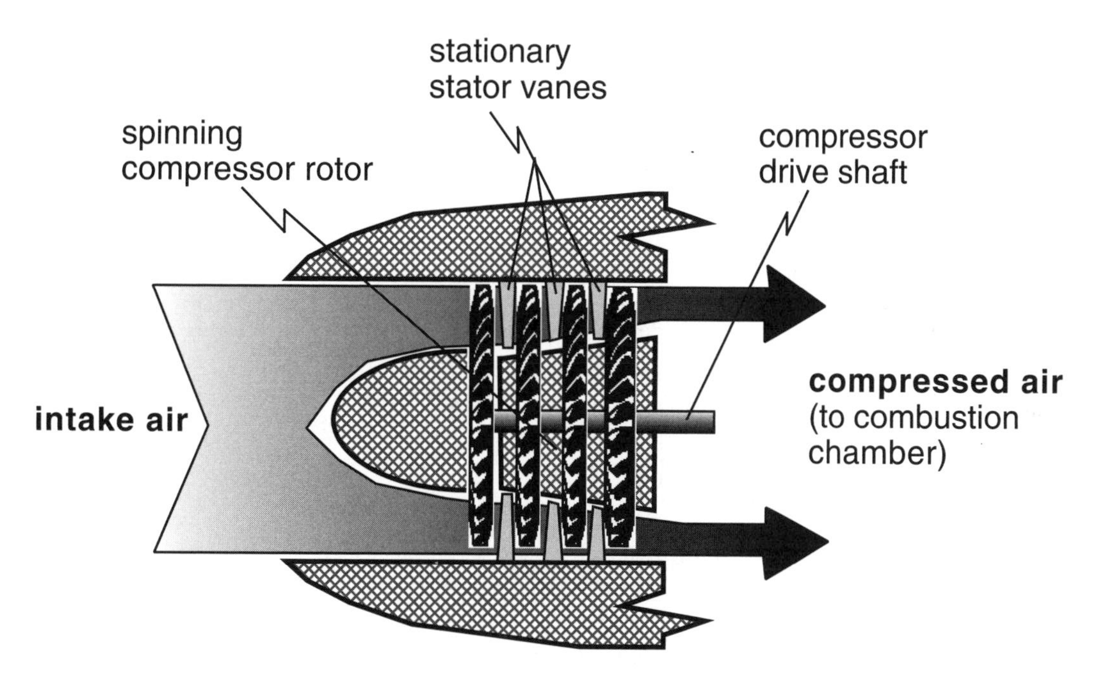

An axial-flow compressor drives air along its longitudinal axis through a progressive series of spinning rotor blades and fixed stator vanes. (By reducing rotational air movement from the spinning blades, stator vanes sustain linear airflow through the compressor and aid compression.) Air pressure increases after each "compressor stage," consisting of one row of rotor blades and the following row of stator vanes.

3.5. Axial-flow compressor. *Dual-axial compressor courtesy of United Technologies Pratt & Whitney.*

Centrifugal-flow compressors are more durable than axial-flow models. Turboprops and most corporate jets use centrifugal-flow compressors because these aircraft are more likely to operate out of unimproved airfields than are large aircraft. Axial-flow compressors, on the other hand, are more efficient. While a centrifugal-flow compressor can normally attain a compression ratio of 10:1, an axial-flow model may achieve ratios of more than 25:1. This means that engines with axial-flow compressors can produce more thrust for the same frontal area, resulting in less drag and more fuel efficiency.

Large jets use the axial-flow type because of the higher compression ratio, which produces higher thrust to weight capability and greater fuel efficiency.

Multistage Compressors

In order to maximize efficiency and power, most modern turbine engines use more than one compressor component, each lined up in sequence behind another. A combination of axial- and centrifugal-flow compressors may be used, or there can be multiples of the same type. In either case, every compressor "wheel" or row of rotor and stator blades acts as one *compressor stage.* Each compressor stage progressively compresses passing gases one step further than the one before it.

Compressor components may be mounted on the same shaft, or on different shafts. An assembly of multiple compressor stages is known as a *multistage compressor,* the number of stages corresponding to the number of compressor components.

Just to clarify this terminology, note that every axial-flow compressor is a multistage compressor. As previously discussed, an axial-flow compressor has the appearance of several rotating wheels, with stators in between. Each "wheel," or row of rotor blades with its stators, is one compressor stage. However, the entire assembly is considered as one compressor, as long as all of the rotating parts are mounted on a common shaft.

The same naming conventions apply to an engine's turbine section. A single-stage turbine has the appearance of a single turbine wheel. A multistage turbine incorporates two or more turbine stages.

Multispool Engines

Many gas turbine engines have two compressors in tandem that are mounted on separate shafts and turn at different speeds. These *dual-compressor* engines deliver higher compression ratios than single-compressor models.

The forwardmost compressor of a dual-compressor engine is known as the *low-pressure compressor.* It is shaft-driven by the rearmost, or *low-pressure, turbine.* The two are connected by the *low-pressure compressor shaft.* Rotational speed of a low-pressure compressor is known as "N_1"; therefore, the low-pressure compressor shaft is also commonly called the N_1 shaft.

The second compressor of a dual-compressor engine is known as the *high-pressure compressor.* It is driven by its own *high-pressure turbine* via a hollow *high-pressure shaft* (or N_2 shaft), which counter-rotates concentrically over the N_1 shaft. As you've probably guessed, N_2 is the speed of the high-pressure compressor.

To again clarify terminology, look once again at the dual-axial compressor shown in Figure 3.5. It incorporates a nine-stage low-pressure axial compressor on one shaft, followed by a seven-stage high-pressure axial compressor on a separate shaft. The two turn separately, at different speeds.

The combination of each compressor, the turbine that drives it, and the shaft connecting them is known as a *spool.* A glance at the separated two spools of a dual-compressor engine (Fig. 3.6) shows the characteristic spool shape of each.

Core Turbine Engine (Gas Generator)

Turbine engines come in many variations. One-, two-, and three-spool models may be found, using any combination of centrifugal-flow and axial-flow compressors. In addition, any number of compressor and turbine stages may be incorporated into each spool. Such engine design details vary significantly by engine type and manufacturer. As much as these differences impact the performance and reliability of competing turbine engines, the basic arrangement and functions are the same. Every turbine engine is built around the same basic core, including compressor, combustor, and turbine sections.

The basic gas turbine engine we've been discussing so far is commonly known as the *gas generator,* or *core turbine engine.* (We will use the terms interchangeably.) In order to most clearly present turbine engine principles, we've pictured variations of the same generic two-spool core turbine engine throughout this book (see Fig. 3.7).

Our "typical" gas generator has a four-stage axial-flow low-pressure compressor, followed by a single-stage centrifugal-flow high-pressure compressor. Its turbine section incorporates a single-stage high-pressure turbine and a three-stage low-pressure turbine. (This arrangement is for example only and is not intended to represent any specific engine.)

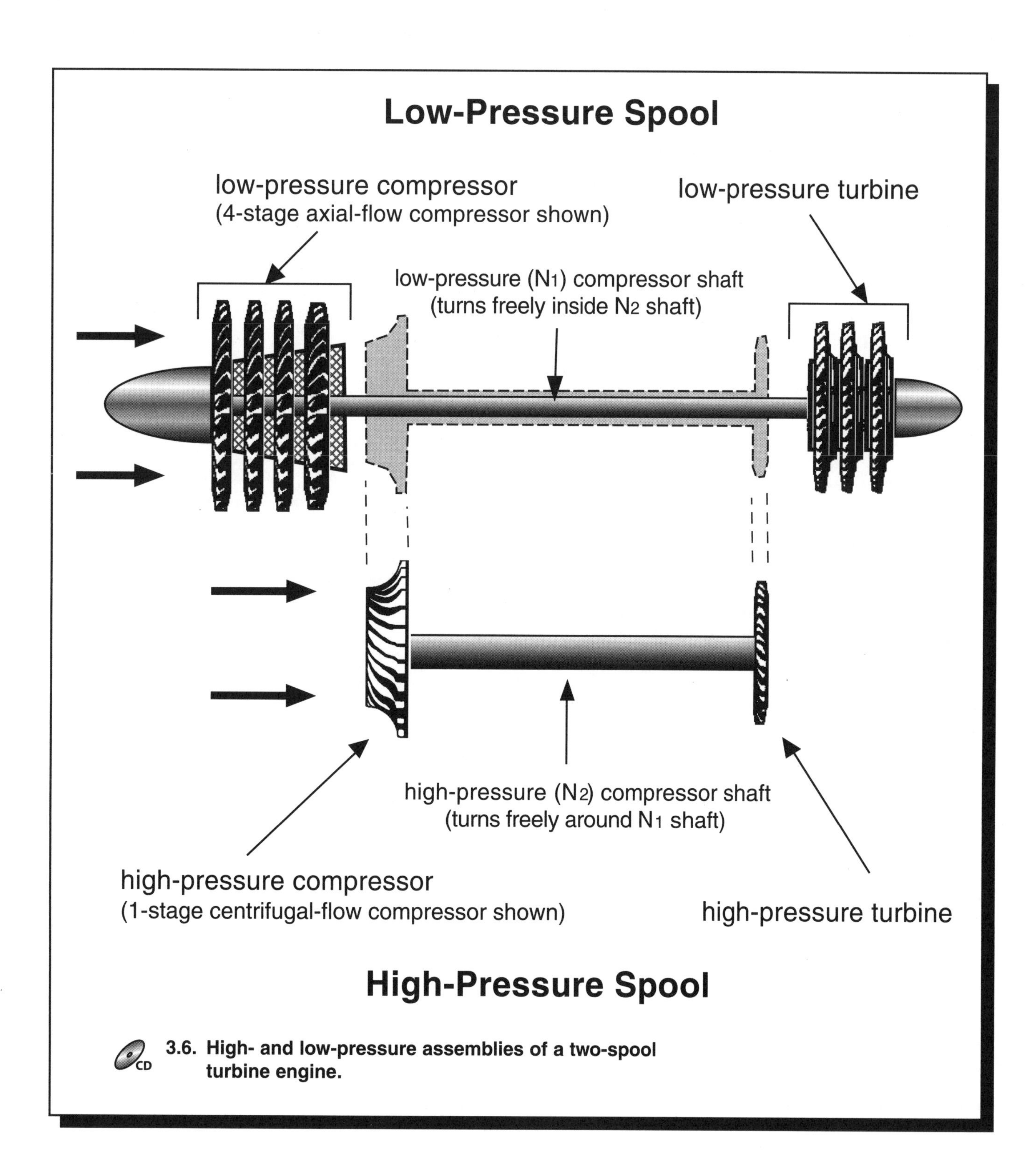

3.6. High- and low-pressure assemblies of a two-spool turbine engine.

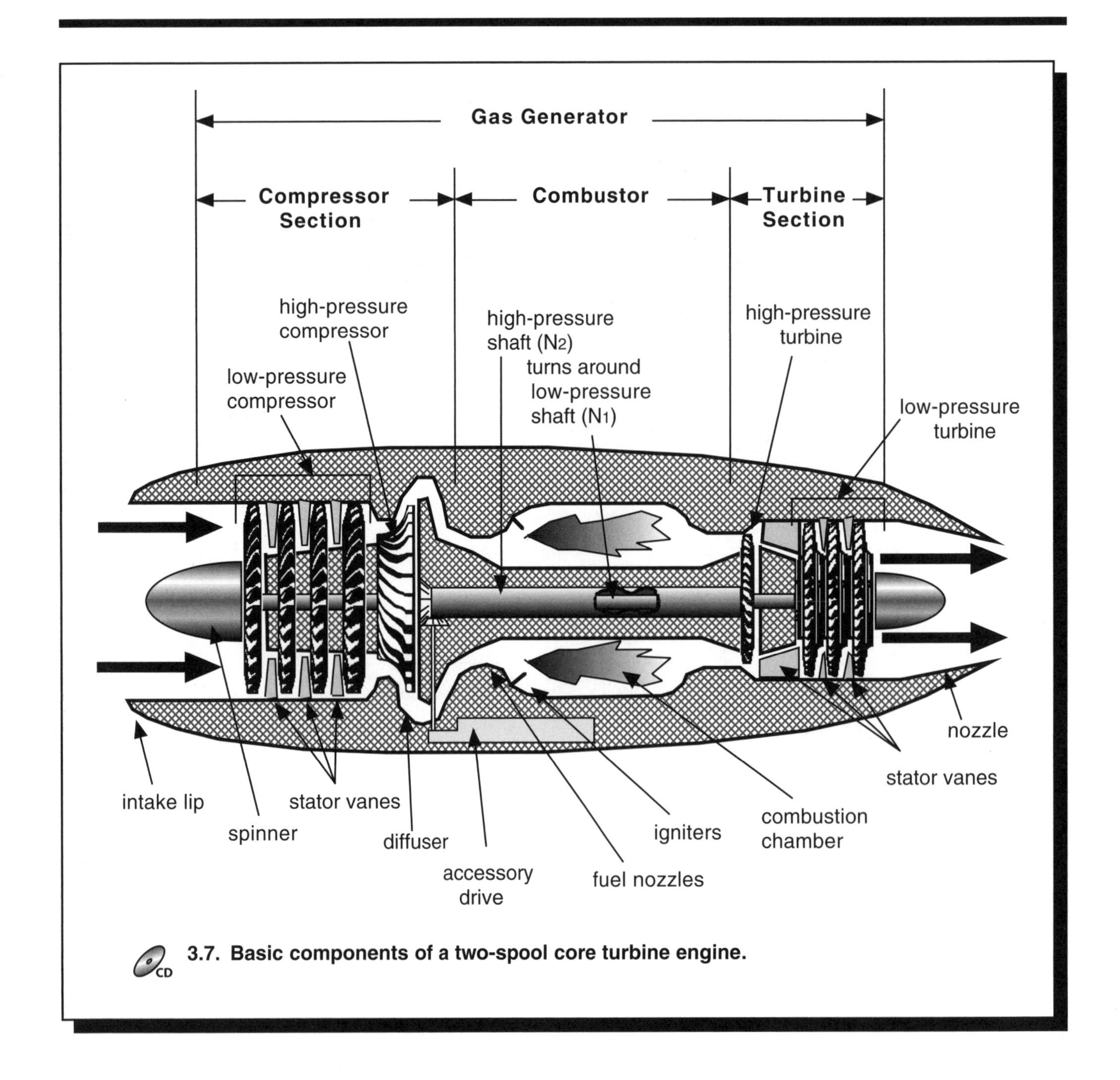

3.7. Basic components of a two-spool core turbine engine.

Turbojets, Turbofans, and Turboprops

Regardless of varying turbine engine types, the basic principles of operation for the core turbine engine are the same. Depending on how the exhaust gases are harnessed, the same basic gas generator may be applied to turbojet, turbofan, or turboprop engines. The biggest conceptual difference between these engines is in how engine output is translated into thrust to propel an aircraft (Fig. 3.8).

Turbojet Engine

The gas turbine engines discussed and illustrated up until now are known as *turbojet engines.* A turbojet engine produces thrust solely by the high acceleration of air as it passes through and exhausts the core engine. Early jet engines were of this type. Turbojet engines, while fairly good performers at high altitudes and high airspeeds, are very inefficient at low altitudes and low airspeeds. They are louder and less fuel efficient than the more modern turbofan engines.

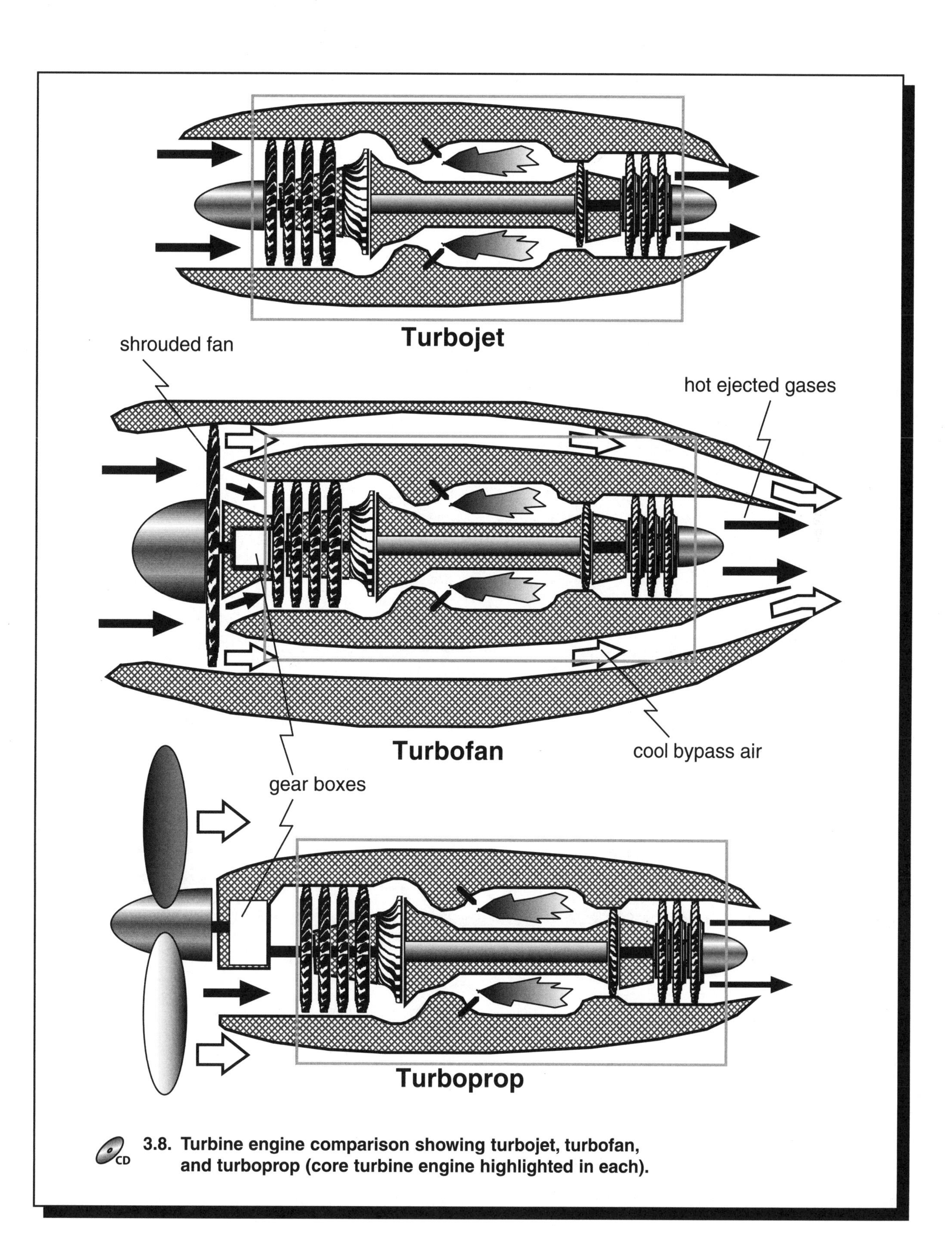

3.8. Turbine engine comparison showing turbojet, turbofan, and turboprop (core turbine engine highlighted in each).

Turbofan Engine

A *turbofan engine* has the same basic core engine as a turbojet. The difference is that a turbofan includes a "shrouded fan" at the front of the engine, driven by the engine's turbine section. (A shrouded fan is essentially a multibladed propeller with a close-fitting nacelle around it.) A turbofan accelerates a larger mass of air than a turbojet. Some of the turbofan's intake air goes through the core engine, while the rest bypasses the engine core around the outside. This *bypass air* generates a great deal of thrust due to the highly efficient design of the turbofan blades.

The ratio of bypass air flowing through the fan section of the engine compared to the amount of air flowing through the gas generator portion of the engine is called the engine *bypass ratio.* For the majority of corporate and airline aircraft designs, the most fuel-efficient engines are those with the highest bypass ratios. (To give an idea of the quantities involved, airflow through the engine is measured in pounds per second.)

Turbofan engines with small amounts of bypass thrust are called *low-bypass engines,* or simply *bypass engines.* Engines with large amounts of bypass thrust from the fan are known as *high-bypass turbofan engines.* At lower altitudes more than 80 percent of a high-bypass engine's thrust comes from the shrouded fan, resulting in greater fuel efficiency.

Turbofans came about because "propellers" (in this case, the shrouded turbofan blades) are more efficient than jet thrust at relatively low altitudes and airspeeds. In essence, a turbofan engine is a compromise between the best features of turbojets and turboprops, the turbojet being best suited to high altitudes and high airspeeds and the turboprop being most efficient at low altitudes and relatively low airspeeds.

Turbojets are very loud because the ejected hot exhaust gases rip into cool surrounding air, creating deafeningly loud wind shear. Turbofans, however, are much quieter due to mixing of the cool bypass air with the hot exhaust gases from the core engine. This helps to insulate and disperse the hot exhaust gases, muffling much of the sound. Virtually all jet engines in production today are turbofans.

Turboprop Engine

A *turboprop engine* simply uses thrust generated by a core turbine engine to drive a propeller. The turbine of a turboprop engine is designed to absorb large amounts of energy and deliver it in the form of torque to the propeller. There are two different ways of doing this.

In a *direct-drive turboprop* engine the propeller is driven directly from the compressor shaft through a reduction gear assembly (Fig. 3.9).

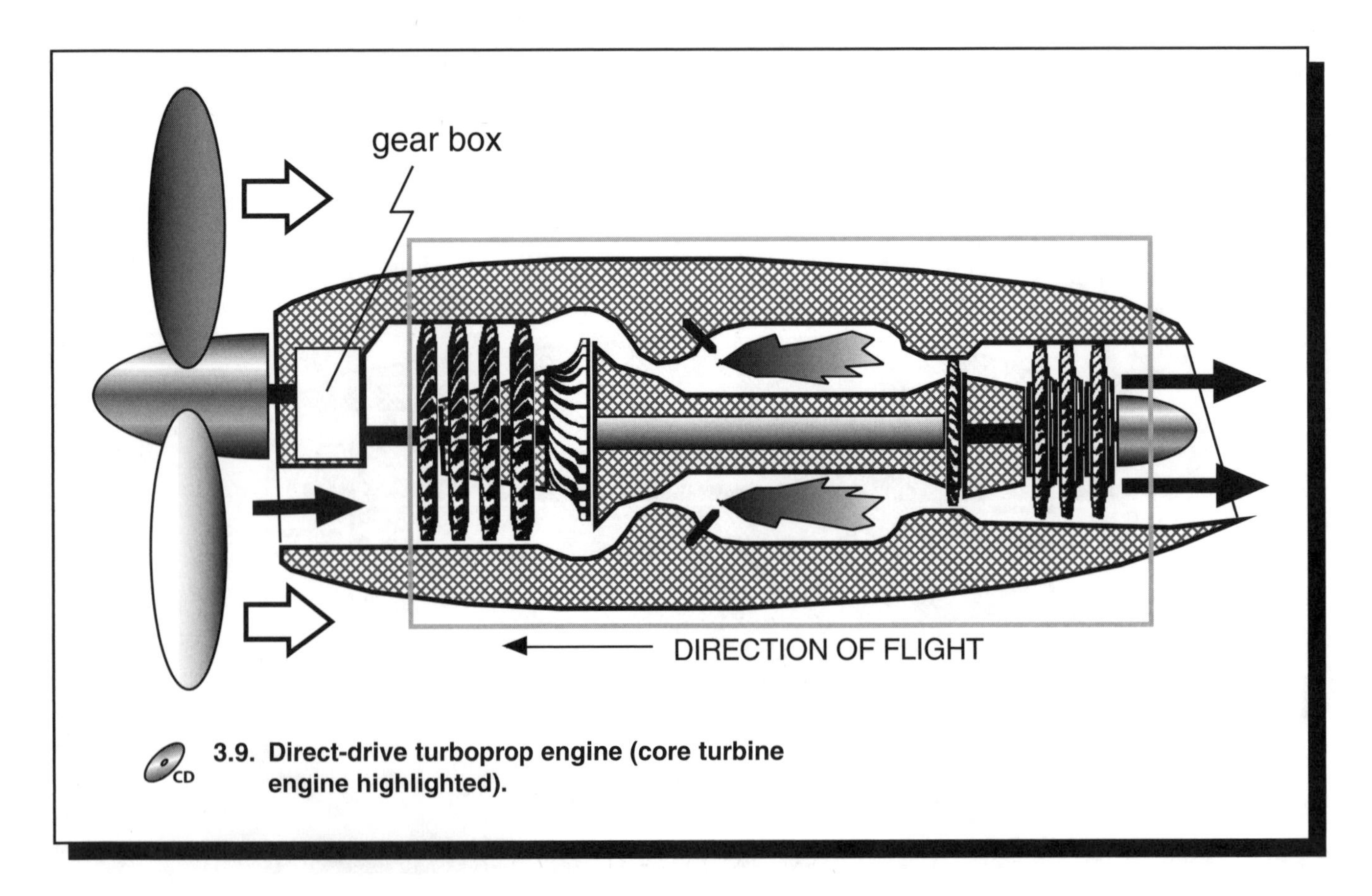

3.9. Direct-drive turboprop engine (core turbine engine highlighted).

In a *free-turbine turboprop* engine the propeller is not connected directly to the engine core. Rather, an additional free-turbine "power section" is added to the basic engine core after the turbine section. High-pressure exhaust gases from the engine core are used to drive the free "power turbine." The power turbine, in turn, rotates a shaft turning the propeller gearbox (Fig. 3.10). This nonmechanical connection between engine core and propeller driveshaft is known as *fluidic coupling.* The same principle is used for automatic transmissions in cars, except that fluidic coupling in turboprop engines is gaseous, rather than liquid.

The difference in design between direct-drive and free-turbine engines can easily be identified after engine shutdown. When the pilot turns the propeller of a direct-drive turboprop by hand, he or she is rotating the engine core as well, meaning the engine is relatively difficult to turn. But with a free-turbine turboprop all that rotates with the propeller is the single power turbine that drives it, meaning the props on these engines spin so freely that they often blow in the wind. Since there is no mechanical connection between propeller and engine core in this case, the engine core does not rotate when the propeller is turned by hand.

Both direct-drive and free-turbine types of turboprops have distinct advantages and disadvantages. Direct-drive engines are designed to operate at more or less fixed rpm. Adding power (by adding fuel) results in increased power output without a significant rpm change. This means that internal aerodynamics of the compressor section, combustion chamber, and turbine section can be engineered closely to the optimum operating range. For this reason, direct-drive turboprops offer more immediate power response and are more fuel efficient. The ground idle of direct-drive engines is high (in the neighborhood of 72 percent of maximum rpm), since the propeller is directly connected to the engine core. Therefore, this type of engine generates lots of noise on the ground. (Propellers are responsible for most "engine noise" on any prop airplane.)

A free-turbine turboprop, on the other hand, has a relatively low noise level at ground idle. Engine core rpm is lower in this type (58 percent on one popular engine), and since there is no mechanical connection between engine core and propeller, the propeller can spin down even slower.

Another significant advantage of some free-turbine engines is that they can be easily disassembled for service, often without removal from the aircraft. The popular Pratt & Whitney PT-6 series free-turbine optimizes this principle via its "backward" installation. Air is drawn to the back of the engine to enter the rear-mounted compressor section,

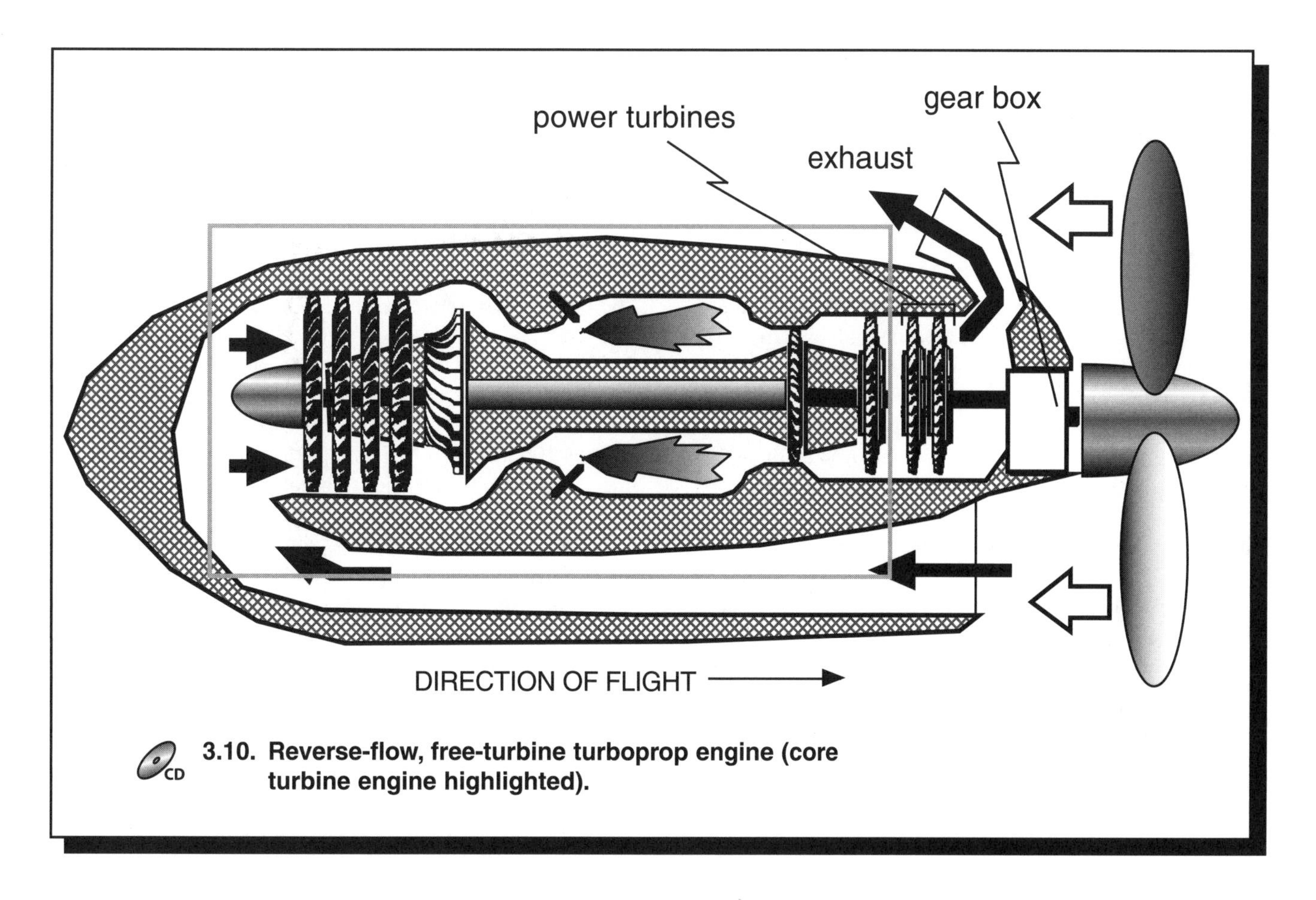

CD **3.10. Reverse-flow, free-turbine turboprop engine (core turbine engine highlighted).**

thereby allowing the power section to directly mount to the propeller gearbox. Since the power section (including the free power turbine) is not physically connected to the gas generator except via the engine housing and plumbing, either section may be removed from the aircraft relatively easily for maintenance.

However, the internal aerodynamics of a free-turbine turboprop must be engineered to operate over a wider range of air pressures and velocities. Therefore a free-turbine engine tends to operate with less optimal internal aerodynamics than its direct-drive counterparts. This makes the free-turbine less powerful by weight and less fuel efficient than the direct-drive type.

It's simple to tell if a turboprop aircraft is a direct-drive type or a free-turbine type. Look at the propellers' blade angle after shutdown. If the props are at a very low blade angle relative to their plane of rotation (that is, in flat pitch), the turboprop is a direct-drive type. If the propellers are in "feathered" position, the turboprop is a free-turbine type (Fig. 3.11).

This difference occurs because direct-drive turboprops have start locks that fix their propeller blades at a 0° blade angle after shutdown. The start locks are metal pins held out by centrifugal force at operating rpm, allowing the blade angle to change. But when rpm is reduced below a certain level at shutdown, the start locks move inward toward the propeller hub and lock the propeller into a low blade angle. When starting the engine, the propellers must be spun up with the engine due to the rigid connection between them, and the flat pitch position lowers air resistance on starting. If the blades were feathered, it would be more difficult to rotate the propeller at start-up.

Free-turbine turboprops do not need start locks, because there's no direct connection between propeller and engine core. The engine can therefore be started easily since propeller mass and drag do not restrict engine core rpm. The propeller begins to turn only once the engine core is producing enough exhaust gases to rotate the power turbine. Therefore, free-turbine turboprop propellers are allowed to move to their "natural" feathered position on shutdown. (See "Turboprop Propeller Systems" later in this chapter.)

Thrust versus Power

If you've been paying close attention, you may have noticed a subtle change in terminology over the past few pages. The output of turbine engines was called "thrust," until we got to turboprops. Then the discussion changed to "power."

This is because the definition of "power," so commonly used for piston engines, does not apply well to pure turbine engines (jets). *Power* is defined as force applied over a distance, per unit of time. But jet engines impart thrust to the air without moving any mechanical parts over a distance. (Although internal engine parts move, thrust is actually imparted to the air by the acceleration of gases.) Accordingly, "jet" engines are rated in pounds of thrust.

Turboprops, however, do generate power in the form of *torque,* or force applied to a propeller over a rotational distance. Therefore, turboprops are rated using power terminology. *Shaft horsepower* (shp) refers to the power delivered by a turboprop engine to its propeller. However, the core-engine exhaust from turboprop engines does produce some jet thrust that augments propeller output. The effect of jet thrust is added to a turboprop's shaft horsepower rating to produce a total power rating known as *equivalent shaft horsepower* (eshp).

Keep this difference in mind when discussing engine management in turbine aircraft. *Turboprop pilots set power or torque. Jet pilots set thrust.*

Engine Operating Parameters

Turbine engine output may be measured and expressed in different ways. In jets *engine pressure ratio* (*EPR,* usually pronounced "eeper") refers to the ratio of engine output pressure to engine intake pressure. Therefore, the greater the EPR, the greater the thrust being produced. (When engine intake pressure and output pressure are the same, meaning EPR = 1, the engine is producing no thrust.)

While EPR is commonly used to set thrust for turbojet and turbofan engines, turboprop power is commonly set and monitored in foot-pounds of torque (or sometimes percentage of maximum torque) delivered to the propeller drive-train.

Other common parameters for turbine engine operation are expressed in terms of rpm's of the various shafts within the engine. As you remember, N_1 normally refers to low-pressure compressor shaft rpm, while N_2 refers to high-pressure power shaft rpm. N_1 and N_2 are often expressed as percentages of a value near maximum rpm, since they represent very large numbers. (However, maximum rpm is not always 100 percent.) N_1 is the primary thrust-setting instrument on some business jet turbofans, rather than EPR (see Fig. 3.12).

As with reciprocating engines, temperatures must also be monitored on turbine powerplants. Depending on engine type and installation, either *EGT* (*exhaust gas temperature*) or *ITT* (*interstage turbine temperature*) is commonly used for this purpose. Fuel flow gauges also provide useful control parameters in turbine engines (see Fig. 3.13).

Turbine Engine Controls in the Cockpit

Thrust or power management is extremely simple for turbine engines. Normally, the primary engine controls on

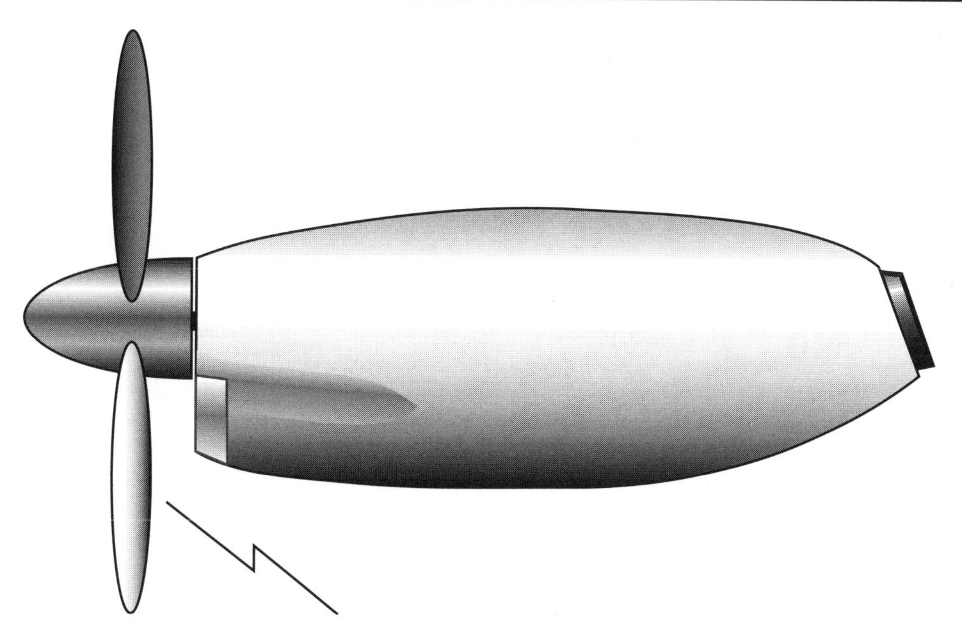

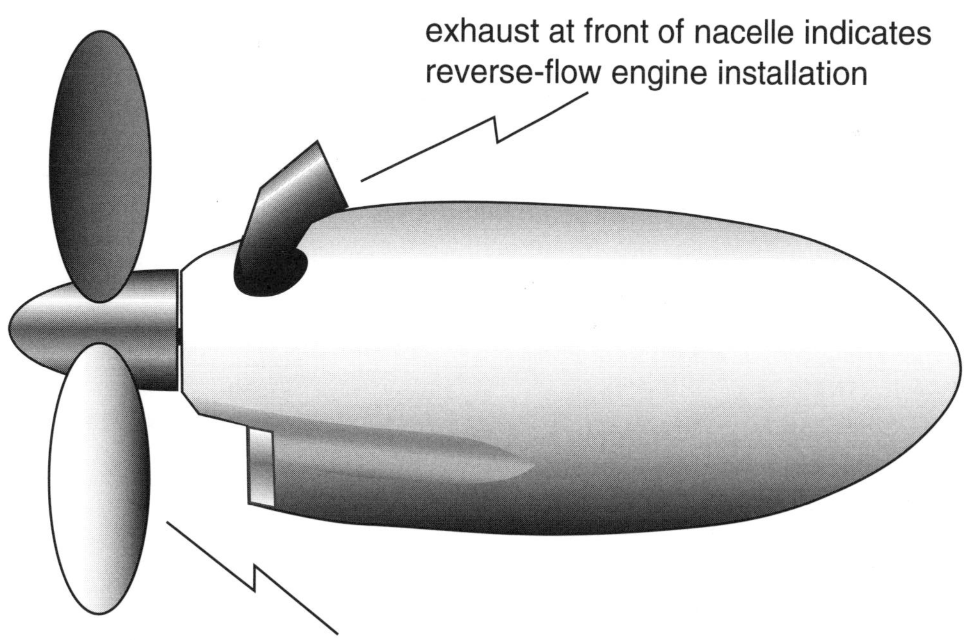

3.11. Differentiating direct-drive from free-turbine turboprop engines after shutdown.

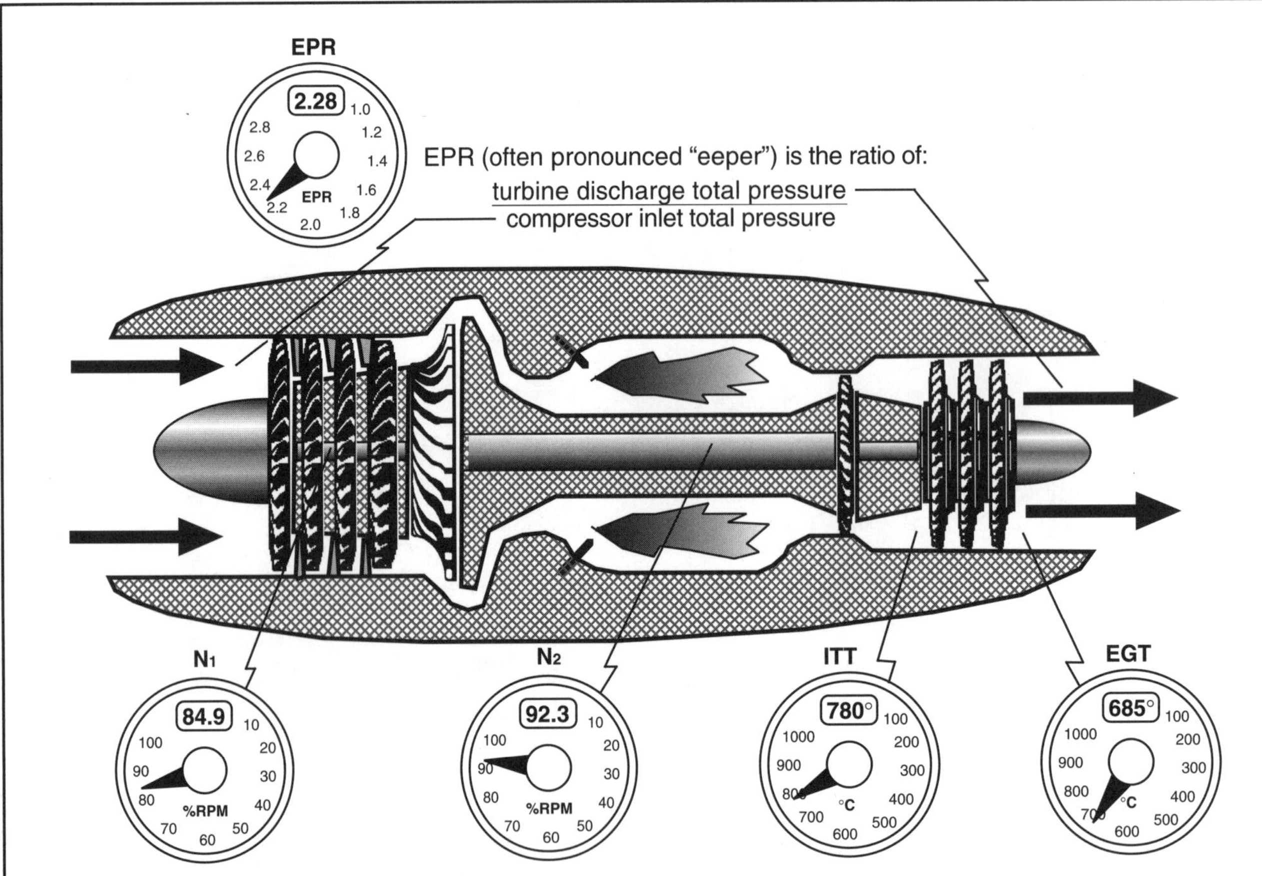

N1: low-pressure shaft rpm. N2: high-pressure shaft rpm. interstage turbine temp. exhaust gas temp.

EPR or N1 indicators are commonly used for setting thrust on jet aircraft. Fuel flow is another measure sometimes used for this purpose. In some aircraft several of these indicators may be used during different phases of flight. For example, while EPR may be calculated and set for takeoff power, fuel flow may be more convenient for targeting airspeeds during approach. Regardless of the measure, thrust is always set using the thrust levers.

3.12. Jet engine powerplant instruments.

jets consist simply of *thrust levers*. A set of *start levers* is used to turn fuel flow to the engines on and off (see Fig. 3.14).

Power levers, or *torque levers,* replace throttles in turboprops, while turboprop fuel levers are commonly called *condition levers*. Most turboprops are equipped with a separate *propeller rpm* control (not unlike those found on reciprocating engines), but in some cases prop control is integrated into the power levers (Saab 2000) or into the condition levers (Dash-8) (see Fig. 3.15).

Turbine Engine Starting

You can see that a turbine engine operates in a beautifully simple manner. Once running, the system is unlikely to stop unless fuel or air supply is interrupted or there's a mechanical failure. Starting, however, is a bit trickier.

You may have wondered at some point, what makes the expanding gases blow out of the proper end of the engine. This is not a silly question. Enough air must be compressed

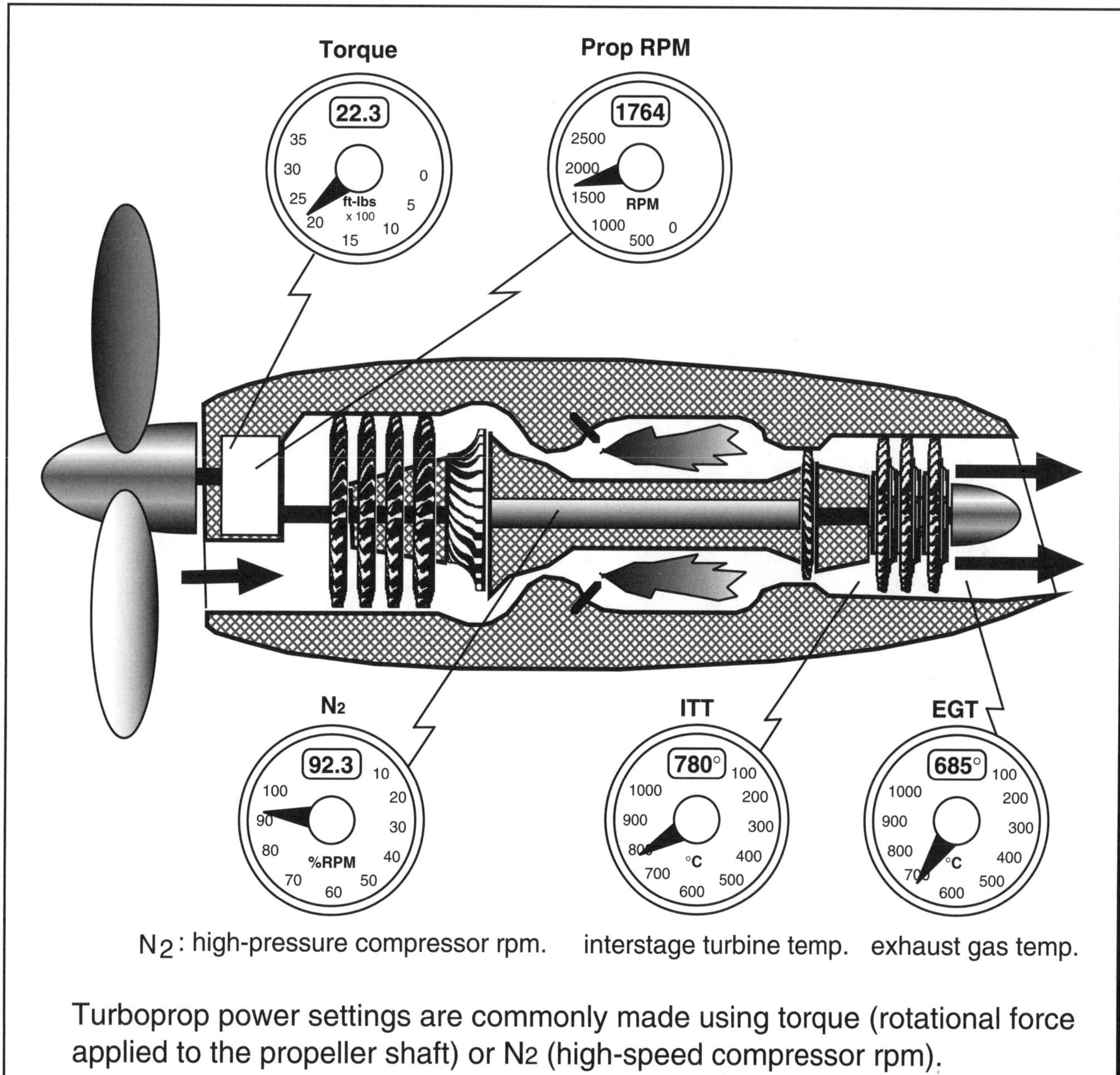

3.13. Turboprop engine powerplant instruments.

into the combustion chamber(s) so that, when fuel is added and ignited, the energized gases from combustion travel aft across the turbine section instead of forward.

Think of the compressed wall of air going into the combustion chamber as sort of an invisible cylinder head and the turbine section as a piston from a reciprocating engine. You can imagine that when you begin the starting sequence of a gas turbine engine it is very important to rotate the compressor section fast enough to build up the wall of air at the forward portion of the combustion chamber. This guarantees that the flow of combustion gases will travel aft once the fuel-air mixture is ignited and that compression will be high enough for a good start.

When these conditions are not met, there's danger of a *hot start.* A hot start occurs when fuel is introduced while compressor rpm is too low, and therefore pressures are

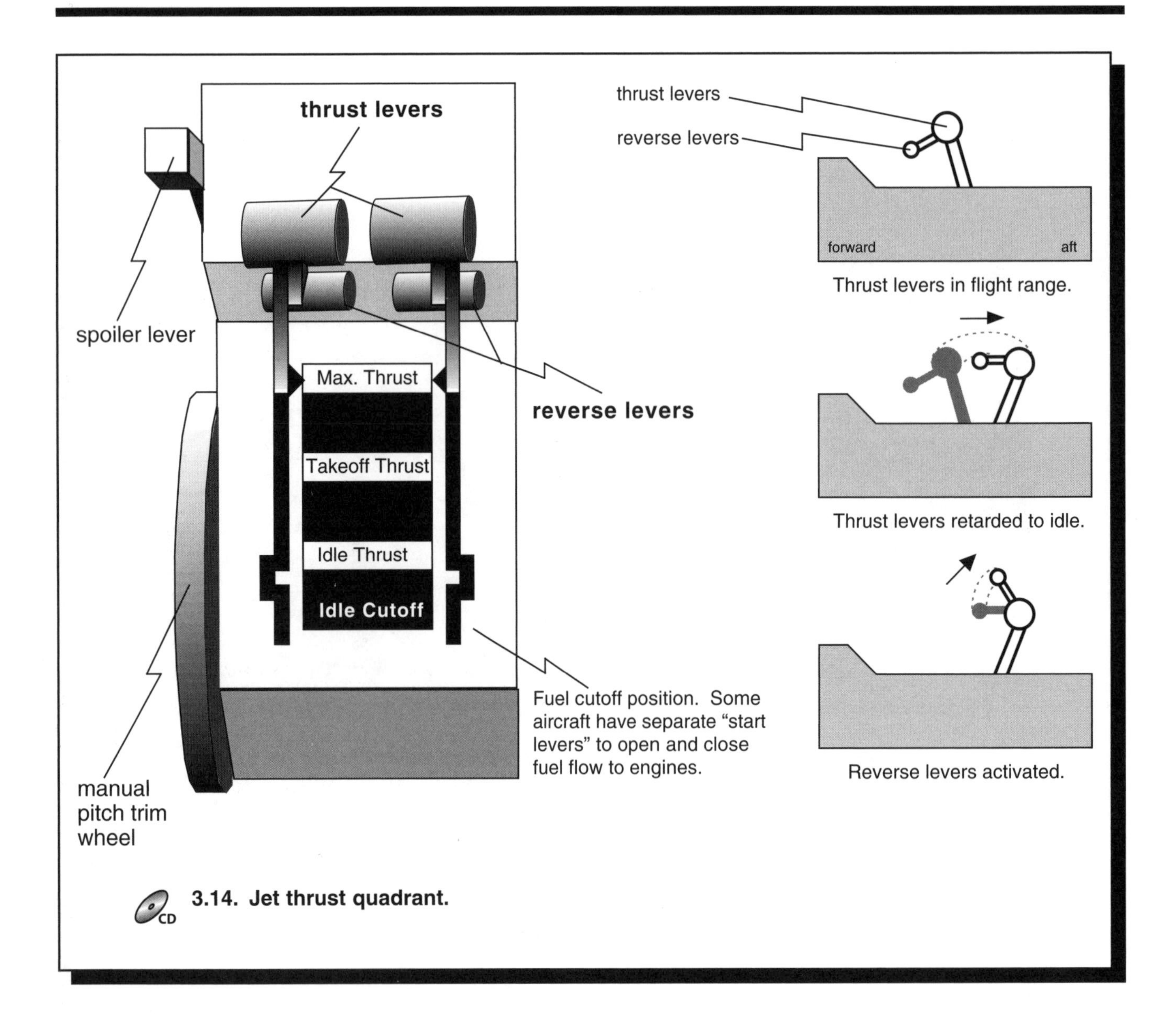

3.14. Jet thrust quadrant.

inadequate in the combustion chamber. In this situation, the fuel burns at very high temperatures in the combustion chamber, with little or no flow aft into the turbine section. If the pilot doesn't recognize this situation immediately through careful monitoring, the engine can overheat and be ruined in seconds. Talk about a bad day!

On small gas turbine engines, the compressor is normally rotated for starting by an electric starter motor. For large turbine engines, high-pressure air drives a pneumatic starter motor. This high-pressure air can come from several sources, such as an APU (auxiliary power unit), ground air source, or pneumatic air cross-bleed from another engine. (More on those in later chapters.) The SR-71 uses a pair of high-performance automotive V-8s to spin up each of its huge engines for starting.

Gas turbine engines use *igniters* to light the fuel-air mixture during starting. Some igniters are glow plugs, similar to those used for the same purpose in diesel engines. (These incorporate a resistance coil through which electric current flows to heat them until they glow steadily.) Other igniters are much like the spark plugs found in reciprocating engines, discharging a steady rhythm of sparks to ignite the mixture.

Once the engine's fuel-air mixture is ignited, the igniters can be turned off. Like a blowtorch, combustion in a turbine engine is maintained simply by injecting fuel continuously into the burning gases of the combustion chamber. The throttle can directly or indirectly control that amount of fuel (depending on the engine manufacturer). The energized gases from combustion then travel aft to turn the power turbines, as previously discussed (see Fig. 3.16).

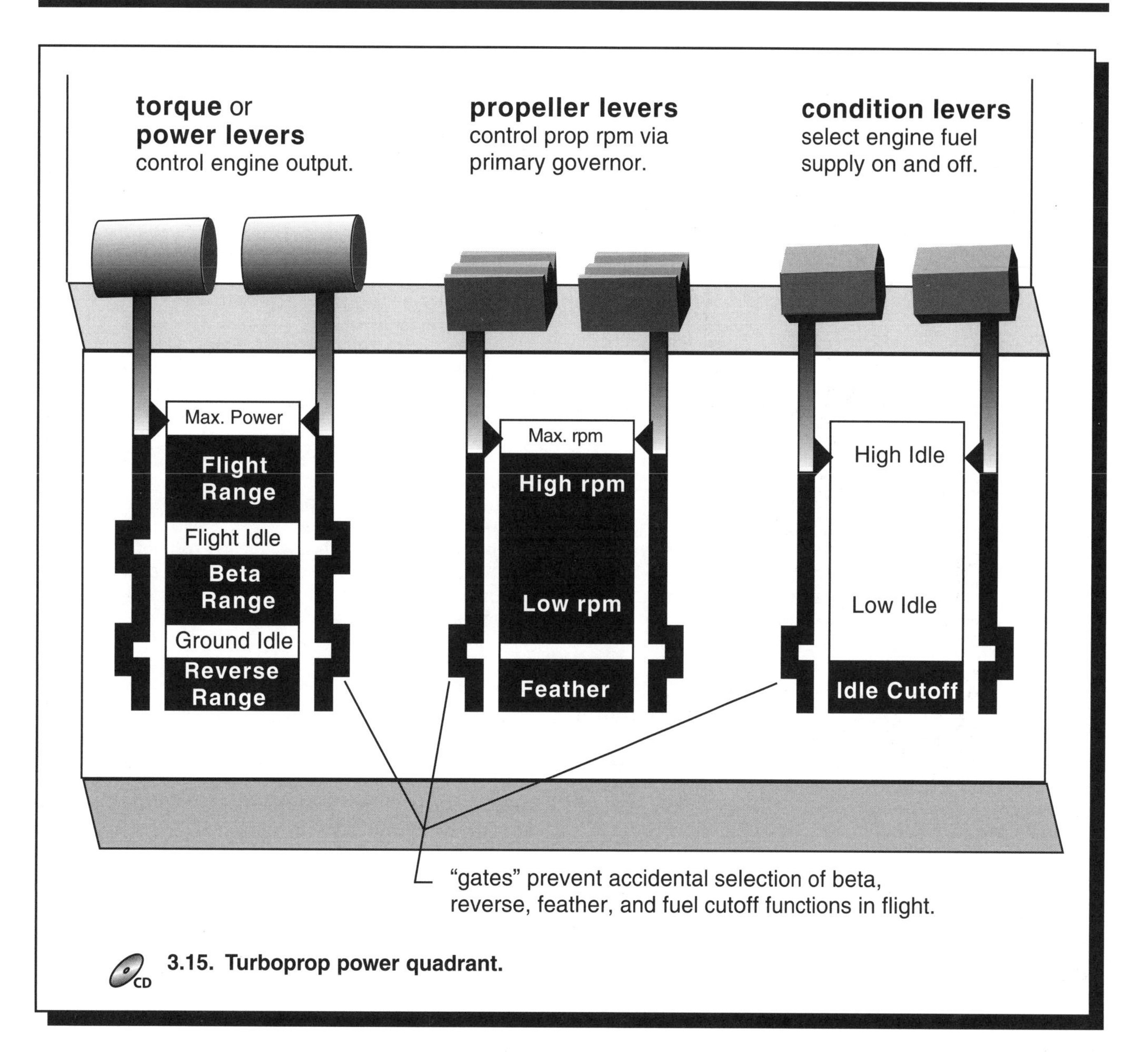

3.15. Turboprop power quadrant.

Turbine Engine Characteristics in Flight

Turbine engines are not particularly sensitive to power changes in the air (especially as compared with turbocharged reciprocating engines) and therefore are extremely easy to manage in flight. While maximum EGT or ITT has to be monitored under certain conditions, you can readily and abruptly change thrust or power in turbine aircraft (say, from cruise to idle) with no adverse effects. Additionally, since the fuel control unit is directly controlled via the power or thrust levers, no "leaning" or "mixture" control of any kind is required.

Power management of turbine engines in flight does vary from that of piston aircraft in several important ways. First, turbine engines take time to *spool up* when power is applied. If left at low power, engine compressor speed is slow. The term *spool up* refers to the time it takes, after power application, for thrust to increase from idle to the selected values. Most of a turbine engine's thrust is developed in the top 10 percent or so of the engine rpm (N_1) range.

Let's say that a jet is coasting along at flight idle (60 percent to 65 percent N_1) when the pilot calls for max thrust. It can take quite a while for the engines to spool up to the 90 percent N_1 range, where most thrust is generated. While adding full throttle in a piston aircraft may result in maximum power within a few seconds, turboprops take a bit longer, and maximum thrust in jet aircraft may not result until five to ten seconds after power application.

Start sequence: electric start

1. Complete Before-Start Checklist.
2. Check battery condition. (For battery starts, 22V minimum voltage must be available to preclude "hot start.")
3. Activate electric starter (or starter/generator) and igniters.
4. Starter spins compressor to specified rpm value.
5. Fuel is introduced and secondary igniters activated.
6. Monitor ITT, rpm, and electrical load as they increase. If any value approaches maximum too rapidly, shut off fuel and abort the start. (Severe engine damage can result from a "hot start.")
7. Electric starter is shut off at predetermined rpm.
8. As temperatures stabilize, reduce power to idle.
9. Complete After-Start Checklist.

Electric starting is common on smaller turbine engines, usually using combined-function electric "starter/generators." It is preferable to use a ground power source or APU for starting, when available. Battery starts require healthy, fully-charged batteries for "cool starts." Repeated battery starts are hard on both batteries and engines.

Start sequence: pneumatic start

1. Complete Before-Start Checklist.
2. Open pneumatic start valve.
3. Pneumatic starter spins compressor to specified rpm.
4. Igniters are activated and fuel introduced.
5. Monitor EGT and rpm as they increase. If either approaches maximum value too rapidly, shut off fuel and abort the start. (Severe engine damage can result from a "hot start.")
6. Start valve closes.
7. As temperatures stabilize, reduce power to idle.
8. Complete After-Start Checklist.

Large turbine engines are started using high-pressure pneumatic power to spin up the compressors. Pneumatic power may be drawn from a ground pneumatic source, from the aircraft's APU, or from another engine that is already running. Pneumatic "start valve" operation must be carefully monitored to ensure that the start valve closes once the engine's compressor is spun up.

Turbine engine starting requires a rather long sequence of events that pilots must carefully monitor. On older aircraft most steps must be accomplished manually, with careful timing. Newer systems automate much of the start sequence, but careful pilot monitoring is always required. Among the most serious malfunctions during starting is the "hot start." This refers to a situation where fuel accumulates in the combustion chamber at an excessive rate, resulting in engine temperatures that can severely damage internal parts. Pilots are trained to monitor engine temperatures and rpm trends, in order to abort starts early when indicated.

3.16. Typical turbine engine start sequences.

For these reasons, being *low and unspooled* in turbine aircraft on approach to landing is absolutely unacceptable, due to the time it takes for the engines to spool up for go-around. (On at least one old foreign military turbojet, spool up time from idle to full power takes twelve seconds. Modern jet engines are required to spool up in five seconds or less.)

Because response to thrust changes can be so slow in pure jet aircraft, power management in these vehicles requires plenty of planning. For example, on final approach in jets all of those high-lift, high-drag devices (flaps and leading edge devices) are extended to slow the aircraft enough so that relatively high thrust must be carried all the way to landing. This aids in go-around situations because, since thrust is already high, spool up time is minimized. Cleaning up the airplane turns out to be faster than waiting for spool up, especially on older turbojet aircraft.

Another challenge arises in jets (less so in turboprops) because turbine engines offer relatively little drag when thrust is reduced to *flight idle.* While the drag of propeller discs will quickly slow a prop aircraft with engines at idle, pure jets tend to just coast. (Glide ratios can exceed 20:1 in modern jetliners.)

That means thrust management to meet descent restrictions in jets can also be challenging, due to the low-drag characteristics of such aircraft. For example, it's difficult in many jets to maintain the 250-knot speed limit below 10,000 feet and to descend at the same time more than 1500 or so fpm. This means that jet pilots must plan ahead on descents and in some cases negotiate with ATC. ("We can hold our airspeed to 280 knots or make our 13,000 altitude crossing restriction, but not both.")

Thrust Reversers

For purposes of both safety and airport size, it's important to stop landing aircraft in the shortest distances possible. Therefore, many turbine-powered aircraft are equipped with thrust reversers to enhance deceleration upon landing. Typically, a little less than 50 percent of an engine's rated takeoff thrust is available in reverse thrust.

Thrust reversers don't impact minimum landing distance as much as you might think. A given aircraft might land and stop in 4000 feet with reversers versus 4500 feet without. In any case, manufacturers are not allowed to incorporate the effects of thrust reversers when calculating aircraft runway length performance. (One reason is that the aircraft must be able to stop within the specified distance in the event of a failed engine, when little or no reverse thrust may be available.)

So why install thrust reversers? They offer a number of valuable benefits. First, it turns out that calculated minimum landing distance for a given aircraft is based on maximum braking. No one ever lands that way, except under extreme conditions. Therefore, the effects of reversers for shortening *routine* landings are significant.

Thrust reversers also tremendously reduce the energy absorbed by an aircraft's brakes. Severe brake damage or even brake fires can result from a maximum energy *rejected takeoff* (*RTO*); thrust reversers reduce that likelihood by taking some of the load off the brakes. Brake and tire wear under routine use is also reduced through use of thrust reversers, thereby lengthening brake and tire life. Proper use of reversers also keeps brakes cool early in the touchdown roll, reducing brake fade and thereby improving brake performance on rollout.

Reversers also improve safety by increasing runway length margins and by imparting an extra measure of control on wet, snow-covered, or icy runways. Under normal runway conditions around 50 percent of an aircraft's stopping power comes from thrust reversers. But under poor runway conditions, such as on wet or icy surfaces, thrust reversers may provide close to 80 percent of the aircraft's stopping power.

Some operators also use thrust reversers to *power back* aircraft from terminal gates, although backing an aircraft under its own power is done only under strictly controlled and approved procedures.

The operating principle of a thrust reverser is simple: engine power is directed forward to slow the aircraft.

Thrust Reversers on Jets

On jets, engine thrust is mechanically directed forward to help bring the aircraft to a stop. Three types of thrust reversers are in common use on jet aircraft: clamshell, cascade-type, and petal door fan reversers. (Within these general types are numerous variations.) The equipment found on any given airplane varies by aircraft model and by installed engine type.

Clamshell reversers are simply metal doors that swing rapidly from the rear of the engine nacelle into the exhaust stream, vectoring thrust forward to slow the aircraft. Aircraft having engines with little or no bypass use this type of reverser, which is simple, effective, and relatively easy to install. Used on most corporate business jets and small to midsized airliners, clamshell reversers may be seen on Citations, Learjets, DC-9s, MD-80s, and 737-200s, to name a few (Fig. 3.17).

Cascade-type reversers are used on aircraft with high-bypass engines. This type of reverser utilizes metal doors to block bypass air, while a thrust reverser nacelle collar slides back to expose cascade vanes that vector thrust forward. While slightly less effective than clamshell reversers, cascade types do not protrude from the nacelles when activated; this reduces ground clearance problems on large underwing-mounted engines. Cascade-type reversers may be seen on 757s, 767s, and 737-300s and -500s (Fig. 3.17).

Petal door reversers are also used on high-bypass engines. They operate in much the same manner as the cascade type, except that pivoting metal doors are used to vector the thrust forward out of the engine nacelle. Petal door reversers may be found on some older 747s.

On jets, thrust reverser cockpit controls are normally installed as secondary, piggyback levers on the thrust levers. These controls are designed to prevent selection of reverse until forward thrust has been reduced to idle. After touchdown, the thrust levers are first pulled aft to "idle"; then the reverse levers are pulled up and aft. Reverse thrust, from reverse idle up to maximum thrust with reversers deployed, is controlled exclusively by the reverse levers. (See Figure 3.14.) As the reverse levers move aft, engine thrust is increased in the reverse thrust configuration.

Reverse Thrust on Turboprops

On turboprops, reverse thrust is accomplished in a relatively simple manner. After touchdown, pilots select reverse thrust by lifting the power levers past a "gate" or detent aft of "idle." (See Figure 3.15.) As the power levers are moved aft, propeller blade angle is gradually changed from positive (flight range) blade angles to negative (reverse range) blade angles. The effect is to first increase drag by "flattening" the propeller blades, then to direct engine thrust forward through negative propeller blade angles of attack. As the power levers move aft in the reverse range, propeller blades are moved to fixed mechanical stops and effectively become a fixed pitch propeller blade

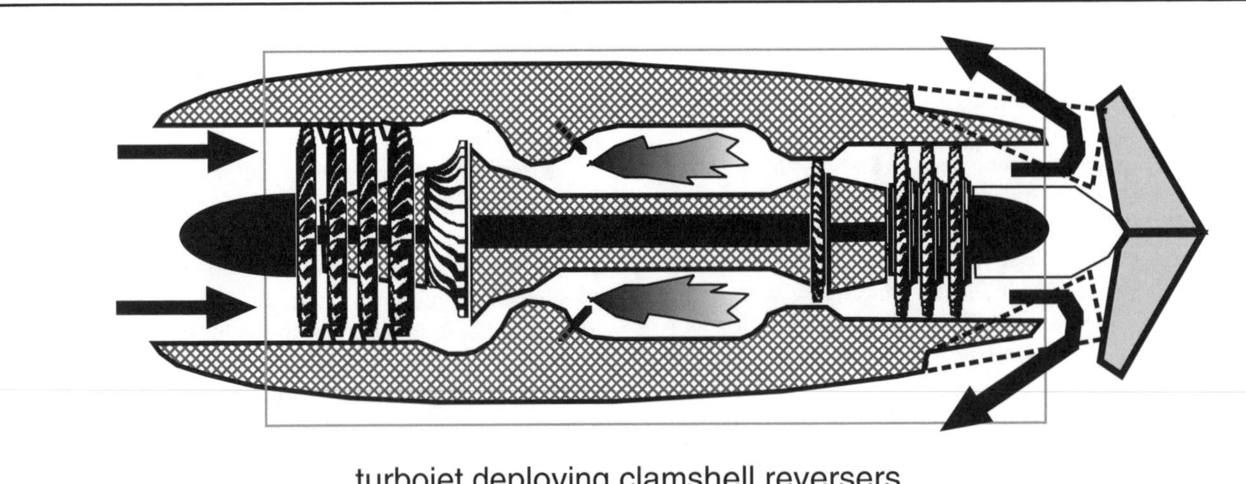

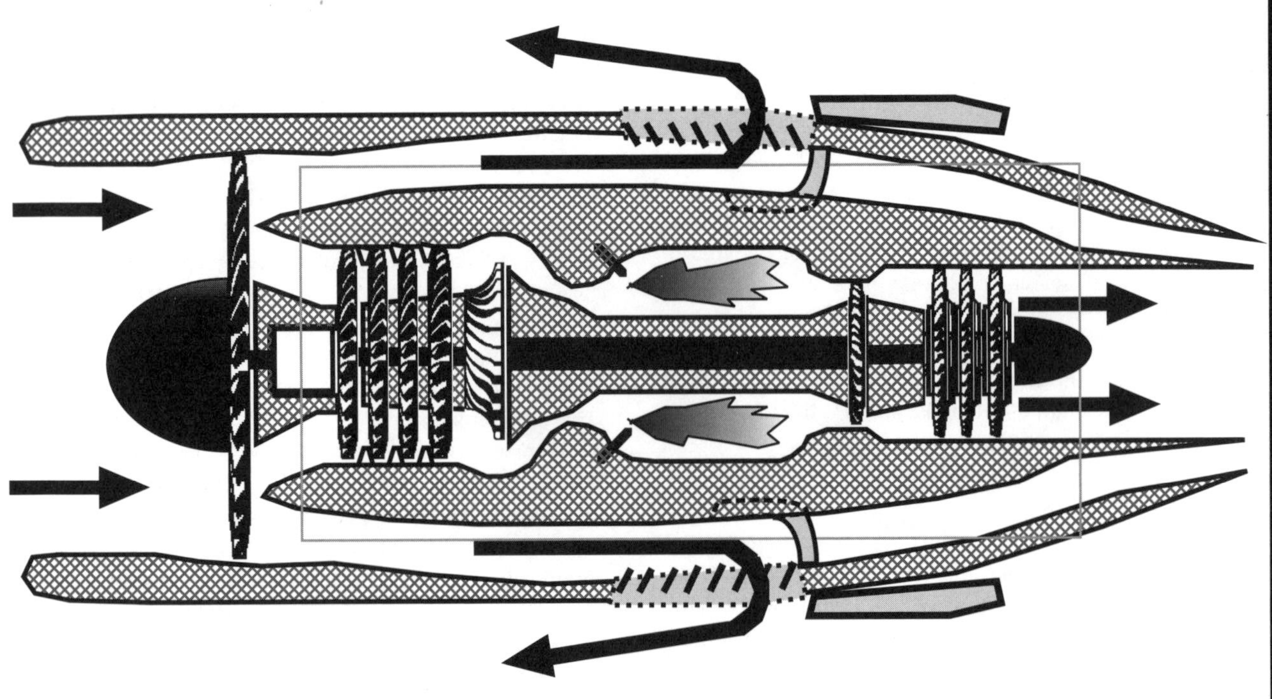

3.17. Two types of jet thrust reversers.

system. Continued aft movement of the power levers increases engine power, and therefore reverse thrust increases. (See also the discussion of turboprop propeller systems in this chapter.)

Use of Reversers

On some aircraft, it would be disastrous (and has been) if one or more thrust reversers were to activate in flight. Many aircraft are equipped with interlock systems (usually tied to landing gear squat switches and the flight idle power or thrust lever position, among other conditions) to prevent inadvertent activation anywhere but on the ground.

Crew training is also important; pilots must learn not to activate reversers too early in the landing sequence. On DC-9s, for example, it's possible to deploy reversers at any time after the thrust levers have been brought to idle. Crews must be thoroughly trained not to activate reversers until nose gear touchdown, especially since reverser deployment shifts weight onto the nose gear.

A few aircraft are certified for thrust reverser use in the air. On some DC-8s, for example, reversers may be used in flight as speed brakes, due to the aircraft's extremely clean (and, therefore, hard to slow) design.

Turboprop Propeller Systems

Turboprop propeller systems are closely related to those found in piston aircraft but have some major differences. Readers with prop flying backgrounds should be ready to dive right in. However, many of our transitioning military readers may have no experience with propeller-driven aircraft. Following this section is "Propeller Supplement for Transitioning Military Jet Pilots." We recommend that pilots having only pure jet experience who expect to operate turboprops, read that section first and then return to this one.

As with constant-speed propellers found on piston aircraft, propeller blade angle on a turboprop engine is varied to maintain constant rpm in flight. Propeller blade angle is controlled by oil pressure. High oil pressure directed to the prop hub pushes the propeller to a low blade angle and therefore high rpm. Reduced oil pressure increases prop blade angle and therefore decreases rpm (see Fig. 3.18).

This relationship is easy (and important) to remember. Think of it this way. Loss of oil pressure is often associated with engine failure. In the event of lost oil pressure due to engine failure, you certainly want the propellers to feather in order to avoid the drag of windmilling props on a dead engine. That's exactly how the system is designed to work. If oil pressure is lost completely, the propeller is pushed by springs and/or compressed nitrogen to the feathered position. The rest of the time engine oil pressure works against the spring system to drive the propeller to the flatter pitch and higher rpm settings required to produce thrust.

Propeller Governors

To accomplish constant-speed operation in flight, the turboprop uses a series of three propeller governors: the primary governor, the overspeed governor, and the fuel-topping governor.

The *primary governor* operates just like the constant-speed version on a piston aircraft. It controls the amount of oil pressure supplied to the prop hub via a valve regulated by spinning flyweights. If prop rpm slows, so do the flyweights. Reduced rotational velocity causes the flyweights to be drawn inward by balancing springs. This, in turn, slightly opens the oil valve to the hub, allowing more pressure to enter the hub. The result is reduced propeller pitch to hold a constant rpm. The primary governor is controlled at the power quadrant by the prop levers. When someone refers to the "governor" on a turboprop engine, it is the primary governor being discussed (see Fig. 3.19).

The *overspeed governor,* as the name implies, protects the engine from propeller overspeed. If the primary governor fails, the overspeed governor prevents excessive propeller speed by increasing prop blade angle automatically at a certain rpm. The overspeed governor is generally a simple relief valve, which dumps oil pressure from the prop hub if rpm exceeds a maximum value.

The *fuel-topping governor* also protects the propeller from an overspeed condition. It does so by reducing the amount of fuel flow to the engine. This reduces engine power and, as a result, propeller rpm. Note that while the overspeed governor simply prevents overspeed at the props' maximum limits, the fuel-topping governor is designed to limit prop rpm to a few percentage points over whatever has been selected by the pilot at the prop levers.

Beta Range

An important concept to understand regarding prop control on turboprops is that of beta range. Basically, in the flight range from maximum power back to flight idle, power levers control engine torque (power), and prop levers control prop pitch and rpm. Similar to piston-powered aircraft, turboprop propellers are advanced for *takeoff* and *climb* and on final approach in order to prepare for the possibility of go-arounds.

Once on the ground, however, it's desirable to manipulate the props further toward flat pitch for slowing and taxiing (this range is known as *ground range* or *beta range*) and then into reverse to shorten landing roll. Control of the props in these regimes is via a *beta valve,* which bypasses the primary prop governor and is controlled by the power levers.

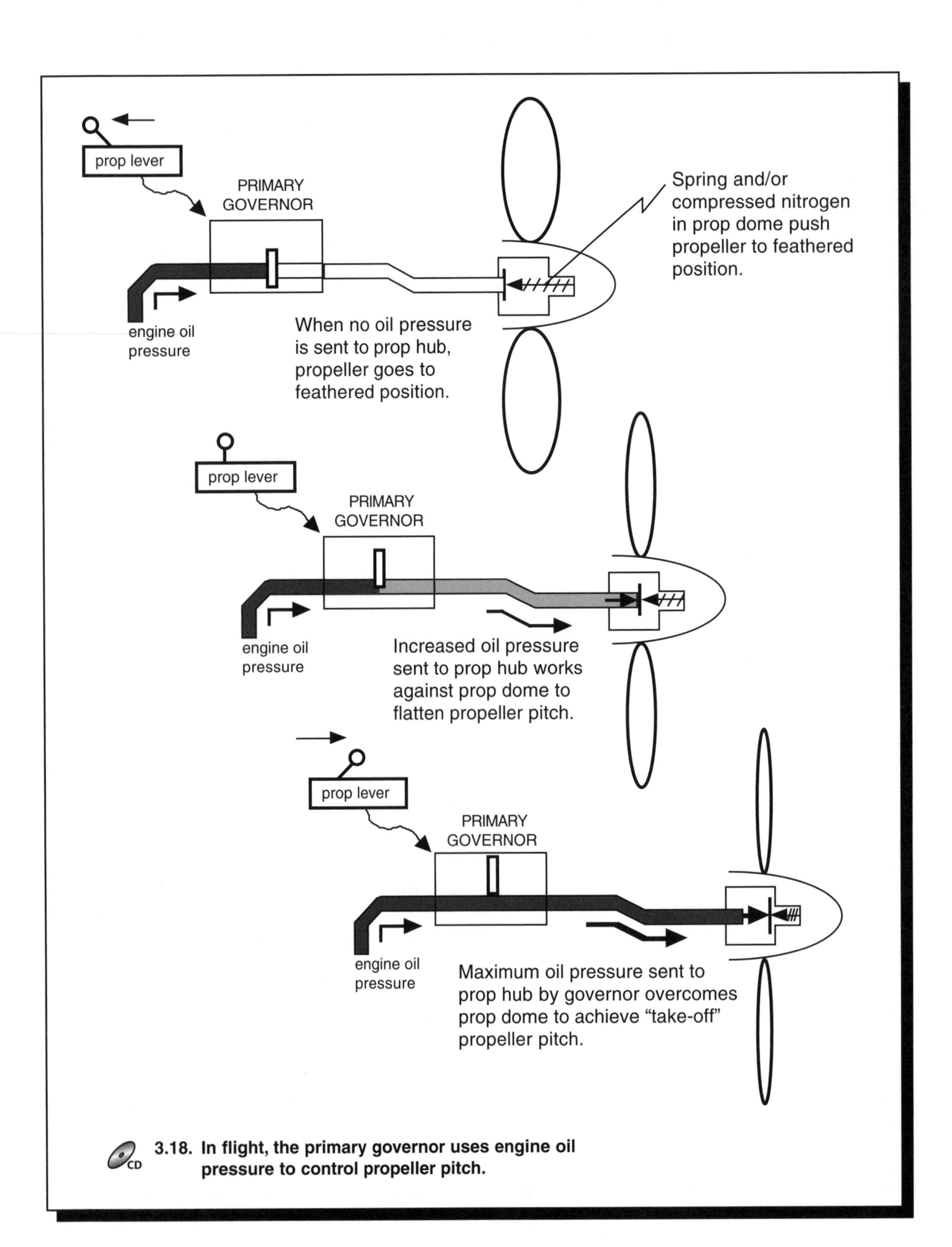

3.18. In flight, the primary governor uses engine oil pressure to control propeller pitch.

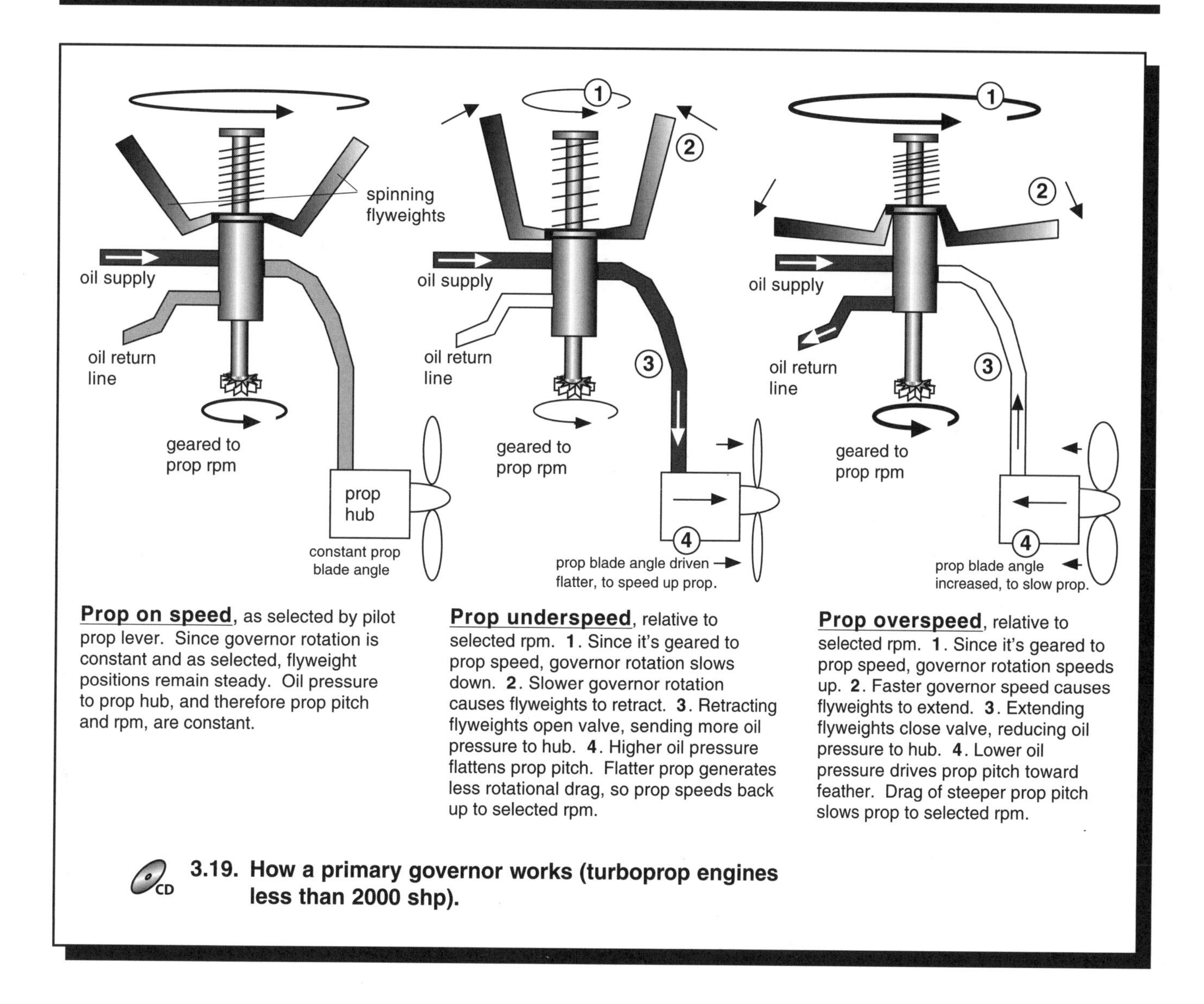

3.19. How a primary governor works (turboprop engines less than 2000 shp).

Think of it this way. In flight, the power levers control only engine torque. In beta range, the power levers control only prop pitch (prop levers remain full forward, and torque remains constant at idle). This, of course, is contrary to the function of the primary prop governors and is why the beta valve exists to bypass that system. In reverse range, the power levers control both torque and props. This allows power to increase again for slowing the aircraft on the ground, while the props are in reverse pitch.

You may wonder why beta range is important for taxiing on turboprop aircraft. Piston engines idle at relatively low rpm, so they don't produce much thrust at ground power settings. However, as you remember, turboprops idle at high rpms, especially direct-drive models. Even at "flight idle" power settings, the props still generate lots of thrust. Beta range allows prop pitch to be reduced to zero-thrust settings, thereby making taxiing easier and lengthening brake life (see Figs. 3.20 and 3.21).

Propeller Auto-Feather Systems

As you remember from your previous multiengine flying, it's critically important to immediately feather the propeller of a failed engine, in order to maintain directional control and any semblance of climb performance. Turboprops are faced with the same problem, further aggravated by the use of very large-diameter multibladed props. Propeller diameters of 12 feet or more are not uncommon on turboprops nor are installations of four to six propeller blades. You can imagine the tremendous drag generated by these huge propeller disks, in the event of engine failure. To address this problem, turboprops are often equipped with

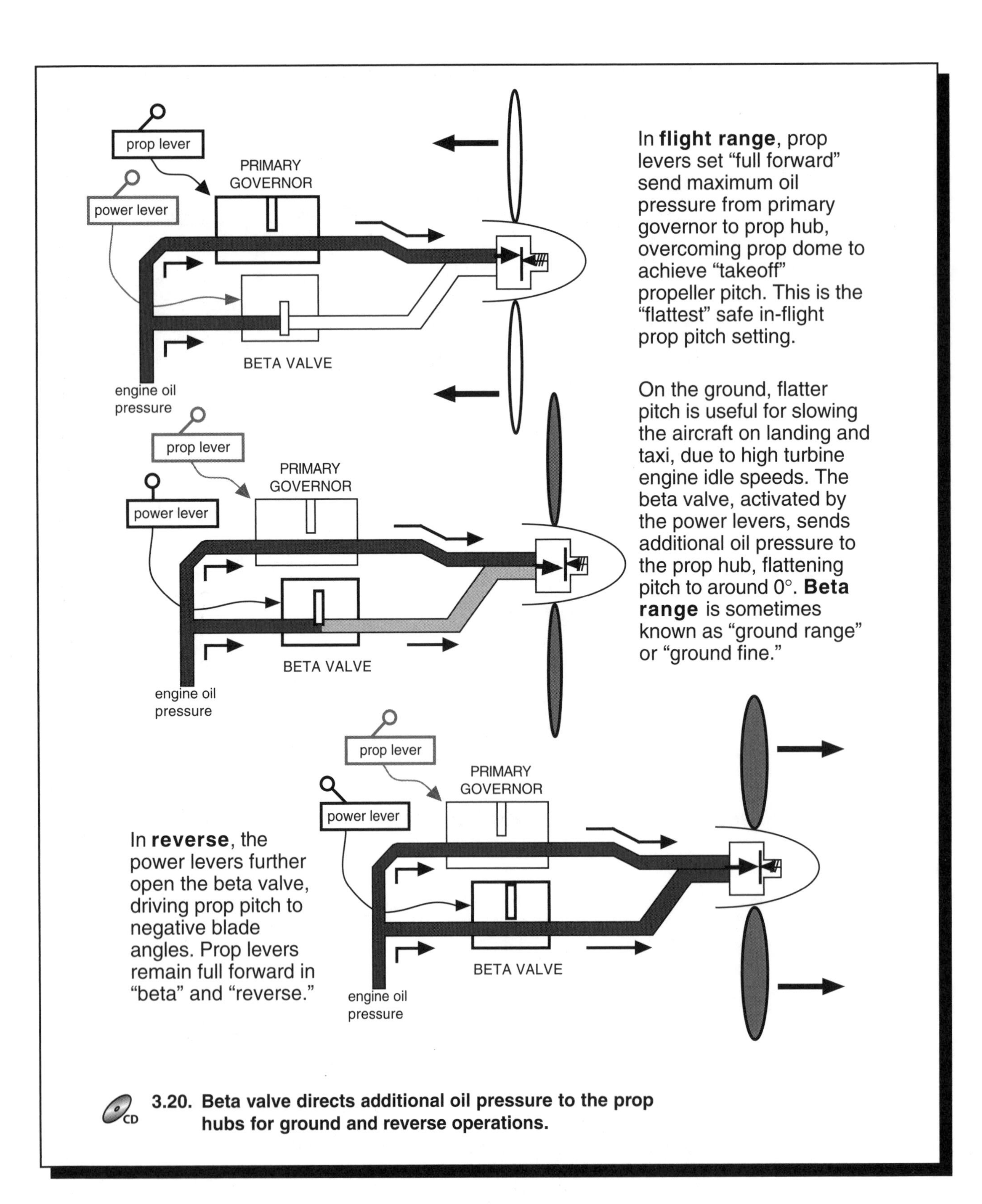

3.20. Beta valve directs additional oil pressure to the prop hubs for ground and reverse operations.

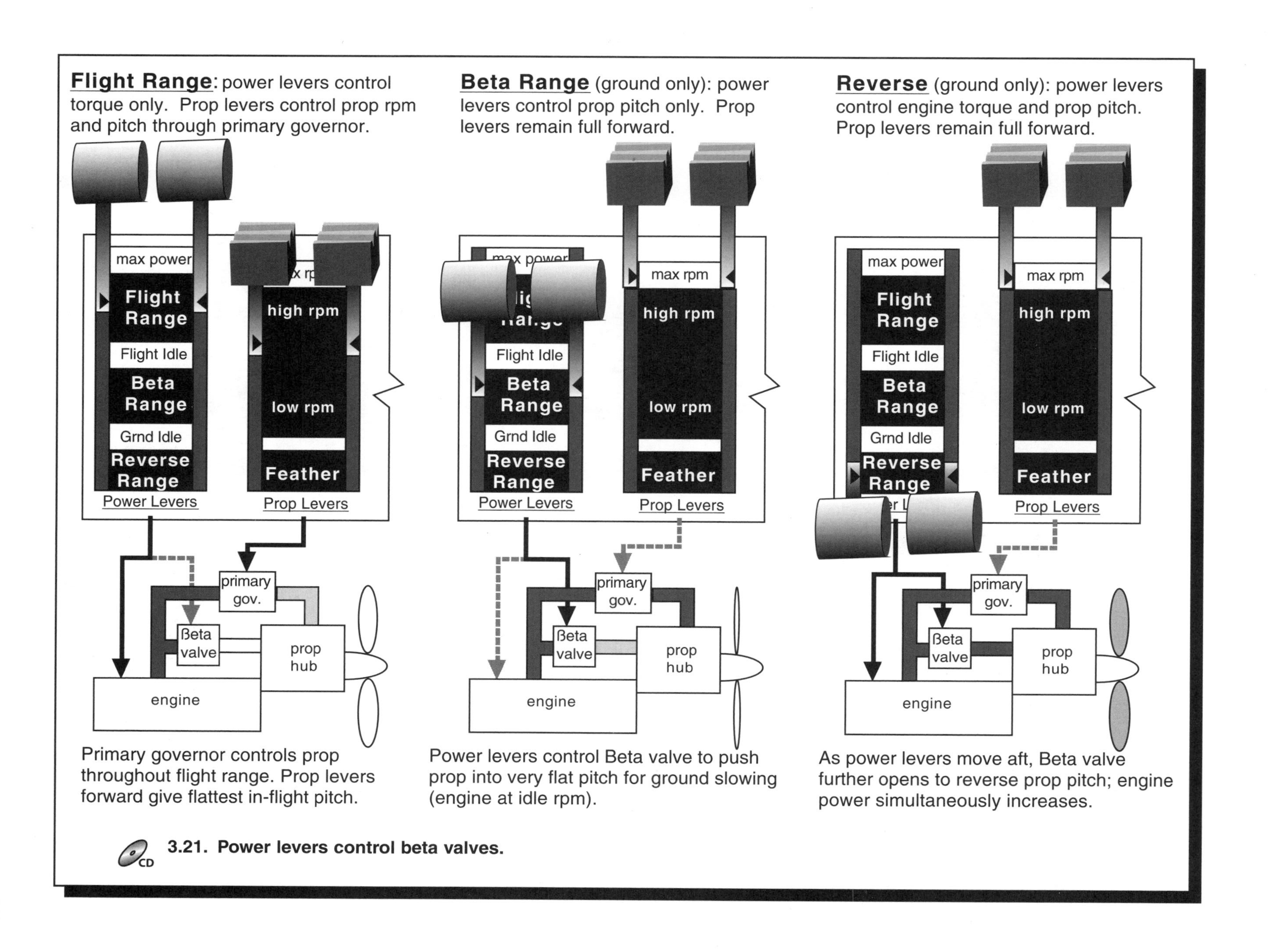

3.21. Power levers control beta valves.

propeller *auto-feather* systems, which sense power loss and automatically feather the propeller of a failed engine.

Propeller Synchronizers and Synchrophasers

As with many multi-engine piston airplanes, multi-engine turboprops are equipped with *propeller synchronization* ("prop synch") systems. These synchronization systems compare and control propeller rpms for purposes of noise reduction. Synchronized propeller operation reduces propeller-induced vibration and eliminates the unpleasant sound of "prop beat" produced by unsynchronized propellers.

There are several types of propeller synchronizing systems, but they fall into two basic categories: the *propeller synchronizer,* which matches propeller rpm between multiple engines, and the *propeller synchrophaser,* which in addition to matching propeller rpm also matches propeller blade positions between engines for maximum noise reduction.

The "brains" of a propeller synchronizer or synchrophaser system is the control unit. The control unit compares electronic signals generated from two different magnetic "pick-ups," one mounted on a stationary location, such as the propeller gear box assembly, and one mounted on the rotating propeller hub backplate.

To control prop rpm, the prop synch control unit signals magnetic synchronization coils mounted on the propeller governor flyweights, thereby magnetically controlling them for precise positioning. In this way, whenever rpm or phase corrections are required, the governor is guided to adjust propeller rpm in a normal manner, by increasing prop blade angle to reduce rpm or decreasing prop blade angle to increase rpm.

In-flight operation of prop synch systems is simply a matter of manually synchronizing the propellers within a predetermined rpm range and selecting prop synchro "ON." The control unit will then match prop rpms between engines, greatly reducing propeller noise levels.

Some prop synch systems require the pilot to manually synchronize the propellers to within a few rpms of each other and then manually select propeller blade phase to the quietest blade position. Other units automatically adjust blade phase position to a predetermined setting that is identified as the quietest during flight testing (see Fig. 3.22).

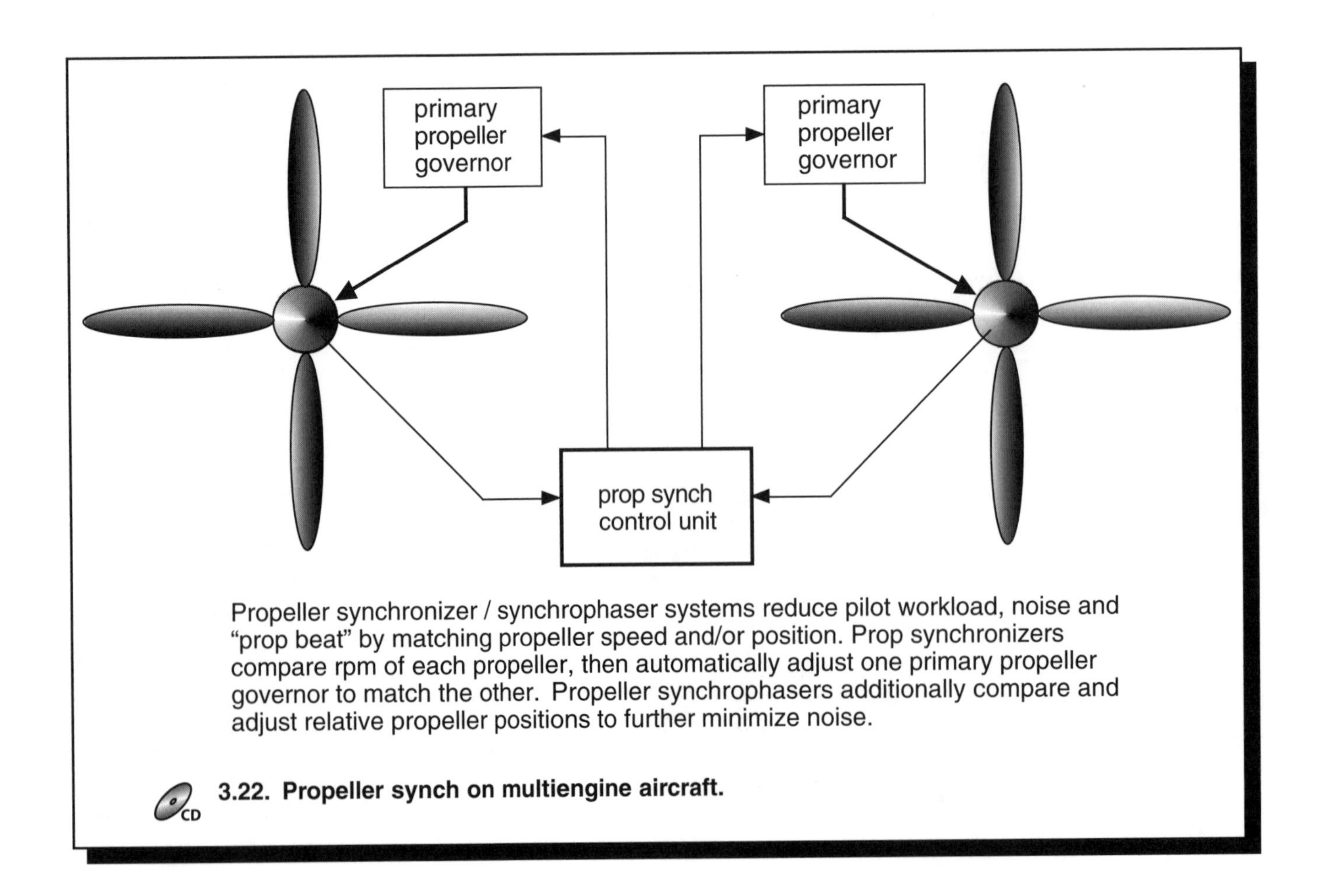

Propeller synchronizer / synchrophaser systems reduce pilot workload, noise and "prop beat" by matching propeller speed and/or position. Prop synchronizers compare rpm of each propeller, then automatically adjust one primary propeller governor to match the other. Propeller synchrophasers additionally compare and adjust relative propeller positions to further minimize noise.

CD **3.22. Propeller synch on multiengine aircraft.**

Propeller Supplement for Transitioning Military Jet Pilots

Whether in fighters, bombers, or tanker/transport aircraft, most recent military pilots have spent their careers piloting pure jet aircraft. Therefore, if that's your background basic propeller theory is probably not familiar to you. Like it or not, the odds are still good these days that your first civilian flying job will be in a turboprop aircraft.

Familiarity with basic propeller characteristics is important, especially in engine-out operations, because these characteristics affect both handling and performance of the aircraft. For this reason, we review some basic propeller theory for those who need it. Pilots should be familiar with these concepts, along with principles of turboprop propeller systems (already discussed; see the preceding), before entering ground school for a turboprop aircraft.

The good news is that propeller systems, although somewhat complex mechanically, are relatively easy to use. To prepare for turboprop training, pilots with only jet experience would benefit by taking even an hour of instruction in any aircraft having a constant-speed propeller. (It need not be a turboprop; piston prop systems are operated in a similar manner.)

Propeller Terminology

Propeller Pitch

Propeller pitch refers to the blade angle of a propeller relative to its plane of rotation. *High, steep,* or *coarse pitch* refers to a large blade angle. *Low, flat,* or *fine pitch* refers to a small blade angle. Flatter pitch is used for takeoff and climb, where acceleration and climb performance are improved by higher rpm and faster acceleration of air passing through the propeller. Steeper pitch is more efficient for cruise flight, where a larger blade angle results in a bigger "propeller bite" and, therefore, power transfer to a larger volume of air at a lower rpm. *Reverse pitch* refers to negative blade angles, meaning those that direct thrust forward to slow the plane in reverse range (see Fig. 3.23).

The *constant-speed propellers* found on turboprops are variable in pitch to provide the performance and efficiency benefits of both high- and low-pitch propellers. Propeller rpm is set by the pilots and then kept at constant rpm by a *propeller governor.* The governor varies blade angle, which in turn impacts rotational drag of the propeller to keep rpm constant. When rpm increases above the value for which a governor is set, the governor turns the propeller blades to a higher angle, thereby increasing load on the engine and reducing rpm. Conversely, when propeller rpm decreases below the value for which the governor is set, the governor decreases propeller blade angle, thereby decreasing propeller load on the engine and increasing rpm.

Windmilling prop refers to a propeller that is at flight pitch settings but is developing no power. In this configuration, the spinning propeller disk creates tremendous drag, often enough to prevent the airplane from sustaining altitude on the remaining engine(s). On some aircraft, it is also difficult or impossible to maintain directional control with a windmilling propeller. *Propeller feathering* addresses this problem by setting propeller blade angle on the failed engine close to 90°, thereby minimizing prop drag. A feathered prop is critical to sustained flight of most propeller-driven aircraft after an engine quits.

Asymmetric Propeller Thrust

An important characteristic to understand, regarding propeller-driven aircraft of all types, is that the thrust produced by any one powerplant propeller is not symmetrical. While a jet engine produces power directly along its axis, aircraft propellers generate forces in both lateral and rotational directions that affect handling. These forces can cause both roll and yaw tendencies. While most pronounced in single-engine aircraft, these effects also arise in twins, especially during takeoff, climb, and engine-out situations. (You fighter jockeys are going to learn to use the rudder in propeller-driven aircraft.)

P-Factor

P-factor (or *asymmetric propeller thrust*) is most noticeable at high angles of attack and high power settings. During flight at high angles of attack, the propeller's plane of rotation is not perpendicular to the relative wind. This causes the descending blade of the propeller to have a greater angle of attack than the ascending blade and a higher relative velocity.

These combined effects give the descending blade significantly more thrust than the ascending blade, thereby offsetting the propeller's center of thrust to the descending blade's side of the propeller hub and causing a yawing moment on the aircraft, which pilots must counter through application of rudder during takeoff and climb (see Fig. 3.24).

In cruise flight the propeller's plane of rotation is nearly perpendicular to the relative wind and therefore thrust is almost symmetrical. On many aircraft, the engines and tail surfaces are installed asymmetrically in order to compensate for any offset propeller thrust in cruise, with the interesting result that rudder in the opposite direction is often required during descent to counter the effects of the asymmetrical configuration.

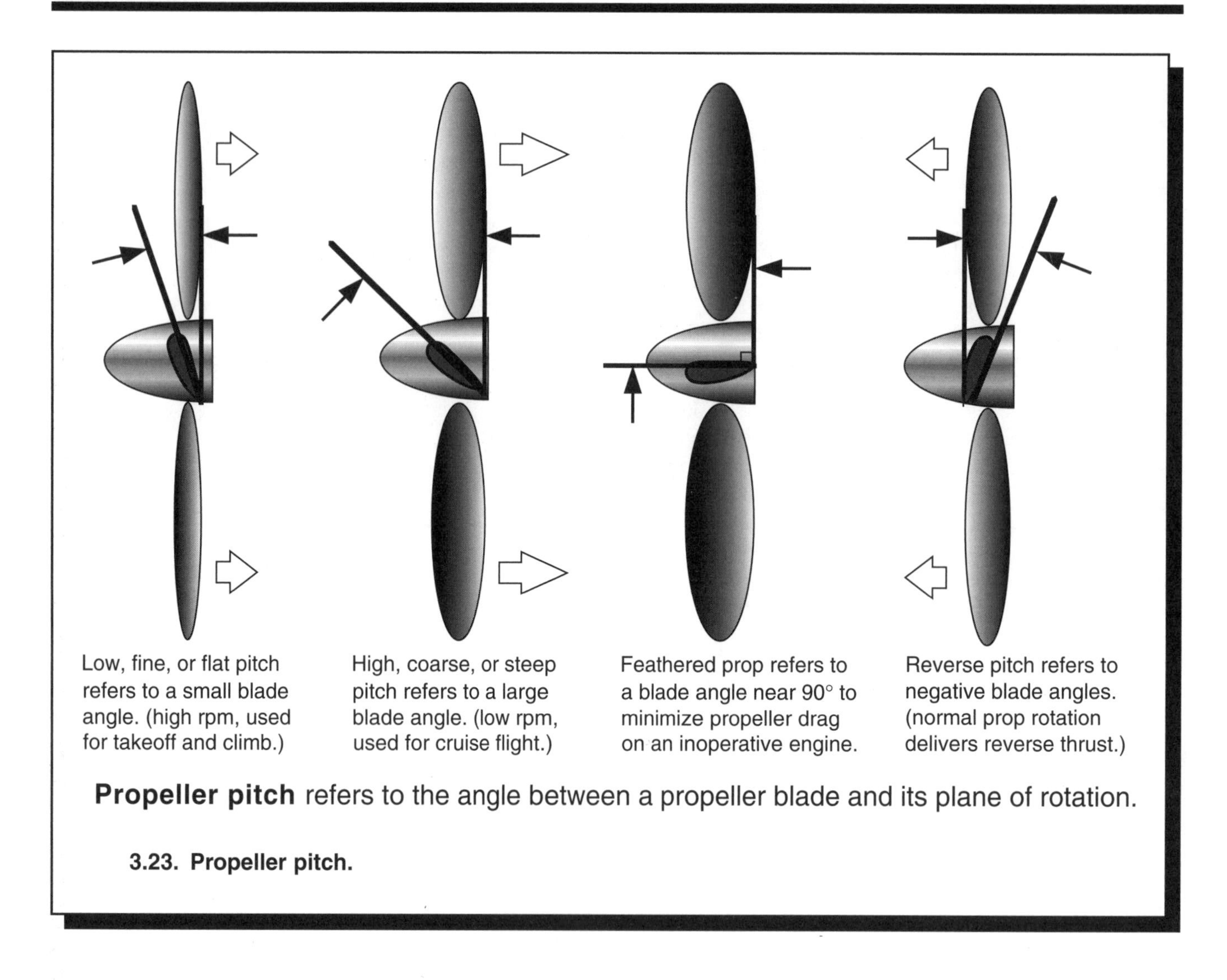

3.23. Propeller pitch.

Torque Effect

We again refer to Newton's third law: for every action there is an equal and opposite reaction. In the case of a propeller, this is called *torque reaction,* or *torque effect.* Simply put, torque effect refers to the tendency for an aircraft to rotate in a direction opposite the rotation of the propeller. (Helicopter pilots are particularly familiar with this effect.) Torque reaction is greatest at high propeller rpm and high power settings and is most noticeable upon rapid power changes.

Propeller Slipstream Effects

Accelerated propeller slipstream: Unlike the relatively clean airflow over surfaces of a pure jet, a propeller can affect the amount and velocity of air flowing over an aircraft's wings and control surfaces. The propeller creates a large area of accelerated airflow over parts of the wing, creating more lift. Accelerated airflow can also increase rudder and elevator effectiveness. Since rapid power adjustments change accelerated airflow, they impact lift and control effectiveness. Therefore, it's important on many propeller-driven aircraft to make smooth power adjustments at slow speeds and low altitudes, such as on short final.

Spiraling propeller slipstream: A rotating propeller imparts its rotation to the airflow it generates. If the propeller is rotating clockwise, for example, it creates rearward airflow rotating in the same direction. In most aircraft this causes asymmetrical airflow to strike the vertical stabilizer and rudder, pushing the tail to one side (see Fig. 3.25).

Critical Engine

Because of all these asymmetrical propeller properties, failure of one engine, on a given aircraft, may be more detrimental than loss of another. An airplane's *critical engine* is the one whose failure would most adversely affect the per-

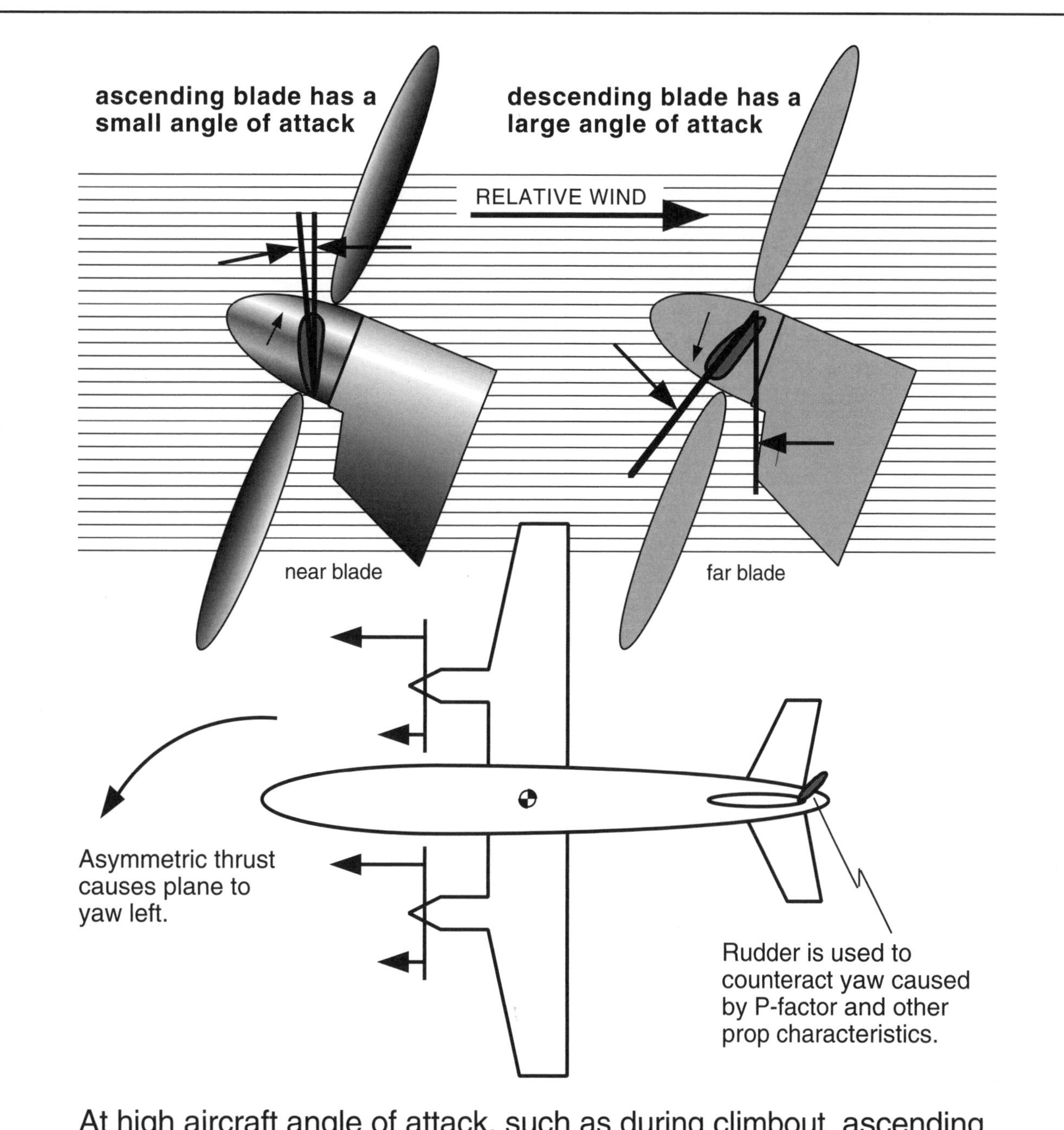

At high aircraft angle of attack, such as during climbout, ascending and descending propeller blades have different angles of attack. The resulting asymmetric thrust must be counteracted with rudder.

3.24. Propeller P-factor. (Figure and comments depict clockwise-rotating propeller installation, as viewed from behind.)

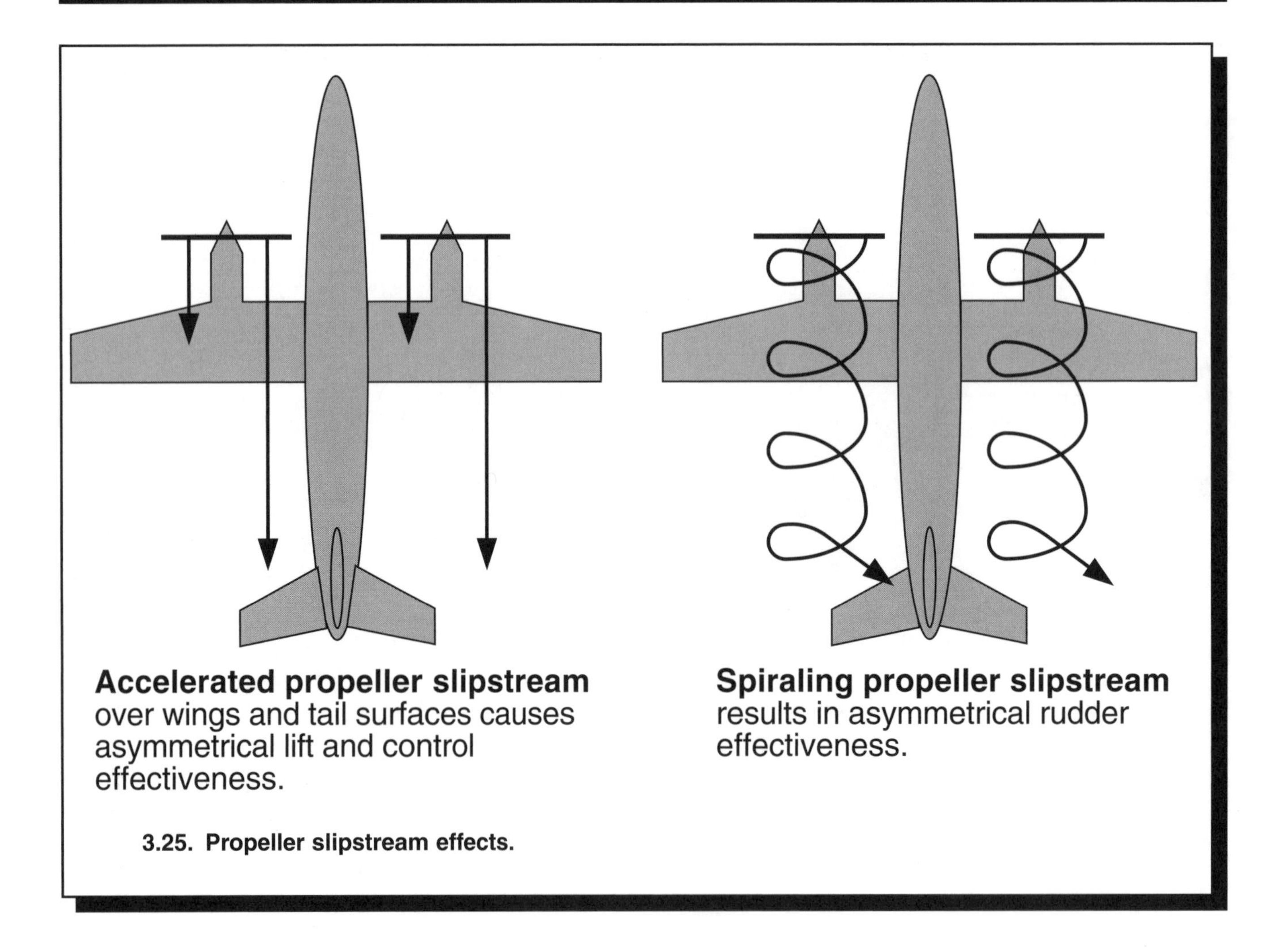

3.25. Propeller slipstream effects.

formance and handling qualities of a given multiengine aircraft. In determining which engine is critical, the manufacturer considers all of the propeller effects previously discussed. (P-factor is generally considered to be the most influential propeller effect.) Engine-out performance and procedures for that aircraft are then developed based on the worst case failure of the critical engine (Fig. 3.26).

Tractor versus Pusher Propellers

Tractor propellers are mounted on the airplane in such a way that their thrust pulls the aircraft forward. Tractor propellers are mounted in front of the engine nacelle with the propeller shaft facing forward. This type of propeller installation is most common because its location in front of the engine nacelle allows it to operate in relatively undisturbed air; therefore, it arguably has highest efficiency.

Pusher propellers are mounted behind the engine nacelle with the propeller shaft facing rearward so that their thrust pushes the aircraft forward instead of pulling it. Although historically pusher propellers have rarely been installed on landplanes, lately there has been renewed interest in using them on canard aircraft designs and for purposes of reducing cabin noise. Both Beechcraft and Piaggio have recently produced multiengine turboprops with rear-mounted engines and pusher propellers, and several other pusher designs are on the drawing board.

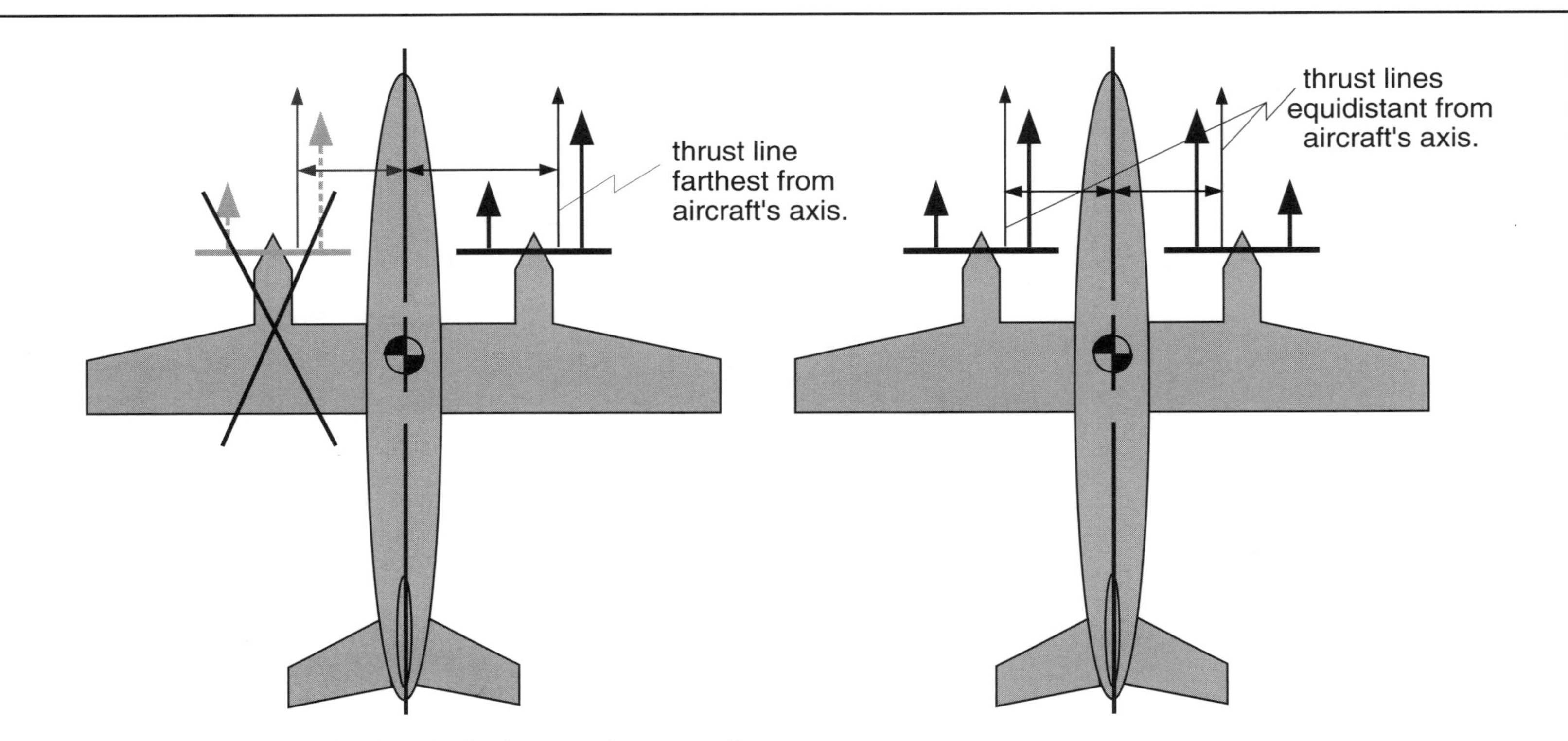

Due to P-factor and other propeller effects, failure of one engine often results in greater aircraft yawing tendency than failure of the other. The engine whose failure most adversely affects flight performance on a given aircraft is known as the "critical engine."

3.26. Critical engine on propeller-driven aircraft.

CHAPTER 4

Turbine Aircraft Power Systems

Basics of Aircraft Power Systems

Almost all the power required to operate a flying turbine aircraft ultimately comes from one source: engine power. (The small amount remaining comes from the pilots themselves in physically actuating the controls.) Having examined how the engines work, let's now consider the systems that transfer engine power to where it's needed to operate the aircraft. For purposes of redundancy and function, there are basically four ways to transmit power around the aircraft: mechanical, electrical, hydraulic, and pneumatic.

While each of the four power systems has unique advantages, a few characteristics and terms apply to all. The four systems are interrelated to provide various protections and features. One power system is often used to control another for specific applications. For example, an electric solenoid valve may be used to direct hydraulic power to the appropriate side of a hydraulic landing gear motor for retraction or extension. Other examples include pneumatic valves switching electric switches, manual levers moving pneumatic valves, and electrically controlled hydraulic cylinders directing gear doors. Any combination is possible.

One purpose of such arrangements is to impact what happens in the event of various types of failures. If electrical power is lost, which way will various valves and switches fail? How about in the case of hydraulic failure? Important switches and valves are designed to perform predictably in the event of a given system failure. The terminology here is "fails-open" or "fails-closed." In ground school, you'll have to learn the failure characteristics of many key control components on the aircraft you fly.

Depiction of Aircraft Systems in Pilot Training

Sometimes ground school gives the impression that pilots must learn every detail of every system, in every airplane flown. While this is true to a certain degree, most systems training is actually on a schematic level. Pilots are taught the major components of each system and then the critical flow of power through it. By understanding conceptually the logical operating sequence of each system and the functions of its major components, a pilot can troubleshoot and solve many problems that occur in flight. This knowledge is also valuable in helping pilots communicate with technicians in the course of normal maintenance and repairs.

The best aircraft systems diagrams for pilots are flowcharts that depict the logic by which power is distributed through each system to key components in that specific design. A pilot's aircraft electrical diagram, for example, does not show the actual physical wiring layout of the system. Most pilots don't have the know-how to read wiring diagrams and, in any case, don't have the time, tools, guts, or legal backing to crawl out onto the wing and fix something in flight.

Therefore, as you examine each systems diagram in training, don't be panicked by the visual complexity. The diagram presents the normal flow of power through each system and what happens to that flow when various problems occur. A sharp pilot understands how faults are contained and power is redistributed under various failure scenarios. That way he or she can do something about problems . . . without crawling out on the wing.

Understanding Aircraft Power Systems: The Reference Waterwheel

Mechanical systems, with their rods, cables, gears, and pulleys, are challenging to engineer and maintain but conceptually simple for most of us to understand. The other three types of power transmission systems are sometimes less well understood by pilots. Therefore, we're going to take a moment here to lay some groundwork that should help in understanding aircraft power systems.

There are no waterwheels in turbine aircraft, but in order to demonstrate some basic principles of power systems, let's consider a simple water power system to which we'll refer during our electrical, hydraulic, and pneumatic system discussions. (Mechanical systems should make sense without this comparison.) If you're already sharp on the basics of electrical, hydraulic, and pneumatic systems, no one will be offended if you skip the waterwheel section.

In the simple waterwheel example in Figure 4.1, a river provides energy to a waterwheel through the force of gravity. One problem with rivers (and most other power sources) is that their flows are not consistent all the time. Therefore, a reservoir is useful so that stored energy is available even when the river's flow is down. (Without it there'd be lots of water power available at some times, and little or none at others.) The reservoir also serves to regulate the flow of water to the waterwheel, making for more consistent operation. The waterwheel itself is useful because it converts the river's energy into mechanical energy in order to drive mankind's gizmos. (For our purposes, "gizmos" represent any type of powered device.)

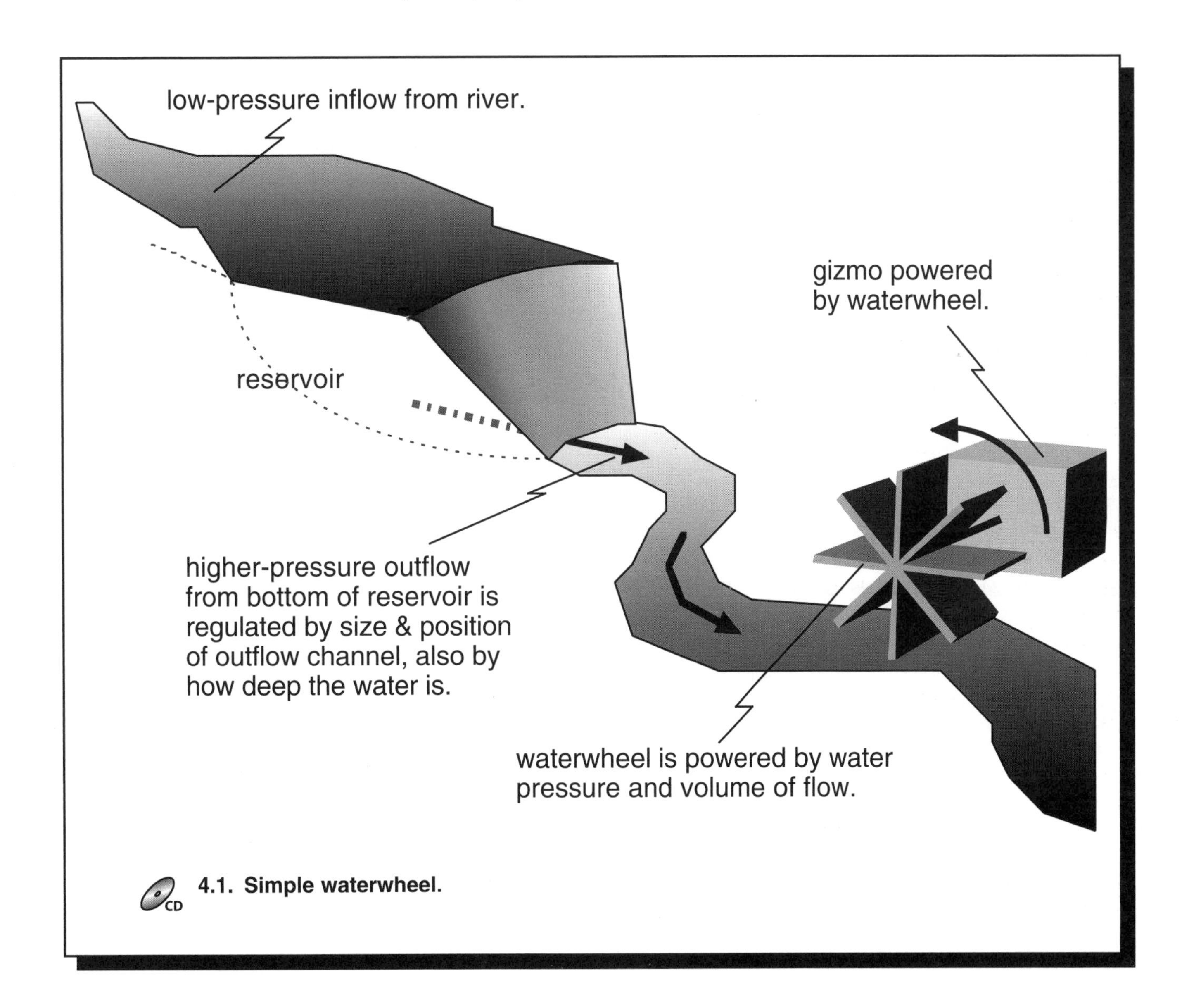

4.1. Simple waterwheel.

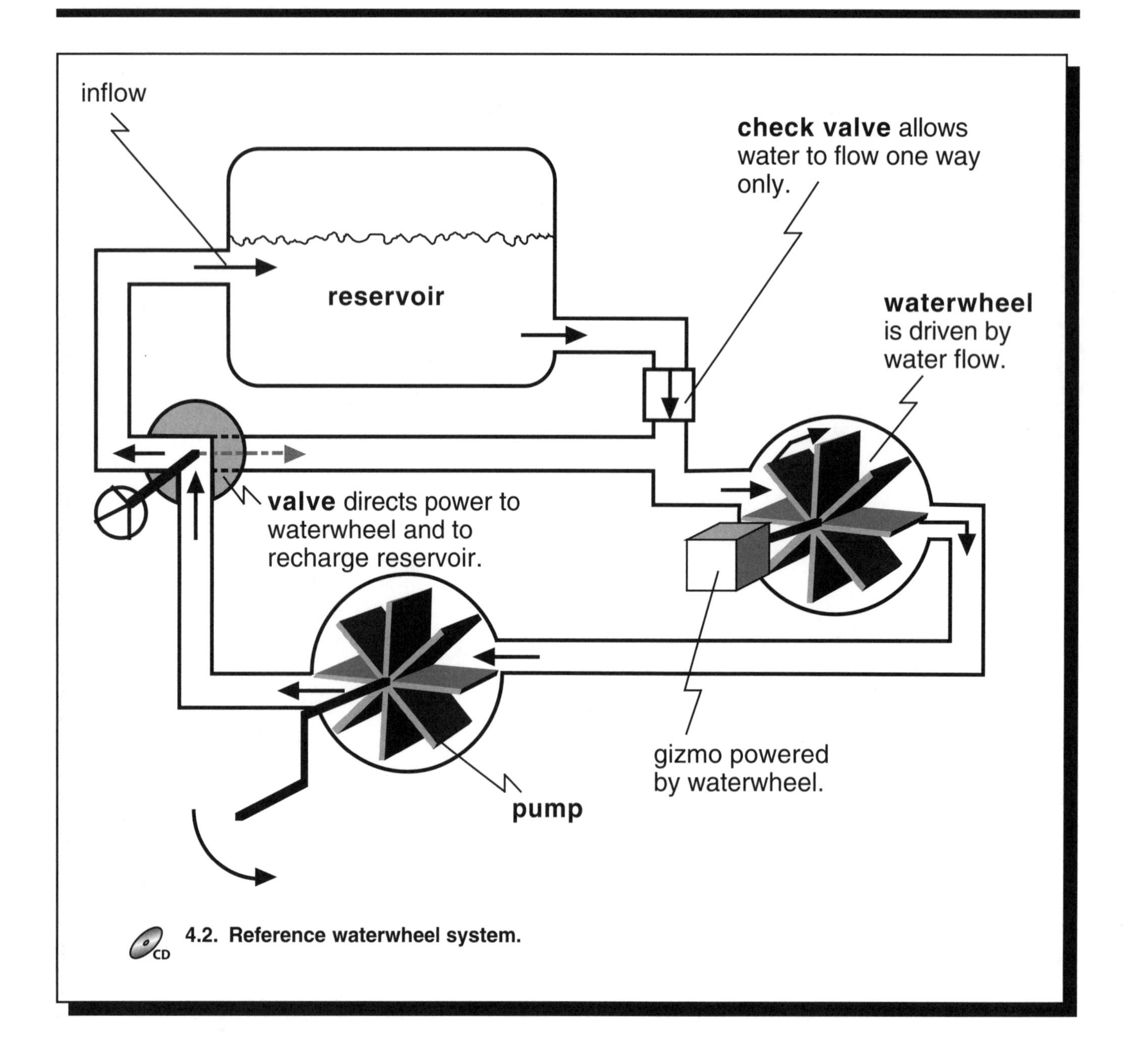

4.2. Reference waterwheel system.

If we do not have a continuously flowing energy source like the river, we can establish a closed system to drive our waterwheel. In this case, the water must be returned to the reservoir, either continuously or periodically. (Otherwise it'd be a one-shot deal.) To accomplish this a pump has been added to the system shown in Figure 4.2. The pump could be a low-pressure pump that simply returns water to the reservoir, or it could be a high-pressure pump that pressurizes the system. Given a strong enough power source (like Hercules or one of the superheroes), the pump could power the whole waterwheel system, without the help of gravity or a reservoir. Note that the waterwheel and the pump are similar in construction. It's just that one device puts energy into the waterwheel system and the other takes it out in a different spot. These two components could swap functions, based upon which one the power is applied to.

Comparing Aircraft Power Systems to the Reference Waterwheel System

As you can see in Figure 4.3, the basic components of every aircraft power system may be compared directly with our reference waterwheel system, though for better understanding a few other observations are in order.

Function	System			
	Reference (water)	Electrical	Hydraulic	Pneumatic
Power Source	Crank	Engine mechanical power	Engine mechanical or electrical power	Engine compressor section
Power Supply	Pump	Generator	Hydraulic pump	Engine bleed air
Energy Storage	Reservoir	Battery	Hydraulic accumulator	Pneumatic accumulator
Power Trans-mission	Pipes	Wires & aircraft structure	Hydraulic lines	High- and low-pressure pneumatic lines
Output Devices ("Gizmos")	Waterwheel	Electric motors, lights, radios, and other devices	Hydraulic motors and cylinders	Pressurization, environmental, gyro air, pneumatic deice
Flow Directional Control	Check valves	Diodes	Check valves	Check valves
Regulation/ Step-down	Flow restrictors	Transformers	Hydraulic regulators	Pneumatic regulators

4.3. Comparison of aircraft power systems with reference waterwheel system.

A reservoir stores water volume, but it also stores potential energy in the form of pressure. In Figure 4.1, the pressure is stored purely in the form of the water's weight above the outlet. (That is, if the reservoir is almost empty, there won't be much pressure at the outlet compared with when it's full.) The system in Figure 4.2, however, may be further pressurized through use of a high-pressure pump. In addition, air can be compressed into the reservoir above the water. Even without much water, air pressure could push a small amount of remaining fluid to power the waterwheel for a short time.

These points are important because each water system's power is tied to two variables: the volume of water being moved and its pressure. A large system like the river and dam in Figure 4.1 needs relatively little pressure to drive the gizmo due to the volume of water flowing. A small system with little water volume requires much greater pressure to turn the same gizmo. As we'll see, these same principles are true in the various aircraft power systems.

Electrical Power Systems

While electrical systems are fundamentally similar for all aircraft, turbine-powered airplanes have the complexity of many electrically powered systems. They fly under extremes of temperature, altitude, and moisture and require extensive cabin environmental systems and passenger amenities. The electrical systems of these aircraft are accordingly complex, sophisticated, and redundant for safety. In comparing a basic electrical circuit to our reference water system, we're reminded that an operating electrical circuit must always form a complete loop. If the electrical circuit is interrupted, or the water channel is blocked, the system ceases to operate (Fig. 4.4).

In aircraft, wires carry electrical power to the devices on each circuit. The "return" of electricity to complete each circuit is normally through the aircraft's metal structure, otherwise known as *ground.*

Every electrical system is composed of one or more circuits containing *power sources, components* powered by each circuit, *control devices* to operate the circuits, and *circuit protection devices* to shield system components from damage due to failures.

Electrical Power Sources

Generator

The engine-powered *generator* in an airplane electrical system corresponds to the pump in our reference water system. Note that the outputs of both pump and generator may be measured using similar parameters: pressure and flow volume. In electrical terms, the generator's output of electrical pressure, or *voltage,* is measured in volts. Its output volume or flow is known as *current* and is measured in amperes (amps).

Generators produce electricity by moving permanent magnets around a coil of wire, thereby motivating electron flow in the coil. They are driven, on modern aircraft, by the engine(s) and sometimes by an auxiliary power unit (APU), if installed. (An APU is a small turbine engine, normally installed in the aircraft's tail, which provides auxiliary power for ground operations, engine starting, and emergency backup. See the APU section later in this chapter.)

Just as a water pump will pressurize a sealed water system to almost its own pressure, a generator drives its associated circuit at virtually its own voltage. Since generators are designed to put out more or less constant voltage, the variable in power required to drive various components is always defined in current (amps). The current required to operate each electrical component relative to the circuit is known as its *load.*

Accordingly, a generator must be rated at adequate amperage to drive all the operating components on its circuit(s). Let's say a circuit has six gizmos on it, each rated at 5 amps. A generator dedicated to that circuit needs to deliver at least 30 amps (5 amps × 6 gizmos) in order to run all six gizmos at the same time.

Of course, on an airplane each generator powers many circuits. Sometimes the generators may not be able to provide enough power for everything that's turned on. An example would be failure of one of the engine-driven generators on a twin-engine plane. While two operating generators will power most everything on the plane at the same time, one alone might not be enough. In that case electrical *load shedding* would be required. (No, nothing has to be thrown out the window!) Enough electrical equipment must be turned off to reduce the electrical load to within the capacity of the operating generator.

A few words are in order regarding generator terminology. On cars and light airplanes, the term "generator" has come to imply a device that generates *direct current* (or DC), while alternators generate *alternating current* (AC). However, in turbine aircraft you'll usually hear such devices referred to as "DC generators" and "AC generators."

Most modern aircraft are electrically powered by AC generators. On large aircraft, in particular, most high-draw electrical devices (such as electric hydraulic pumps and windshield heat) are AC powered. AC is converted to DC to power DC systems. A few aircraft have both AC and DC generators installed, but in most cases when you hear the term "generator" on a modern turbine aircraft, assume that it's an AC generator.

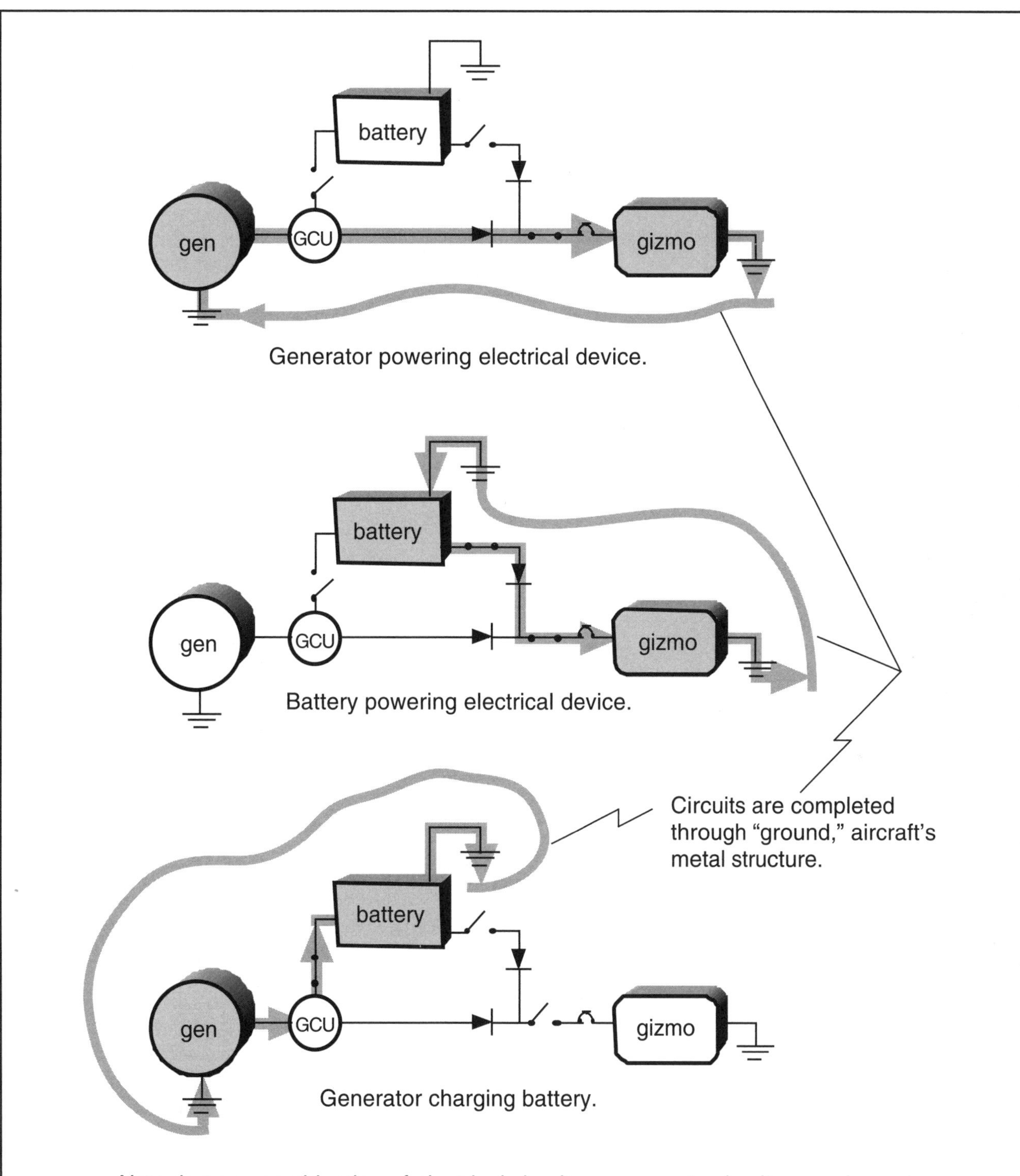

Note that any combination of electrical circuits can operate simultaneously, and that return portions of each circuit are never shown in electrical diagrams.

4.4. Basic electrical circuits.

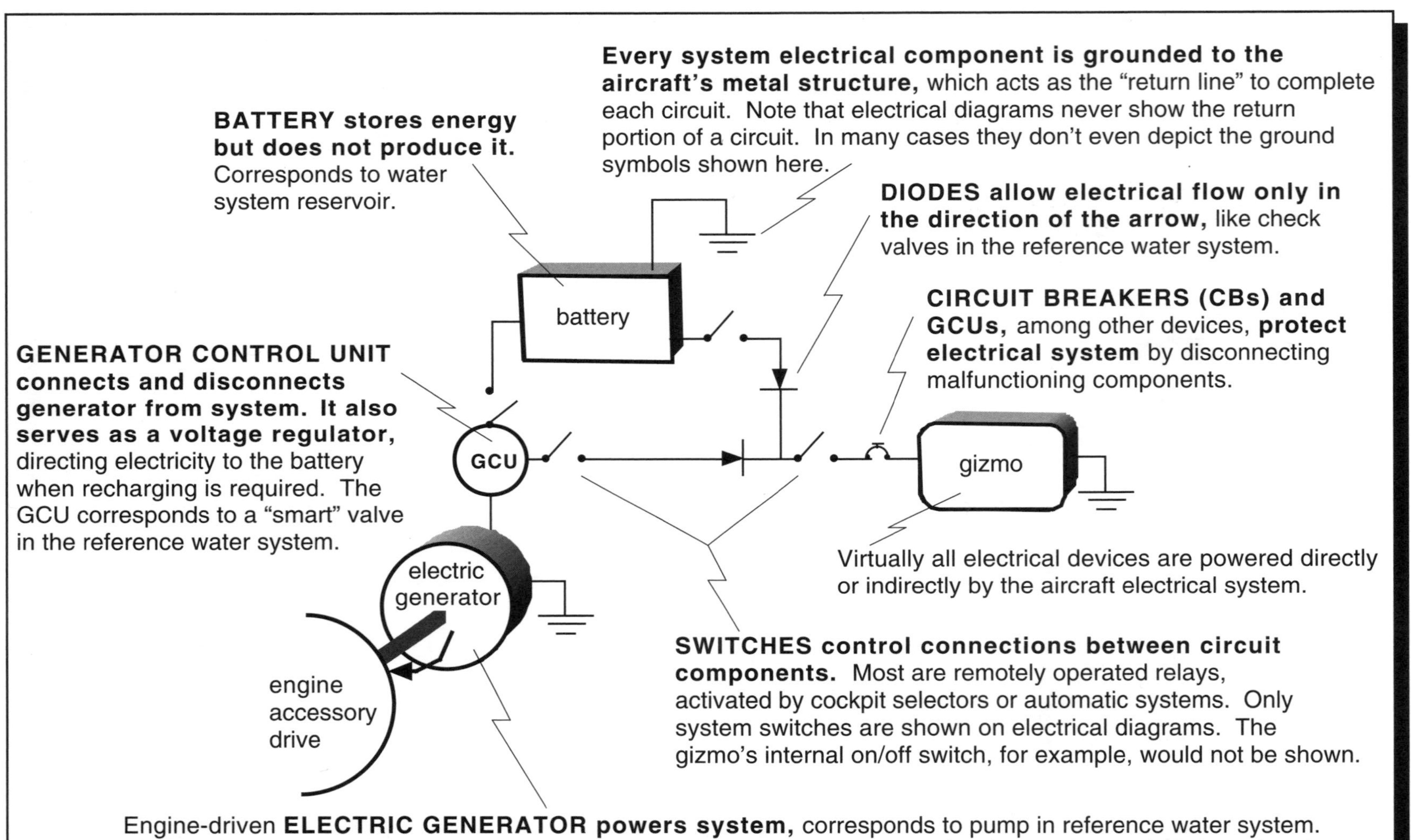

Every system electrical component is grounded to the aircraft's metal structure, which acts as the "return line" to complete each circuit. Note that electrical diagrams never show the return portion of a circuit. In many cases they don't even depict the ground symbols shown here.

BATTERY stores energy but does not produce it. Corresponds to water system reservoir.

DIODES allow electrical flow only in the direction of the arrow, like check valves in the reference water system.

CIRCUIT BREAKERS (CBs) and GCUs, among other devices, **protect electrical system** by disconnecting malfunctioning components.

GENERATOR CONTROL UNIT connects and disconnects generator from system. It also serves as a voltage regulator, directing electricity to the battery when recharging is required. The GCU corresponds to a "smart" valve in the reference water system.

Virtually all electrical devices are powered directly or indirectly by the aircraft electrical system.

SWITCHES control connections between circuit components. Most are remotely operated relays, activated by cockpit selectors or automatic systems. Only system switches are shown on electrical diagrams. The gizmo's internal on/off switch, for example, would not be shown.

Engine-driven **ELECTRIC GENERATOR powers system,** corresponds to pump in reference water system. Motors and generators are so similar in construction that the same unit can often be used to perform both functions. On many smaller turbine aircraft, single unit **starter/generators** serve double duty as both electric generators and engine starter motors. When electricity is applied from an outside source, the unit acts as a motor. When turned mechanically, the same unit "pumps" out electricity.

4.5. Basic electrical components.

Note that an electric motor is simply the opposite of a generator. In a motor, electricity channeled through a coil creates an electromagnet, which works against fixed magnets to turn a shaft. In many smaller turbine installations, each engine's generator is also used as a starter motor. With this type of *starter/generator* the difference is simply whether (1) the device is being driven electrically by a battery (or other power source) to spin the engine or (2) the generator is being driven mechanically by the engine to produce electricity (Fig. 4.5). You can see that this is no different than our reference water system, where the waterwheel and pump are more or less interchangeable, depending upon where the power is applied (see Fig. 4.2).

Battery

The *battery* is best thought of as a power reservoir that stores electrical energy in chemical form. Like the water system reservoir, or like a spring in a toy, energy has to be sent to the battery from another source (the airplane's generator). The battery can then act as a power source itself, until it runs out of stored energy. Once charged, a battery will deliver to its circuit, for a limited period of time, a voltage and current. Accordingly, batteries are rated in volts and amp-hours.

People are often confused as to why battery and generator voltages aren't the same in a given system. Think of our water system: in order to fill the reservoir the pump must generate more pressure than that of the reservoir. Otherwise the reservoir could not be refilled. For the same reason, in an electrical system generator voltage must overcome battery voltage in order to charge the battery. Accordingly, battery voltage is always a little below generator voltage for a given system. (The battery is normally rated at a little over 24V in a 28V system.)

Battery amp-hour ratings simply indicate how many amps the (fully charged) battery is designed to put out over what period of time. If you're interested in learning how long your battery will last after a generator failure, simply add up the loads, in amps, of the operating electrical components and divide into the battery amp-hour rating. For example, a fully charged 30-amp-hour battery will theoretically supply 30 amps for one hour, 15 amps for two hours, or 60 amps for one-half hour. To conservatively determine component loads, just add up the amperage ratings stamped on the circuit breakers (CBs) of the operating components. You'll likely be amazed at just how little battery reserve is available!

Thirty available minutes of operation under battery power is required by the regulations. However, that's based on properly reduced electrical loads and on a new, fully charged battery. You can see that in the event of total generator failure, it takes skill and maybe some luck to promptly identify battery reserve, descend, and land before losing the entire electrical system.

In addition to storing energy, batteries have another handy characteristic. A battery acts as a shock absorber in a circuit because (again like a spring or a reservoir) it can absorb a surge or "spike" of electrical energy that might otherwise damage components of the circuit.

The generator(s) and battery ultimately power, through many circuits, all of the electrically powered devices and systems of the aircraft, including lighting, avionics, electric motors, and many flight instruments. While circuits may normally be powered by either the generator or the battery, the battery does not put out enough power over time to be useful as the sole electrical source. In fact, on large aircraft many electrical devices can't even be operated when only battery power is available. Generators effectively power everything in flight, with the battery acting only as a reservoir and shock absorber. The same is often true at the gate: ground power from an AC outlet, an APU, or a *ground power unit* (*GPU*-a generator or battery cart) runs the aircraft systems and keeps the battery charged.

BATTERY-POWERED ENGINE STARTS

A few other points about turbine aircraft batteries merit discussion here. Most smaller turbine aircraft engines are started using electric starters or starter/generators. It takes *a lot* of electrical power to start a turbine engine. For only one-half second or so, an average electrical turbine engine start requires almost 2000 amps! Therefore, a *battery start* requires a very healthy battery. Since weak batteries can lead to a damaging hot start of the engine, battery condition must be checked before each attempt. (Weak batteries cause hot starts because they fail to "spin up" engine compressors adequately. See "Introduction to Gas Turbine Engines" in Chapter 3.)

Typically, minimum battery voltage for a battery start is around 22V on a 28V system. Otherwise, the pilot must request a ground power unit (GPU) for assistance in starting. Many turbine operators use GPUs for starting whenever they're available, in order to reduce wear and tear on their expensive engines and batteries.

NI-CAD VERSUS LEAD-ACID BATTERIES

Two different types of batteries are commonly found in turbine aircraft. Smaller turboprops often have the same basic type of *lead-acid battery* found in piston aircraft and cars. Increasingly, turbine vehicles these days are equipped with *ni-cad batteries* (nickel-cadmium). Ni-cads are much more expensive than lead-acid models but have some distinct advantages. They put out sustained voltage over a longer period of time, whereas lead-acid batteries tend to

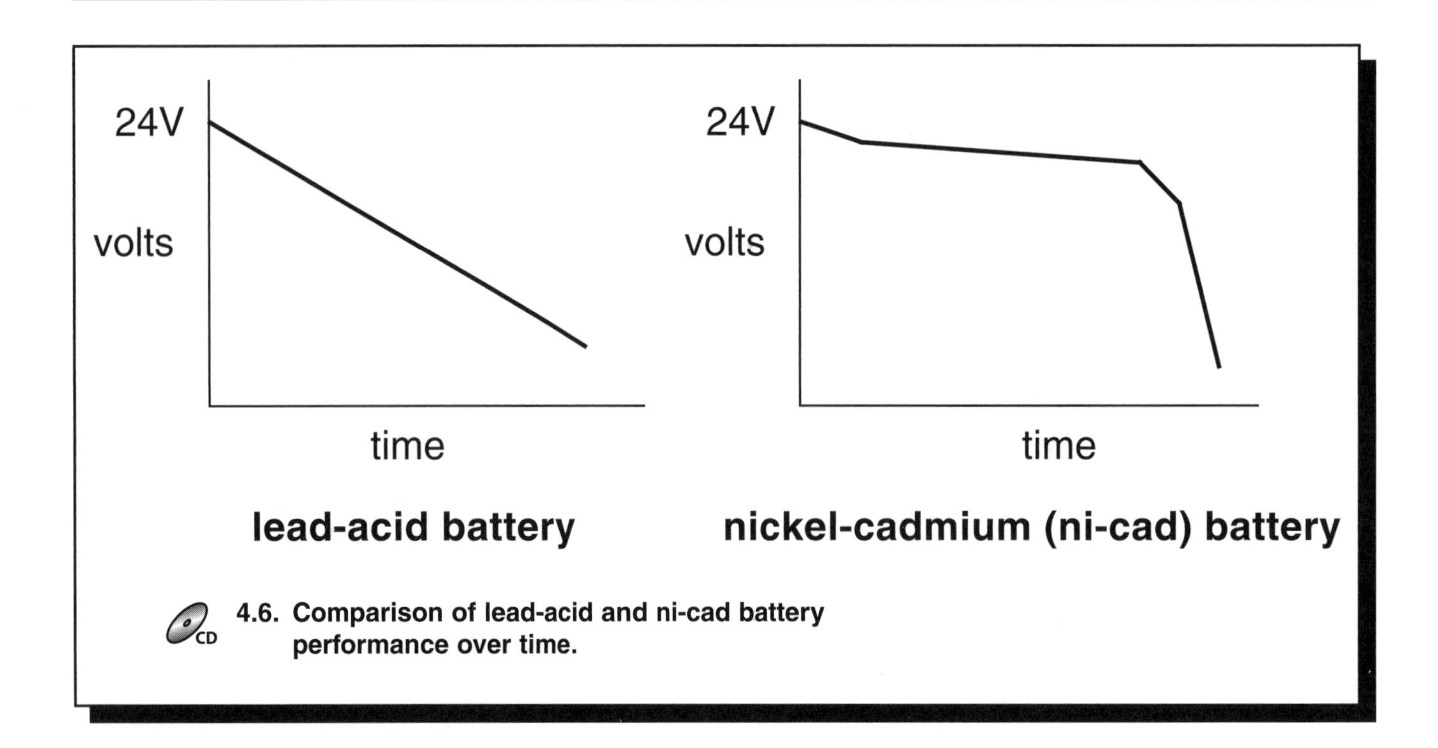

4.6. Comparison of lead-acid and ni-cad battery performance over time.

drop off more quickly. Ni-cad batteries recover quickly from heavy usage. And finally, they put out lots of current. Since a weak battery can lead to a damaging hot start of the engine, many operators find ni-cads are well worth the extra investment (Fig. 4.6).

Ni-cad batteries have a few disadvantages, however, in comparison with lead-acid batteries. One is their *memory* characteristic. If a ni-cad battery is kept nearly fully charged all the time, it tends to lose some of its ability to meet more demanding situations. For example, if GPUs are used for every start over a period of months, the battery gets used to providing only limited power and then recovering over a short time span. It may not then be able to fully power a battery start. When this happens, the battery must be removed by maintenance and *deep-cycled.* This refers to the process of fully discharging the battery and then slowly recharging it on a trickle charger. (Use a rechargeable razor or flashlight? These have ni-cad batteries, too, in case you'd like to test the principles of battery memory and deep-cycling.) You can see that it's good preventative maintenance to make battery starts, at least periodically, in order to exercise the battery.

Finally, ni-cads are sometimes subject to *thermal runaway* (sometimes called *runaway battery*). Thermal runaway occurs when excessive current is drawn from and then replaced to the battery. The battery overheats and begins to self-destruct. Unless attended to early, overheating sometimes cannot be stopped until the battery is completely physically destroyed. Some aircraft have battery temperature indicators in the cockpit for identifying this condition; others may only be tested electrically by the pilots. In any case, thermal runaway can be a very serious problem. The procedure is to immediately isolate the battery from all other circuits and to make a prompt landing. Runaway batteries can potentially damage other electrical components, cause fires, and damage airframes due to release of internal battery chemicals. In some cases, a runaway battery can, under extreme circumstances, burn its way through its compartment and fall out of the aircraft.

Control Devices

Control devices on each electrical circuit meter electricity to operate the aircraft's systems and components. *Manually operated switches* range from simple on/off toggles for lights and radios to aircraft and avionics master switches that control electrical supply to many different items at once. *Rheostats* adjust electrical flow to variable devices such as instrument lights. Remote mechanical switches are used for such purposes as *limit switches* (for example, to turn off the flap motor when flaps are fully retracted) and "squat switches" (to confirm, when closed, that the landing gear is down and weight is on the wheels).

Relays and solenoids are remotely or automatically controlled switching devices built around electromagnets. In each device, a small amount of current is sent through a coil to move a switch or shaft.

Relays serve to remotely control electric circuits carrying large amounts of current. (Newer technology solid-state devices are replacing relays for many applications.) A solenoid is also an electrically powered remote control device, but is designed to move a shaft over a short distance when powered. Solenoids are used to remotely operate hydraulic and pneumatic valves and small mechanical devices, as well as other switches.

Relays, solenoids, and associated devices are very important in modern aircraft because of the trend toward computerized systems. They allow remote electronic control of all types of aircraft systems, whether hydraulically, pneumatically, or electrically powered.

Generator Control Units (GCUs)

Turbine aircraft generators are normally controlled by *generator control units* (GCUs). These multifunction devices fill the role of voltage regulators, directing generator current to the battery when necessary for recharging. They also provide circuit and generator protection by disconnecting the generator from the system when electrical abnormalities occur. (GCUs often incorporate relays to switch generators on and off line.)

Examples of other GCU functions include overvoltage protection, load paralleling relative to other generators, and electrical protection at engine startup (see Fig. 4.5).

Electrical Conversion Devices

Most larger turbine aircraft these days require a variety of different types of electrical power. Once again consider our reference water system. The combined amount of water pressure and flow required to drive a waterwheel is a function of the waterwheel's "load" on the system. If the waterwheel is tiny, and turning only itself, little pressure and flow are required. If, on the other hand, the waterwheel is large and driving tons of machinery, lots of power will be required. The same is true of electrical components. While a navigation light or radio draws little power from the system to operate, consumption by an electric flap motor may be tremendous. To properly address these varied needs, different types of electrical power are required. Many general-usage items are driven off 28V DC (direct current) power. AC (alternating current) power often drives high-draw items on larger aircraft.

A number of electrical conversion devices are required to match the varying power requirements of aircraft electrical components with the available sources. *Transformers* are used to step system voltage (usually down) for specific applications. *TRUs* (transformer-rectifier units) convert AC generator output to DC.

Some electric flight instruments, such as gyros and some engine instruments, are commonly driven by low-power 26V AC. *Inverters* convert DC power to AC power for this purpose. (Incidentally, AC-powered nongyro instruments fail in a different manner than DC-powered instruments. The old saying is "AC lies; DC dies." While a DC meter goes to 0 when depowered, an AC needle freezes right where it was when power was lost.)

Circuit Protection

Electrical Faults

There are basically three types of electrical faults that we worry about in aircraft (Fig. 4.7).

An *open* is an uncommanded interruption of electrical supply to or in some electrical component or system. When an open occurs, the affected device or system ceases to operate and has little or no effect on other components outside the affected one. Some opens are used in our favor for system and component control. On/off switches, circuit breakers, fuses, solenoids, and various other devices may be used to intentionally open circuits, thereby turning electrical components off by depriving them of electricity.

A *short,* on the other hand, occurs when electricity is allowed to take a "shortcut" through or around a component or system. The flow of electricity goes directly to ground, without the normal component resistance to check the flow. The problem here is not only failure of the affected component or system, but also the drain of electricity out of the system through the short. If unchecked, a short is likely to damage or ruin virtually every component on its circuit. Furthermore, it may draw more electricity than the system power source(s) can produce, disabling the power source(s).

In order to understand the effect of a short on a power source, imagine riding a bicycle uphill. You're pedaling hard, with constant power, when suddenly the chain breaks. For a moment there you'll put out a whole lot more energy than you planned upon. What's more, if you're not smart enough to stop pedaling, you'll work yourself to death without keeping that bike rolling.

This is basically what happens to a power source if there's a short circuit. There's a certain amount of electrical resistance provided by the lights or radios or whatever that power source is driving, and if the electricity takes a shortcut, that resistance goes away. If the power source is not isolated immediately from the short, it will quickly fail due to its inability to produce an infinite amount of electricity.

A third type of electrical fault we'll call *logic failure.* This occurs when some component (often computer controlled, but not necessarily) basically gets the wrong idea about what it's supposed to do. For example, on one popular commuter aircraft, landing gear system operation is controlled by a central computer. On early models, when pilots selected "gear down," the gear sometimes extended right through the gear doors, which had not opened first! The

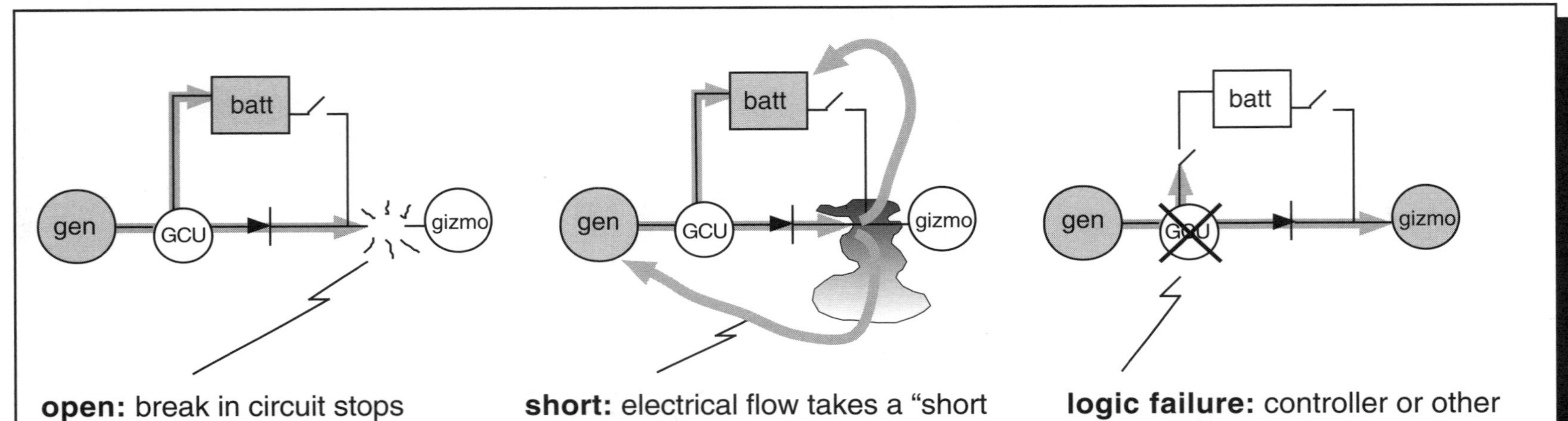

open: break in circuit stops flow of electricity. Affected components don't work. Here the gizmo has lost its electrical supply, due to a broken wire or bad connection.

short: electrical flow takes a "short cut," instead of following its planned circuit. In this case power is traveling through spilled water to the aircraft's metal structure and then back to generator and battery. The gizmo is unpowered, and all related circuits and components are threatened.

logic failure: controller or other "smart" device malfunctions, so circuit doesn't operate properly. Here the GCU has lost its voltage regulation function. It can't signal the battery charging relay to close, so the battery can't be recharged.

4.7. Electrical faults.

central computer had malfunctioned in a manner that threw off the gear operation sequence.

Bus Bar Systems

The key safety feature of sophisticated aircraft electrical systems is redundancy. Multiple power sources are used, each protected from the failures of others and each capable of driving a wide (and overlapping) variety of equipment. This is done through use of an *electrical bus bar system,* meaning the aircraft's electrical system is carefully organized into a bunch of separate but interconnected circuits. Bus bar systems allow important circuits to be isolated from one another and to be supplied by alternate power sources in cases of failure.

A bus bar system is set up so that each power source supplies one or more specific buses. Each bus may be thought of as an electrical manifold, with a variety of items hooked up to it for power. The buses are interconnected via circuit protection devices of one kind or another so that in the case of a fault on one bus other buses are instantaneously disconnected to prevent involving them, too. In addition, a bus that has lost its normal power source due to a failure can often be alternately powered by another source, through another bus (Fig. 4.8).

Let's look at how electrical buses are set up to provide protection through redundancy. If two important aircraft safety systems back each other up, each will be located on a separate bus so as to prevent simultaneous loss of both due to an electrical problem.

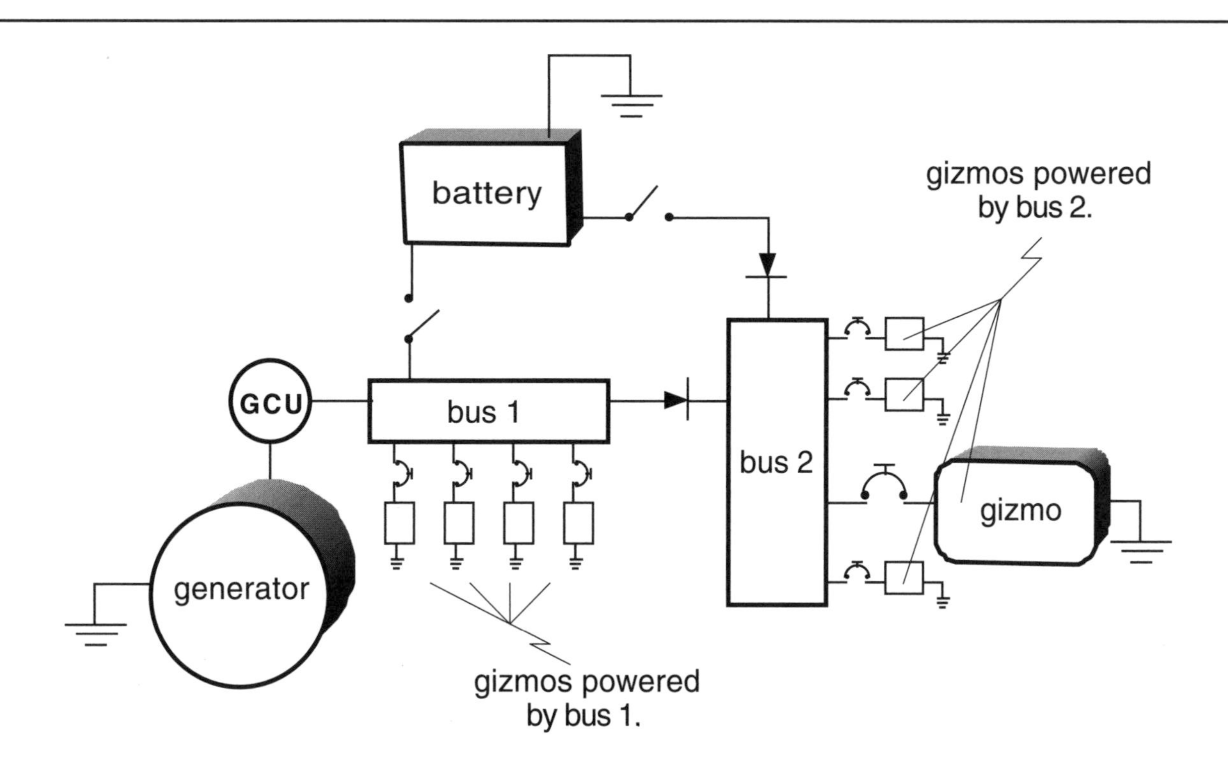

An aircraft electrical system supplies power to a variety of electrical buses, each of which acts as a manifold to supply various electrical components. Since each is separately powered, malfunctioning buses can be isolated from the rest of the electrical system, thereby protecting the system. Even if components on a failed bus are unusable, backup components powered by other buses can keep operating. Redundant systems are always powered by different electrical buses for this reason.

4.8. Electrical buses.

Each engine-driven generator, for example, normally drives its own "generator bus," on which are normally found the highest-draw items on the aircraft. Redundant items will be split between the buses. The pilot's windshield heat and copilot's quarter-window heat might be powered from the left generator bus, while the copilot's windshield and pilot's quarter-window heat are on the right generator bus. That way, if either generator bus fails during icing conditions, somebody should be able to see out the windshield for landing.

Items like built-in fire extinguishers and emergency lights are usually powered off a "hot battery bus." That way they are powered for use even when no generator power is available and all switches are off.

A "battery bus" is normally powered off the aircraft master switch so that radios, lighting, and a few other items can be operated off the battery when the engines aren't running. You can see why electric windshield heat wouldn't be on that bus. It'd drain the battery in minutes flat, and who needs it when sitting on the ramp, anyway?

Most aircraft systems are set up so that the battery bus(es) can be powered by the generators should the battery fail in flight. Some generator buses can also be selectively powered by the battery, but not for long. (The electrical demands are too high.)

Circuit Protection Devices

A variety of devices are used to protect system components from damage due to failures. Resettable *circuit breakers (CBs)* "pop" to disconnect individual components that are drawing too much current. (You'll need to know the procedures for resetting them in each aircraft you fly. In some aircraft, pilots are not to reset CBs except in emergencies. More commonly, the CB is reset one time only, then left alone. Incidentally, the CB pops when its housing expands due to heat. If it hasn't cooled down it won't reset, even if the component is okay. Wait a few moments before attempting to reset.) (See Figure 4.5.)

Current limiters and *fuses* also may be used to open circuits that are drawing too much current. They may isolate specific components or one bus from another. Unlike CBs they are not resettable and usually must be changed by maintenance after opening their circuits. Most fuses blow relatively instantly, but in some cases slow-blow fuses or current limiters are used where a higher-than-normal current must be accommodated for short periods of time (during engine start cycles, for example).

Diodes correspond to check valves (one-way valves) in a water system; they allow electricity to flow only one way through a circuit, thereby protecting "upstream" components from electrical flow in the wrong direction.

Electronic circuit protection devices include a wide variety of automatic and computer-controlled sensors and switching systems. Among the more common ones are the previously discussed *generator control units (GCUs)*, which protect generators and their associated buses from faults. *Hall effect devices* are sometimes used to protect against one-way surges in current, usually between buses. Some aircraft, such as the Dash-8, have sophisticated computerized *bus bar protection units,* which electronically monitor and protect the entire electrical bus system from faults.

Bus ties are switches or relays used to connect or disconnect buses from one another. They serve to isolate failed buses from working ones and may be manually or automatically actuated, depending on the specific aircraft installation. Bus ties are also used to reroute electrical power to buses that have lost their normal power sources.

Reading an Airplane Electrical Diagram

The first question that comes to mind when examining an electrical diagram is "What happened to the circuit?" (Everyone wonders about this question but dares not ask it.) The answer is that an electrical diagram shows only the flow of power from its sources out to each of the powered components. The "return" portion of each circuit back to its source is not shown, but it's definitely there (Fig. 4.9).

One reason is that the schematic is easier to read without all those return lines. The other reason we've already discussed; in the metal aircraft that most of us fly, electrical current returns to its sources largely through "ground," meaning the vehicle's metal structure. (If the connection of a component to the aircraft's structure is poor, the circuit will not be complete, and the component won't operate properly. This situation is known as a "bad ground.")

Reading aircraft electrical schematics is not too difficult, if you keep a few things in mind. Remember that our key consideration, as pilots, is in the flow and distribution of electricity to operate key systems (see Fig. 4.10).

To reinforce the concept of reading flow of electrical power through a system, look at Figures 4.11 and 4.12, which show the flow of power through the same electrical system under two different situations: fully powered in flight and operating under battery power only.

As you can see, various electrical buses are powered or unpowered, depending on the scenario. By now you should be asking yourself, "What the heck is on all those buses, anyway?" Your aircraft *POH (Pilot's Operating Handbook)* or *AFM (Aircraft Flight Manual)* will have, in some form, a listing of the items on each electrical bus. Most aircraft fuse and circuit breaker electrical panels are also organized by bus. Figure 4.13 shows a partial listing of what might likely be found on the buses of our fictitious electrical system. Note which electrical items are unpowered, and therefore inoperative, when only battery power is available.

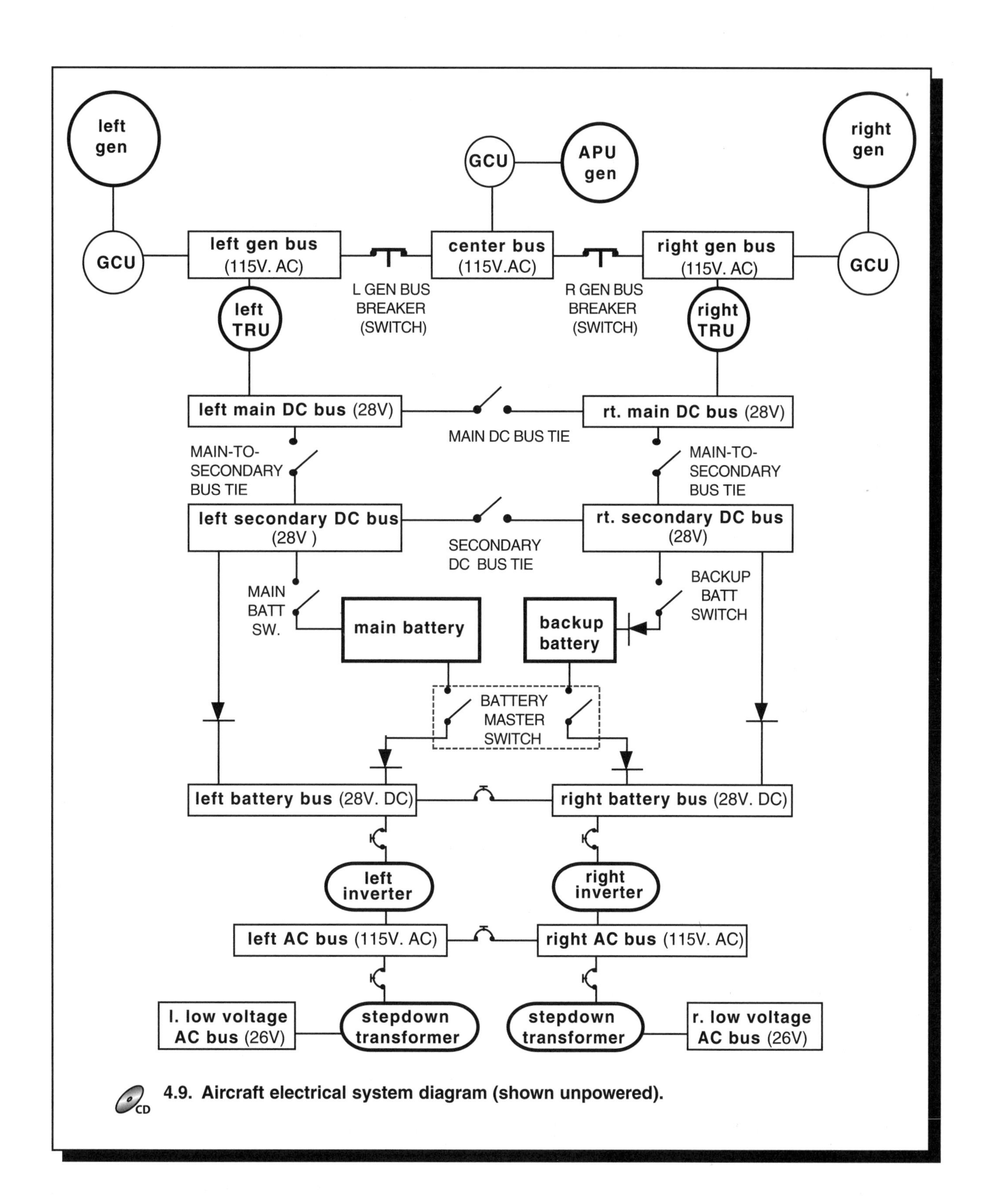

4.9. Aircraft electrical system diagram (shown unpowered).

Always start at the power source, and follow the flow of electricity through the system

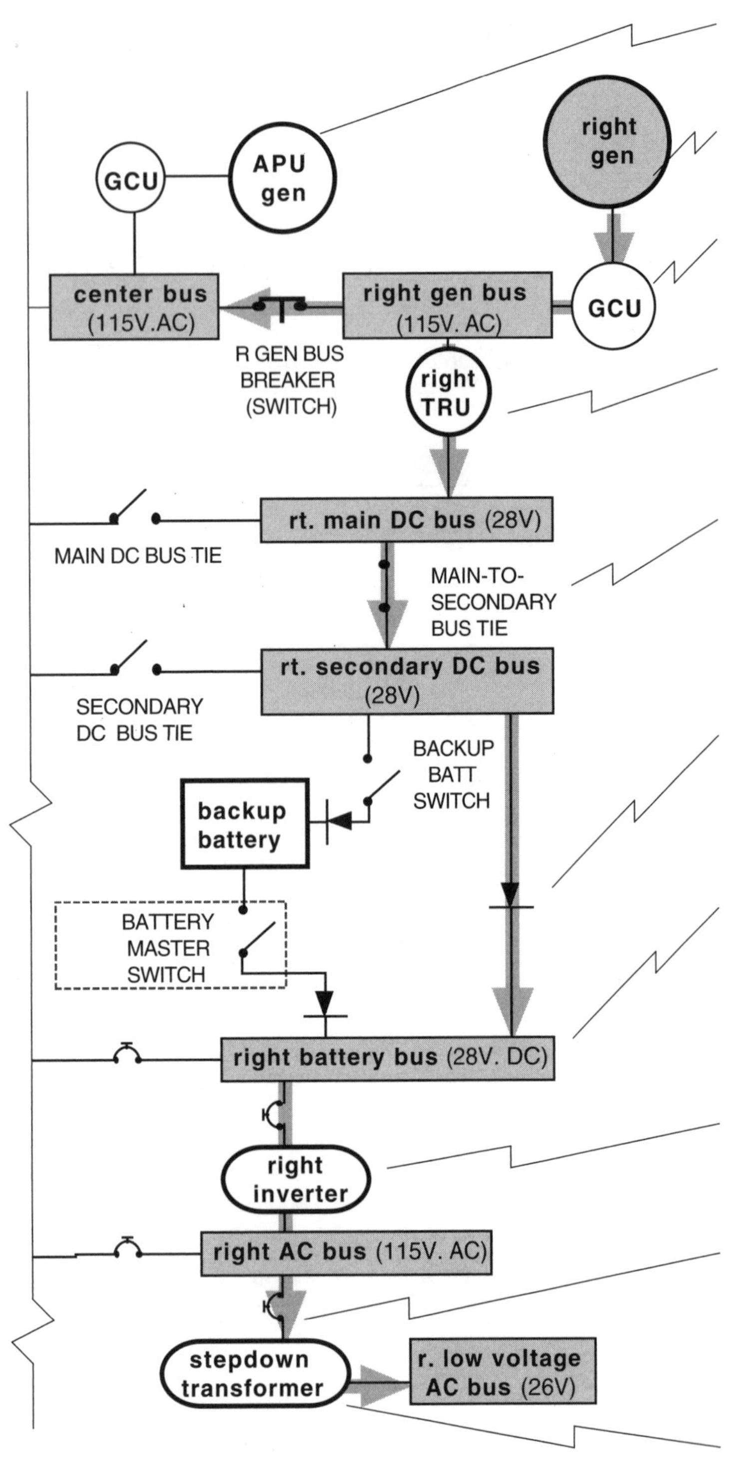

Electric generator driven by APU.

Electric generator driven by aircraft engine.

Generator control unit ("GCU") controls generator connection to system.

Transformer rectifier unit ("TRU") converts AC power to DC power (one direction only).

Bus ties are relays that disconnect buses or reroute power in the event of faults. Some are manual, some are automatic, and some are both.

Diode acts as one-way "check valve" in electrical system, allows electrical flow only in direction of arrow.

Electrical buses are independently powered electrical "manifolds." Electrical devices are powered by different buses, in order to provide circuit protection and alternate power sources for backup systems.

Inverter converts DC power to AC power (one direction only).

Circuit breaker ("CB") disconnects bus from circuit in the event of a short or other fault drawing high current.

Transformer changes electrical voltage (one direction only).

4.10. Reading an aircraft electrical system diagram.

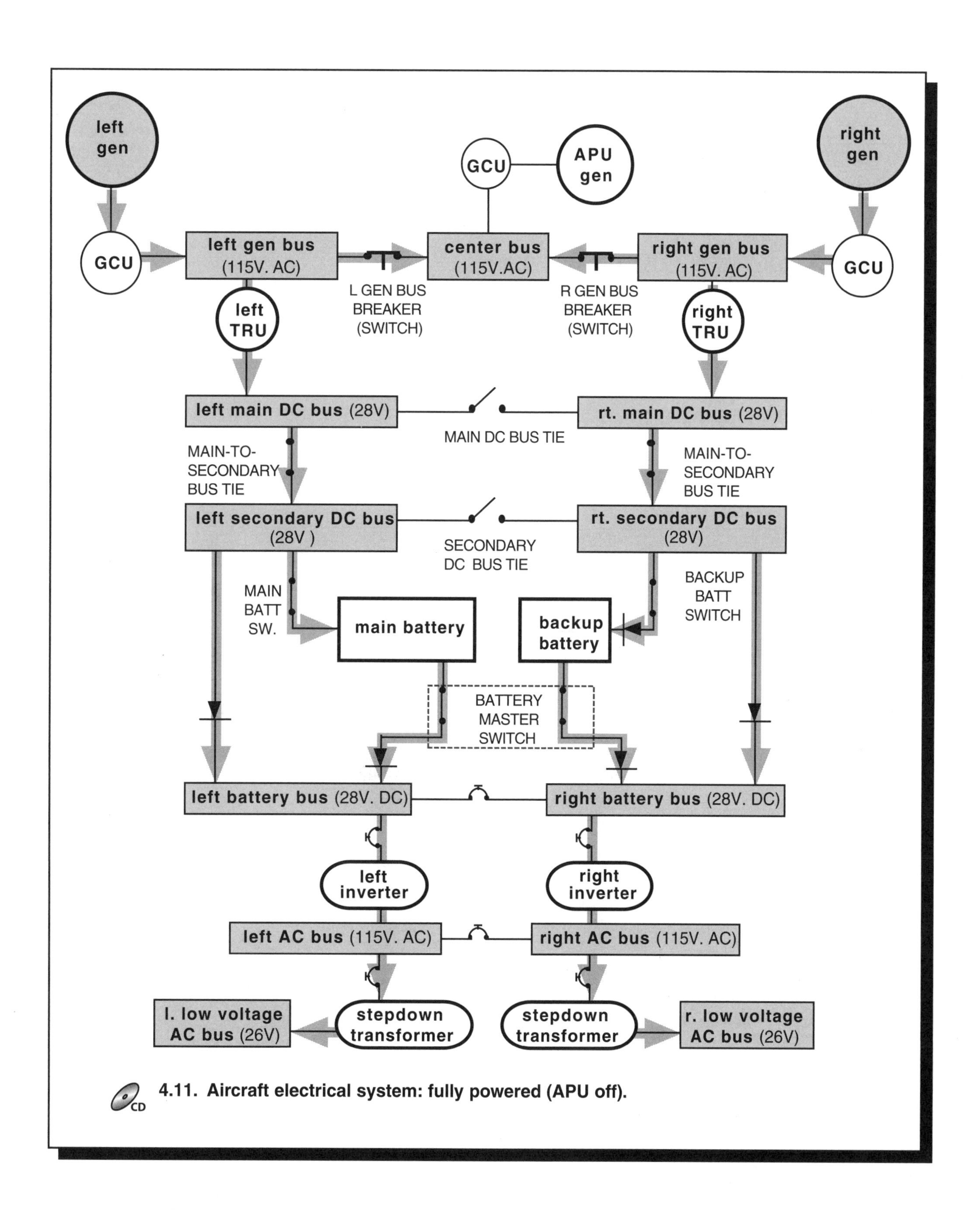

4.11. Aircraft electrical system: fully powered (APU off).

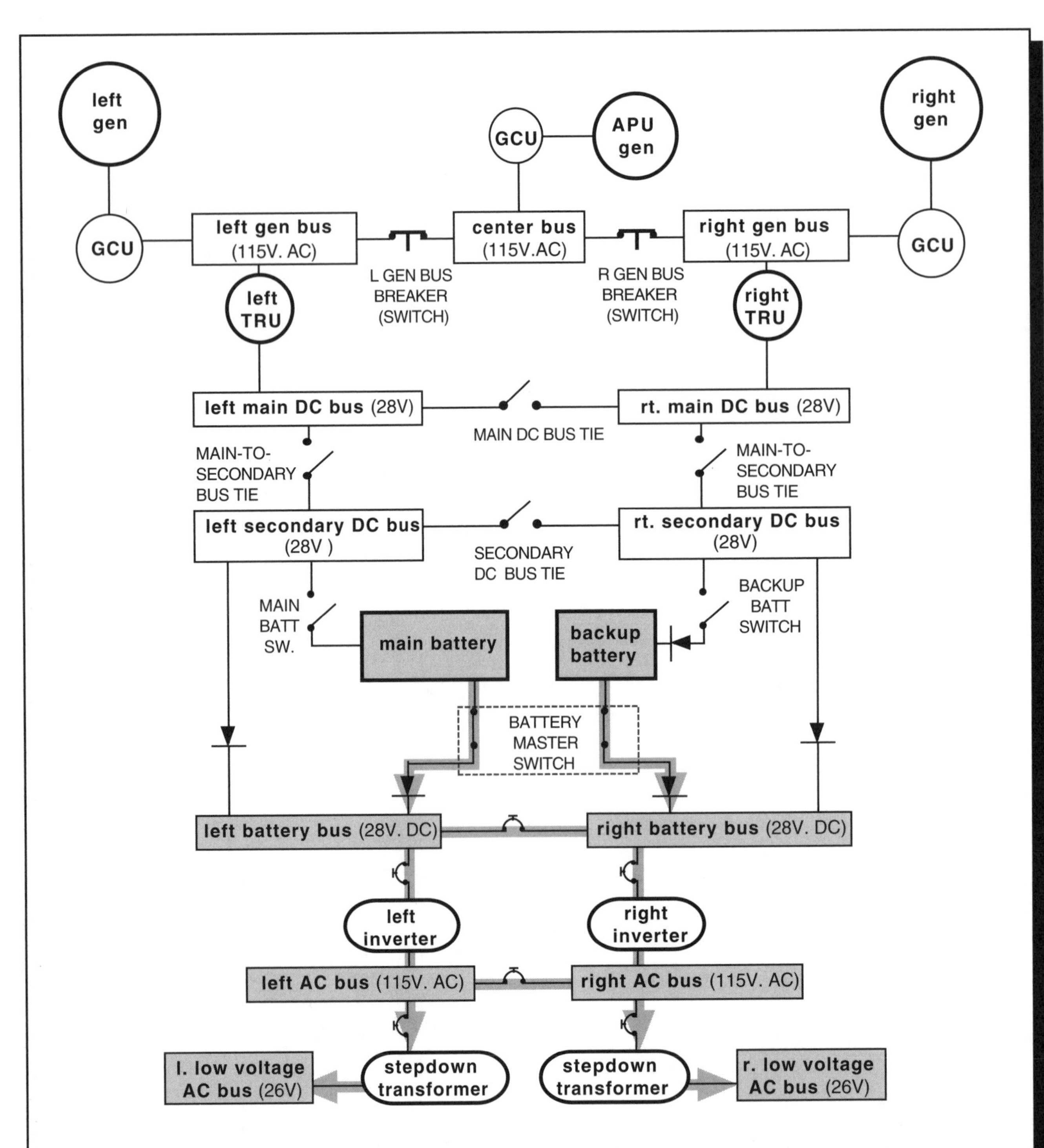

System is shown with battery master switch on, only. The batteries are powering the "battery buses," "AC buses," and "low voltage AC buses." The rest of the DC buses can also be battery powered by closing the main battery switch and some bus ties. (But not for long; that's lots of electrical load.) It is not possible to battery-power "center" or "gen buses" on this system.

4.12. Aircraft electrical system: battery power only.

L. Generator Bus (115V AC)
L. windshield heat
R. 1/4 windshield heat
L. electric hydraulic pump
Prop deice-inboards
Galley power

L. Main DC Bus (28V DC)
L. stall warning
Standby cabin pressurization
L. engine thrust reverser
R. windshield wiper
Inboard brake antiskid
Nosegear steering control

L. Secondary DC Bus (28V DC)
L. DC radio bus
FO's instrument panel lights
L. landing light
Prop deice controller- inboard
Emergency exit lights & charger (also self-powered by battery pack)

L. Battery Bus (28V DC)
R. engine fire extinguisher
L. pack valve
Master caution and warning lights
Fire detection systems, loops 1L & 2R
R. engine fuel shutoff valves
L. fuel quant. & temp. indicators
Captain's advisory display
L. DC engine instruments

Hot Battery Bus (28V DC, always powered)
(Note: not shown on aircraft electrical diagram.)
Fire extinguishers
Firewall shutoff valves (fuel & hydraulic)
P.A. system
Emergency intercom
Dome light

L. AC Bus (115V AC)
L. pitot heat (Captain)
L. radio bus
L. fuel aux. pump

L. Low-Voltage AC Bus (26V AC)
L. air data computer
L. ADI, HSI, ADF & RNAV, R. altimeter
Air data computer comparator
Hydraulic quantity indicator

R. Generator Bus (115V AC)
R. windshield heat
L. 1/4 windshield heat
R. electric hydraulic pump
Prop deice-outboards

R. Main DC Bus (28V DC)
R. stall warning
Auto cabin pressurization
R. engine thrust reverser
L. windshield wiper
Outboard brake antiskid

R. Secondary DC Bus (28V DC)
R. DC radio bus
Captain's instrument panel lights
R. landing light
Prop deice controller- outboard
Ground spoilers control
No smoking & seatbelt lights

R. Battery Bus (28V DC)
L. engine fire extinguisher
R. pack valve
Warning horns
APU ignition and starter
Fire detection systems, loops 2L & 1R
Engine ignition
L. engine fuel shutoff valves
R. fuel quant. & temp. indicators
Fuel crossfeed valve
FO's advisory display
R. DC engine instruments
Roll spoilers control
Flight recorders
Radio altimeter
RMI
Stby. attitude indicator

R. AC Bus (115V AC)
R. pitot heat (FO)
R. radio bus
R. fuel aux. pump

R. Low-Voltage AC Bus (26V AC)
R. air data computer
R. ADI, HSI, ADF, L. altimeter
Altitude alerter
Standby gyro

4.13. Distribution of electrical equipment by bus.

Troubleshooting

When troubleshooting, you'll evaluate your aircraft's electrical system by tracing electrical flow from each power source through the electrical buses. (Most of this process will be guided by checklists, but it sure helps to understand what's going on.) The effects of failures are considered using steps similar to the following:

1. Execute the appropriate abnormal checklist.
2. Identify the operating power source(s), and trace the flow of power from that source through operating buses of the system.
3. Determine which electrical buses are unpowered as the result of the failure(s).
4. Assess what you can do as pilot (such as closing bus ties) to power any of the inoperative buses. (Keep in mind the power capacity of remaining sources, as well as likely causes of the failure.)
5. Identify which important electrical devices remain on unpowered buses, and plan a course of action to complete the flight without them.

To illustrate this process, let's evaluate the failure depicted in Figure 4.14, using the preceding steps:

1. There is a fault on the right generator bus. Call for and execute the Generator Bus Failure Checklist. The bus has been isolated from the rest of the electrical system, along with the right generator.
2. The left generator continues to operate and is available to power the system, as are the batteries.
3. Only the right generator bus remains unpowered.
4. All appropriate pilot actions have been completed. The right generator bus cannot be repowered. Check that the left generator is producing adequate electrical power to operate all currently operating systems. Pilots should also be alert to any signs that the failure is adversely affecting operating buses.
5. Check Figure 4.13 to see which electrical components are on the right generator bus, and therefore unpowered and unavailable for use. Plan course of action for safe completion of flight, with reference to appropriate checklists.

To develop your troubleshooting skills, you may wish to make a few copies of the unpowered electrical diagram figure (for your own use only, of course). Simulate some failures, and using a colored pen or marker, trace the resulting flow of electrical power. Identify which buses can be powered under your scenarios and which cannot. Then consider what electrical equipment won't work and how you'd safely complete the flight. *This is exactly what you'll be doing in the electrical portion of your ground school.* Keep in mind that this troubleshooting process is similar for most aircraft systems.

Hydraulic Power Systems

The extensive hydraulic systems found in large aircraft may be totally unfamiliar to you, if you've been flying smaller recent-model reciprocating engine aircraft or light turboprops. The only hydraulic systems generally found in those smaller vehicles are simple brake systems and, in some aircraft, self-contained hydraulic "power pack" systems for landing gear extension and retraction.

Hydraulic systems are based on the simple principle that fluids are flexible but noncompressible. Assume you have a long tube full of fluid. If force is applied to the fluid at one end of the tube, that force is transmitted, with virtually no loss, to the other end (Fig. 4.15).

This can be illustrated, at the simplest level, by considering a basic brake system like that found in a light turbine twin (or even a Cessna 152, for that matter). The power transmission requirement is straightforward. The pressure created by the force of your foot must be transmitted to the disc brakes at the wheels, in order to stop the airplane.

This could have been done using cables and levers, but as early auto manufacturers learned, cables stretch and mechanical linkages wear over time. Combine this with mechanical friction, and much of the force applied by your foot is lost before it ever gets to the brakes. Besides, it takes a heavy and complex linkage to route mechanical force around all of those corners between your foot and the brakes.

The solution is hydraulic brakes. A simple piston in the brake master cylinder under your foot applies force, through a frictionless and flexible brake-line, to a similar piston in the slave cylinder at the brakes. The slave cylinder, in turn, translates the hydraulic pressure back into mechanical force to squeeze the brake caliper.

Furthermore, by controlling the sizes of the master and slave brake cylinders and pistons, designers can establish proportional mechanical advantage. (That's why you can stop an airplane weighing many thousands of pounds, using just the pressure applied with your feet.) (See Figure 4.16.)

Benefits of Hydraulic Power in Large Airplanes

In larger airplanes, hydraulics provide a powerful but relatively lightweight transmission method. Instead of using high-draw electric motors and heavy mechanical drivetrains to power each flap, landing gear, and spoiler, one hydraulic pump can transfer electric or engine power

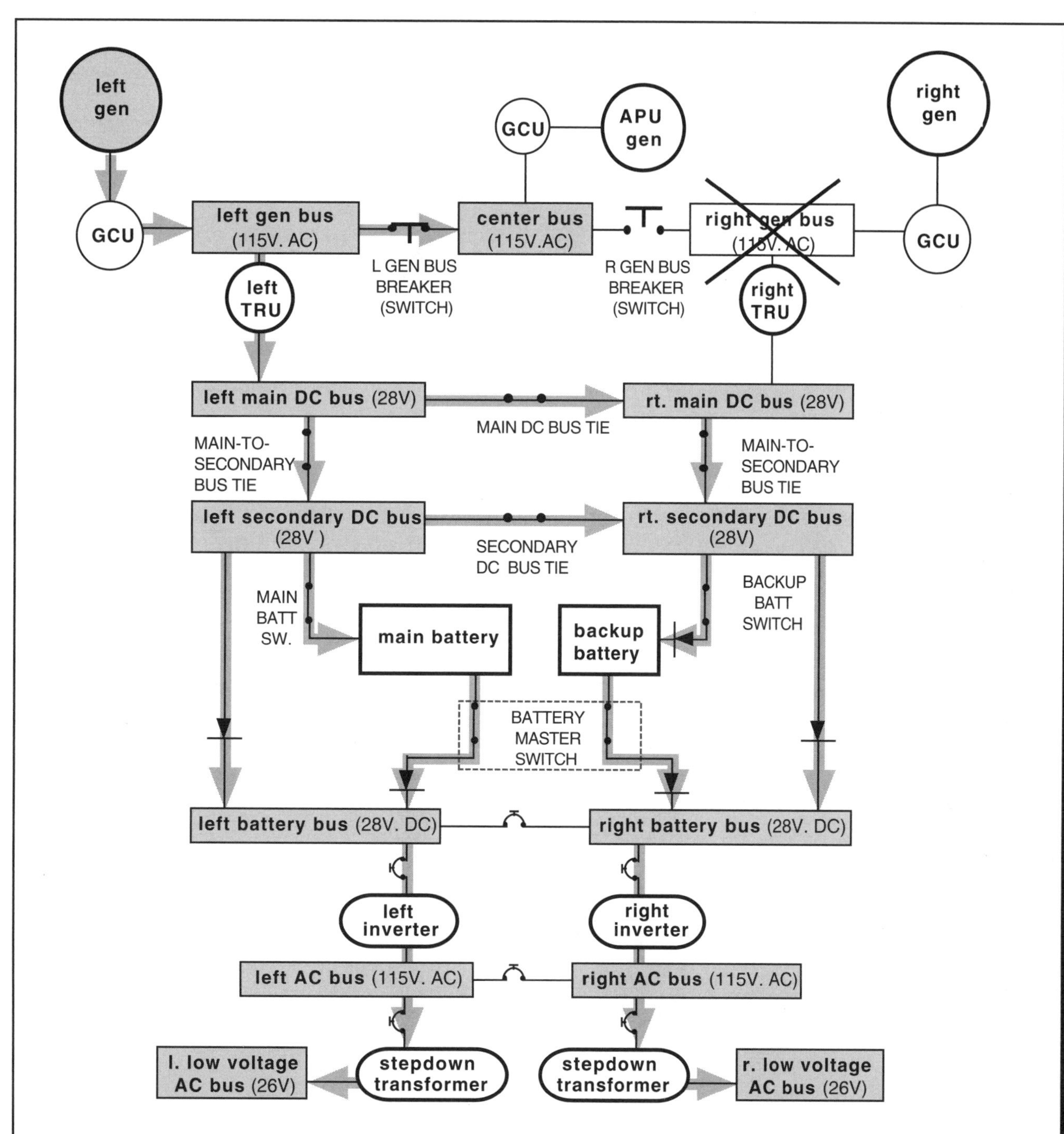

Right generator bus has been isolated due to fault. Pilots would be notified by caution or warning light, plus loss of various electrical components. Right GCU has taken right generator off line. Left generator is now powering all buses except for the "right gen bus." Any components on that bus are unpowered, and plans must be made to complete the flight without them. Course of action begins with pilot call for "Generator Bus Failure Checklist."

4.14. Aircraft electrical system: right generator bus fault.

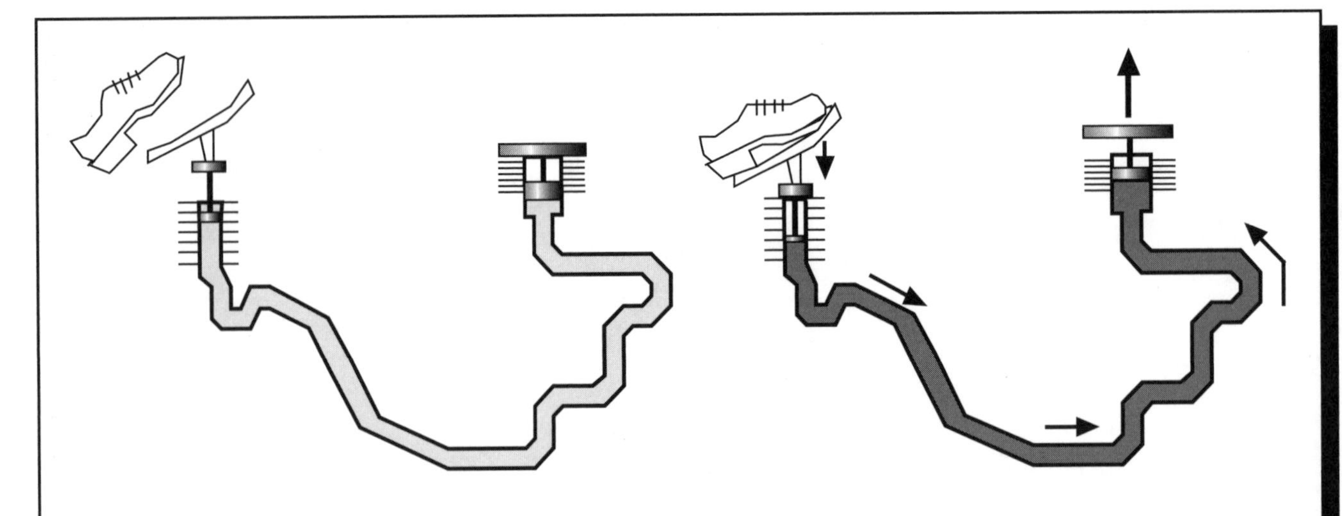

 **4.15. Hydraulic properties: power transmission.**

through a hydraulic system to run everything. Hydraulic power is especially valuable for heavy-duty applications because it can be drawn directly from engine power, so as not to tax aircraft electrical systems. Electric motors draw tremendous current when used for heavy-duty, intermittent operations. (That's why all the cockpit lights dim when you cycle electrically powered landing gear in a small aircraft.)

For these reasons, hydraulic power has proven more reliable and less maintenance intensive for heavy-duty applications than traditional electromechanical drivetrains. An added benefit is that since fewer and smaller electric motors are needed, the aircraft electrical systems can be lighter and last longer, too. Finally, an engine-driven hydraulic system provides another power source for redundancy on critical systems.

Hydraulic Systems and Components

Large aircraft hydraulic systems are much like our reference waterwheel system, which is a hydraulic system, itself. The differences are in the details. Aircraft hydraulic systems operate at very high pressures, usually around 3000 psi (pounds per square inch), and sometimes at high temperatures. Hydraulic fluids are specially formulated to withstand these conditions without vaporizing. This is because the key feature that makes a hydraulic system work is the noncompressibility of fluids. If the fluid vaporizes in the line, it becomes a gas with totally different properties, which can cause loss of transmission efficiency or even a "vapor lock" in small lines. (Incidentally, hydraulic fluids are often highly caustic; avoid getting them on your skin or in your eyes.) (See Figure 4.17.)

Most turbine aircraft are designed with two or more completely separate hydraulic systems. For standardization, these hydraulic systems are given letter or number designations (e.g., A System and B System or #1 System and #2 System.) These separate systems share the hydraulic workload and are designed so one hydraulic system can back up another in case a system pump fails or one or more systems lose hydraulic pressure or fluid. Some aircraft incorporate dedicated standby hydraulic systems for use when a primary system fails (see Fig. 4.20).

Hydraulic Pumps and Components

Hydraulic pumps are generally rotary pumps, geared directly off the aircraft engines or driven electrically. These pumps convert the rotary motion of the power source to hydraulic pressure and flow, which is then delivered to mechanical systems around the aircraft via

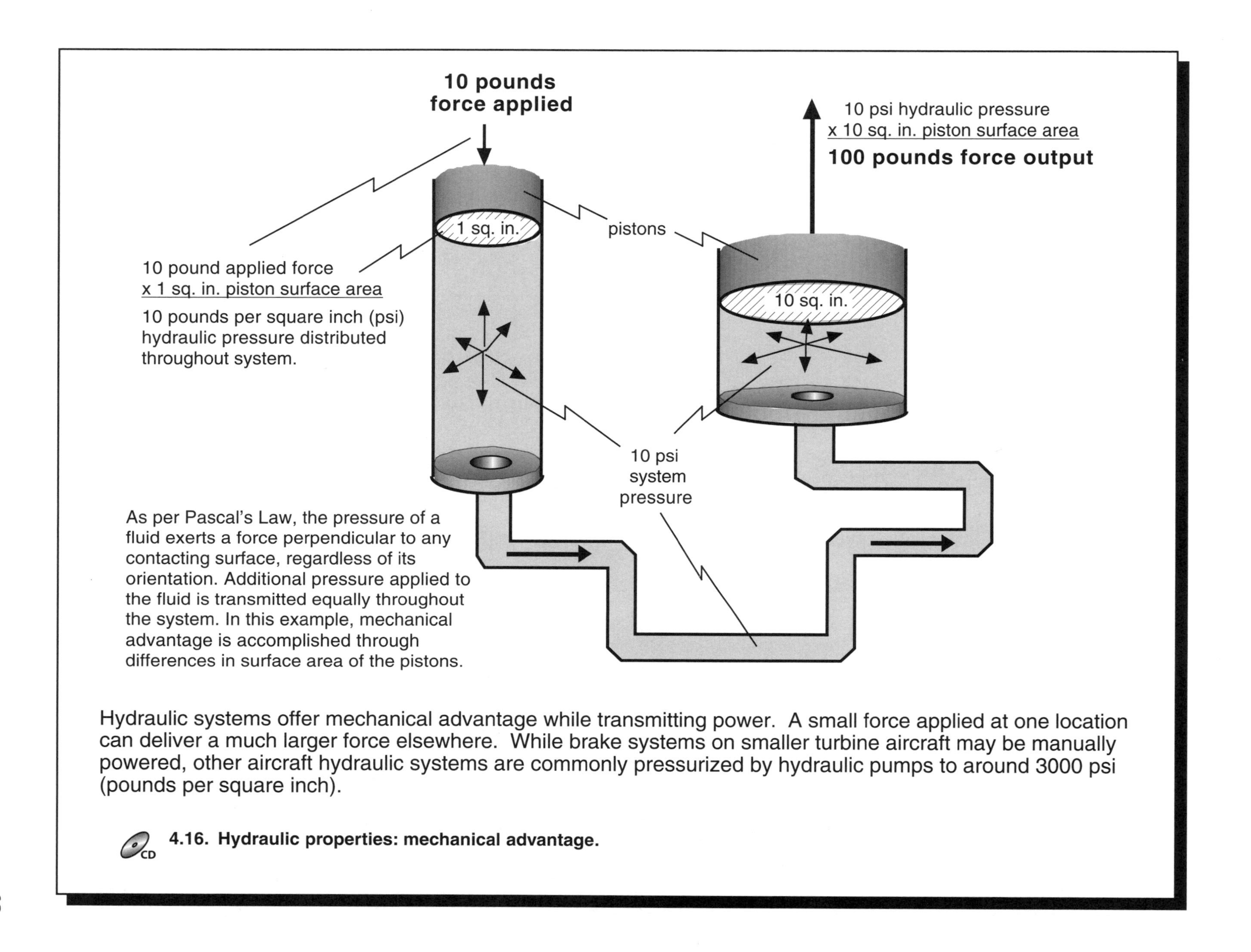

Hydraulic systems offer mechanical advantage while transmitting power. A small force applied at one location can deliver a much larger force elsewhere. While brake systems on smaller turbine aircraft may be manually powered, other aircraft hydraulic systems are commonly pressurized by hydraulic pumps to around 3000 psi (pounds per square inch).

CD **4.16. Hydraulic properties: mechanical advantage.**

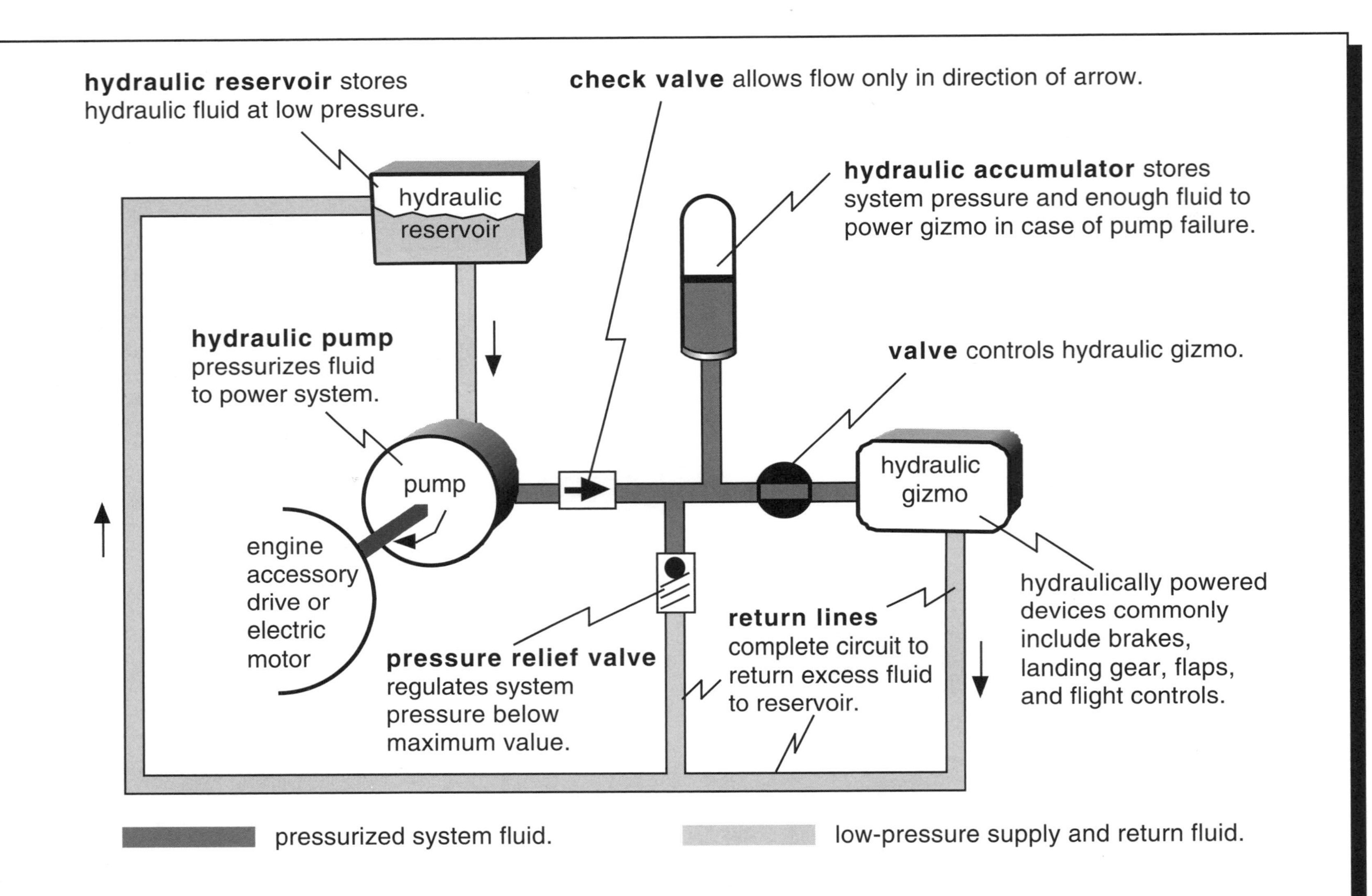

Hydraulic systems accomplish their work primarily through pressure, rather than flow. Therefore, fluid movement through the system is minimal. The hydraulic fluid is used more like a flexible metal rod pushed through a tube, rather than the flowing water of our reference water system.

4.17. Basic hydraulic components.

hydraulic lines. Pumps may be designed for continuous use or for periodic operation.

On large aircraft, one engine-driven pump is normally installed per powerplant, along with one or more electrically driven pumps for redundancy. An airplane's hydraulic systems are usually interconnected but with isolation valves installed to separate them. Each pump alone can normally power most or all of the hydraulic demand of the aircraft. Some aircraft have *standby hydraulic pumps* installed to back up the main pumps in case of failure.

Hydraulic motors are relatively small units (compared with electric motors) that convert hydraulic power back into mechanical power. They are normally rotary impeller units (basically "pumps run backward"), which convert hydraulic pressure and flow back into rotary output to turn shafts operating, for example, flaps or landing gear.

Hydraulic cylinders use pistons to translate hydraulic pressure into linear mechanical movement. These are used for many purposes, brakes being among the most obvious. Hydraulic cylinders are also used to power control surfaces, gear doors, air stair doors, and other devices with relatively short travel.

Hydraulic lines (flexible and rigid versions) deliver hydraulic power from the pump to the hydraulic motor.

Valves direct the flow of hydraulic fluid, and therefore power, to where it's needed. For example, in the case of hydraulic landing gear, valves can direct hydraulic flow/pressure to one side of a hydraulic landing gear motor for retraction or to the other side to reverse the motor's direction for extension. Remember that the valve needs to get the message somehow from the cockpit gear handle in order to properly direct the power. This actuation requires power in itself to change the position of the valve.

Hydraulic Reservoirs and Accumulators

Hydraulic reservoirs are required to retain adequate hydraulic fluid to operate the aircraft systems, as well as some reserve to allow for leakage. Usually, low-pressure air is applied to reservoir vessels to minimize foaming of hydraulic fluid (see Fig. 4.17).

Hydraulic *accumulators* store hydraulic pressure in order to provide backup for key operations in the event of a pump failure. They normally consist of a sealed pressure container with a diaphragm or piston installed. System hydraulic pressure compresses nitrogen or "dry air" (and sometimes a spring) to store energy for a short application of power in the event of a pump failure. Accumulators may also be used to control power fluctuations or surges in the system, much like a battery in an electrical system (see Fig. 4.18).

Hydraulic Backup Pumps

Aircraft hydraulic systems, as we've discussed, come in a variety of designs. Common to all is the need to have some sort of backup system for providing hydraulic pressure to operate critical systems, such as flight controls or landing gear, when the primary source of hydraulic pressure has failed. These backup systems range from simple to complex, depending on the relative complexity of the hydraulic system, itself.

Smaller turbine-powered aircraft usually come equipped with a hydraulic hand pump, which is used as a backup method to provide hydraulic pressure to extend the landing gear when the primary pump fails. Most hydraulic hand pumps are either single-acting pumps (fluid moves during power stroke but not during return stroke) or double-acting pumps (fluid moves during both strokes). During flight training in such aircraft, you will be required to manually extend the landing gear with the hydraulic hand pump. Many hand pumps require more than 100 strokes to extend the landing gear, meaning that emergency extension takes a lot of physical work and a good deal of time. So don't attempt manual gear extension on final approach with these types of systems; climb to level flight at safe altitude before performing the emergency extension (see Fig. 4.20).

In larger turbine aircraft having engine-driven pumps as their primary source of hydraulic pressure, electrically driven hydraulic standby pumps are used to supplement the primary pumps and serve as emergency backup. Generally electrically driven hydraulic pumps can produce the same pressure as the engine-driven pumps but with much less fluid volume. Therefore, when operating high-load devices, such as landing gear or flaps, using standby pumps, it may take much longer to operate these systems than with the normal engine-driven pumps. Plan accordingly!

Hydraulic power transfer units, found on more complex hydraulic systems, are hydraulically driven pumps that utilize pressure from one hydraulic system to pressurize the other system. Typically there is no exchange of fluid between systems and electrical power must be available to operate the unit. Hydraulic power transfer units are normally used as emergency backup to the engine driven hydraulic pump (see Fig. 4.20).

Air turbine hydraulic pumps or *air turbine motors* (*ATM*) use a large volume of compressed pneumatic system air from an operating engine, auxiliary power unit (APU), or ground air source to spin an air turbine geared to a hydraulic pump (See Fig. 4.20. See also "Pneumatic Power Systems" and "Auxiliary Power Units" later in this chapter).

Ram air turbines (*RAT*) can also be used to produce emergency in-flight hydraulic pressure; these work by

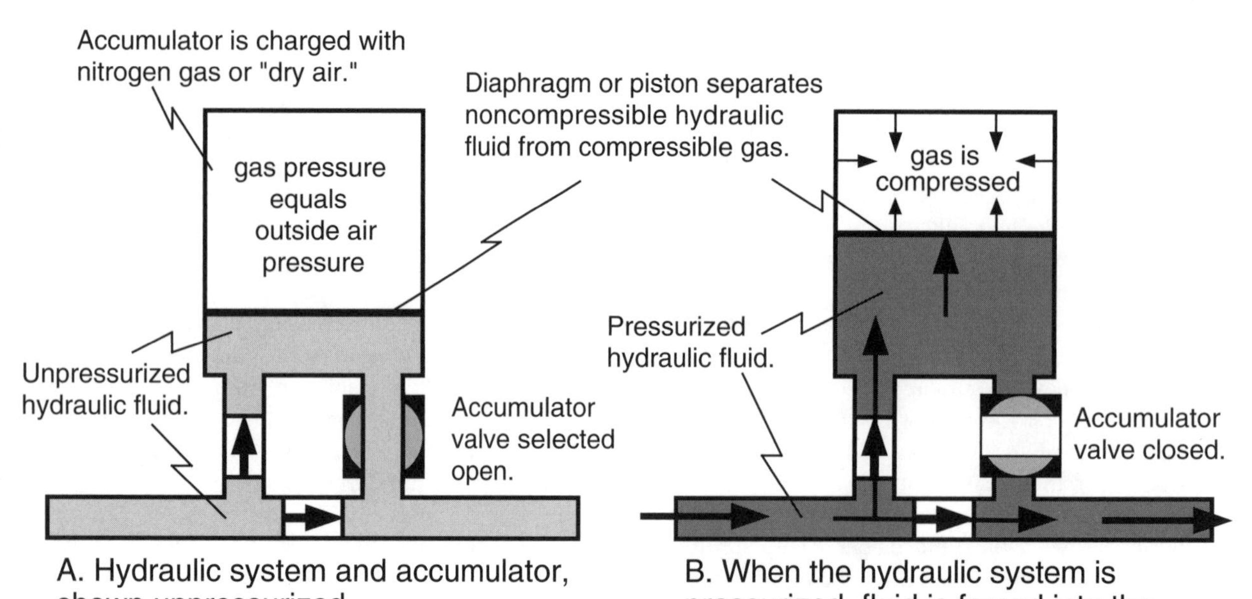

A. Hydraulic system and accumulator, shown unpressurized.

B. When the hydraulic system is pressurized, fluid is forced into the accumulator, compressing the gas to match system hydraulic pressure.

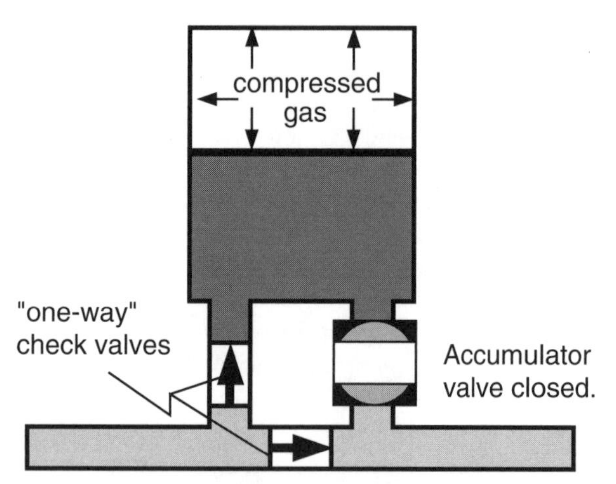

C. If system hydraulic pressure drops, the accumulator remains pressurized.

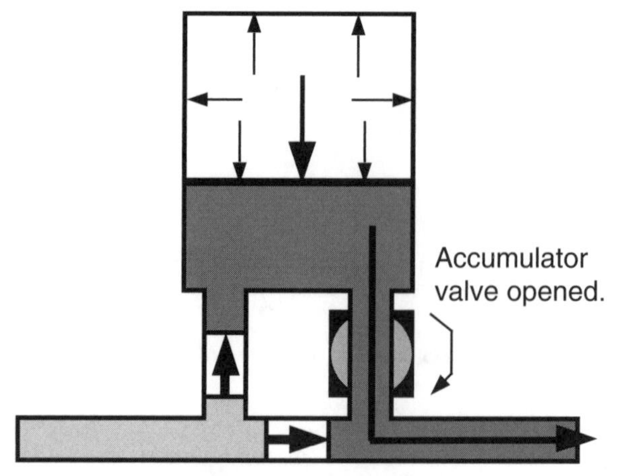

D. The pressurized accumulator can momentarily power the hydraulic system when the accumulator valve is opened.

Hydraulic accumulators store system pressure as backup power sources for critical systems in the event of hydraulic pump failure. (They store only enough pressure for limited or one time use.) This type, with selectable accumulator valve, would likely power emergency flap or landing gear extension. Other types, such as brake accumulators, may simply provide backup pressure and have no selectable valves.

4.18. Hydraulic accumulator.

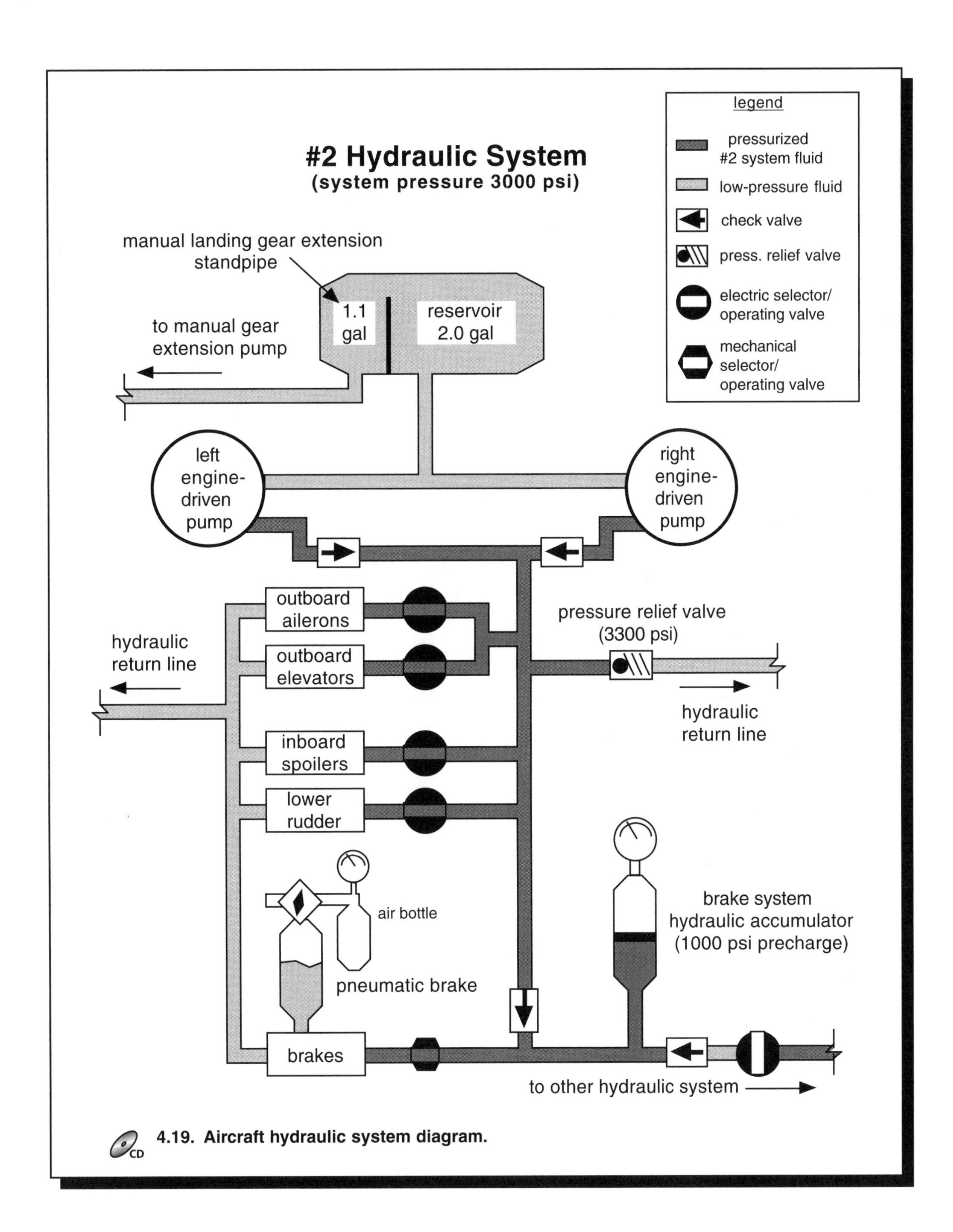

4.19. Aircraft hydraulic system diagram.

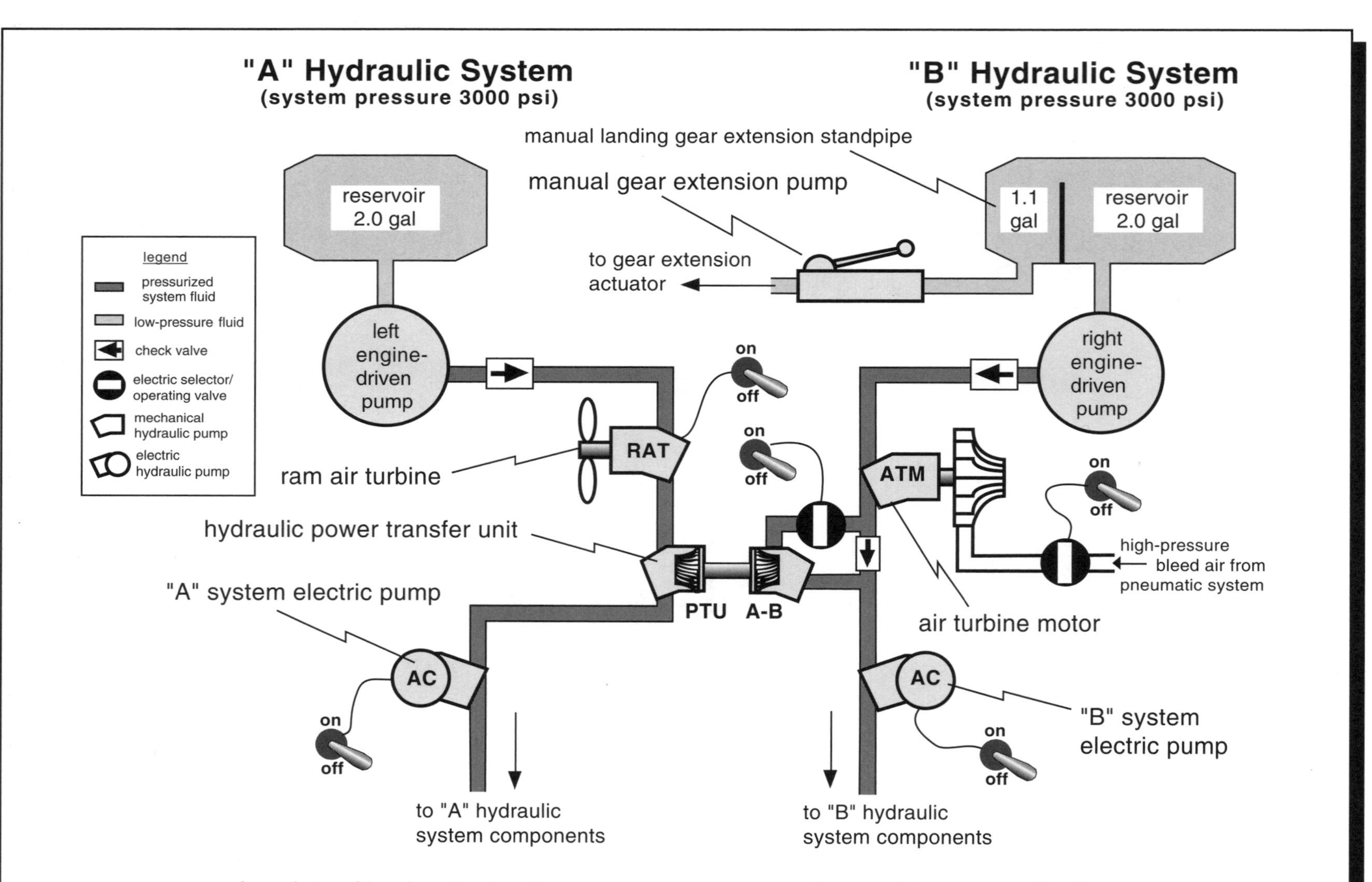

A variety of backup hydraulic pressure sources may be used to power critical hydraulic system functions should the primary engine-driven pumps fail. Depending upon the specific aircraft system, one or numerous backup pumps may be installed.

4.20. Backup hydraulic pumps.

extending an air turbine into the outside airflow. Spun by ram air pressure, the turbine, in turn, drives a hydraulic pump, thereby providing emergency hydraulic pressure. RATs are normally found on aircraft having only hydraulically actuated flight controls with no mechanical linkage between cockpit controls and flight control surfaces. In case of complete hydraulic system failure, the RAT then supplies necessary hydraulic pressure to actuate the flight controls. In order to produce the required hydraulic pressure to do this, the aircraft must be flown at a high enough airspeed to generate sufficient air pressure to turn the air turbine. (For example, the Lockheed L-1011 Tristar requires a minimum airspeed of 160 KIAS be maintained when using the RAT.) RATs are typically installed in the underside of the aircraft fuselage, where they can be easily extended by gravity alone. RATs may be extended manually or automatically depending on the installation (Fig. 4.20).

Hydraulic System Characteristics

One interesting characteristic of hydraulic systems is that most of the work is done with relatively little movement. Continuous-operation systems are kept constantly pressurized by hydraulic pumps, but fluid movement occurs only when something hydraulic is operating. Even then, fluid flow is minimal when small actuators and cylinders are being powered. In these cases, the work is done more by pressure than by flow. Compare a hydraulic system to a flexible rod being pushed through a tube—small movement at the driving end of the system is transferred hydraulically to push a faraway button, to move a valve, or to deflect a control surface.

Since hydraulic pressure is kept bottled up between the pump and devices it powers, hydraulic return lines generally carry little pressure or flow back to the reservoir. Exceptions occur when rotary hydraulic motors are powered to operate high-load devices such as landing gear and flaps. Then, both hydraulic flow and pressure are significant.

Pneumatic Power Systems

Pneumatics provide yet another method for transmitting engine power to various aircraft systems. In this case, the medium for power transmission is compressed air. Since air, being a gas, is compressible, pneumatic power is far less efficient than hydraulic power for heavy-duty jobs. On the other hand, pneumatic systems are much lighter than hydraulic systems, need little maintenance, and require no special fluids.

On piston airplanes, pneumatic power comes from pressure or vacuum pumps driven mechanically off the engines. These systems are basically pretty simple. Rotary pneumatic pumps correspond to the pump in our reference water power system. The ready source of atmospheric air eliminates the need for any type of reservoir, and control is relatively simple through a series of valves. Pneumatic systems in piston aircraft typically operate gyro instruments, pressurization, and deicing boots.

Turbine-powered aircraft use pneumatic systems for the same types of applications, plus many more. The reason is that turbine engines are essentially giant pneumatic pumps. As you remember, the engine's gas generator compresses huge amounts of air to support combustion. It's a relatively simple matter to draw this "bleed" air from the engines and use it to power all sorts of things (see Fig. 4.21).

High-pressure bleed air is drawn from the compressor section of a gas turbine engine. Some systems draw bleed air from two or more "stations" on each engine, yielding different pressure and temperature outputs. (These differing bleeds are sometimes identified as "high-pressure bleed air" and "low-pressure bleed air." In this book, the terms "high-pressure bleed air" and "bleed air" are used interchangeably to refer to all such variations.)

High-Pressure Bleed Air

High-pressure bleed air has many applications straight out of the engine, including engine and wing thermal anti-ice. It is also used, via some combination of air cycle machines (ACMs, or Packs) and heat exchangers, for cabin pressurization, heating, and cooling. (See "Pressurization" and "Environmental Systems" in Chapter 5.)

On large turbine aircraft, high-pressure bleed air also powers the engine starters. Large engines such as the P&W JT8D are started pneumatically, as an alternative to the heavy electrical loads and motors required for electrical starting of turbine engines.

Typically for engine starting, the aircraft APU is started electrically to provide bleed pneumatic power, or else a ground pneumatic power source is used. When the captain calls for starting, a *start valve* is opened, sending compressed air to spin up a small turbine in the pneumatic starter. This, in turn, spins up the engine core compressor. Fuel is introduced, and when N_2 (high-pressure compressor rpm) reaches a predetermined value, the start valve is closed, and the starter turbine is disengaged from the engine's compressor.

Low-Pressure Air

Some high-pressure bleed air is metered through a *pressure regulator* to much lower pressures, normally around 18 psi. This *low-pressure air* (sometimes called *instrument air*) corresponds, in pressure and in applications, to the

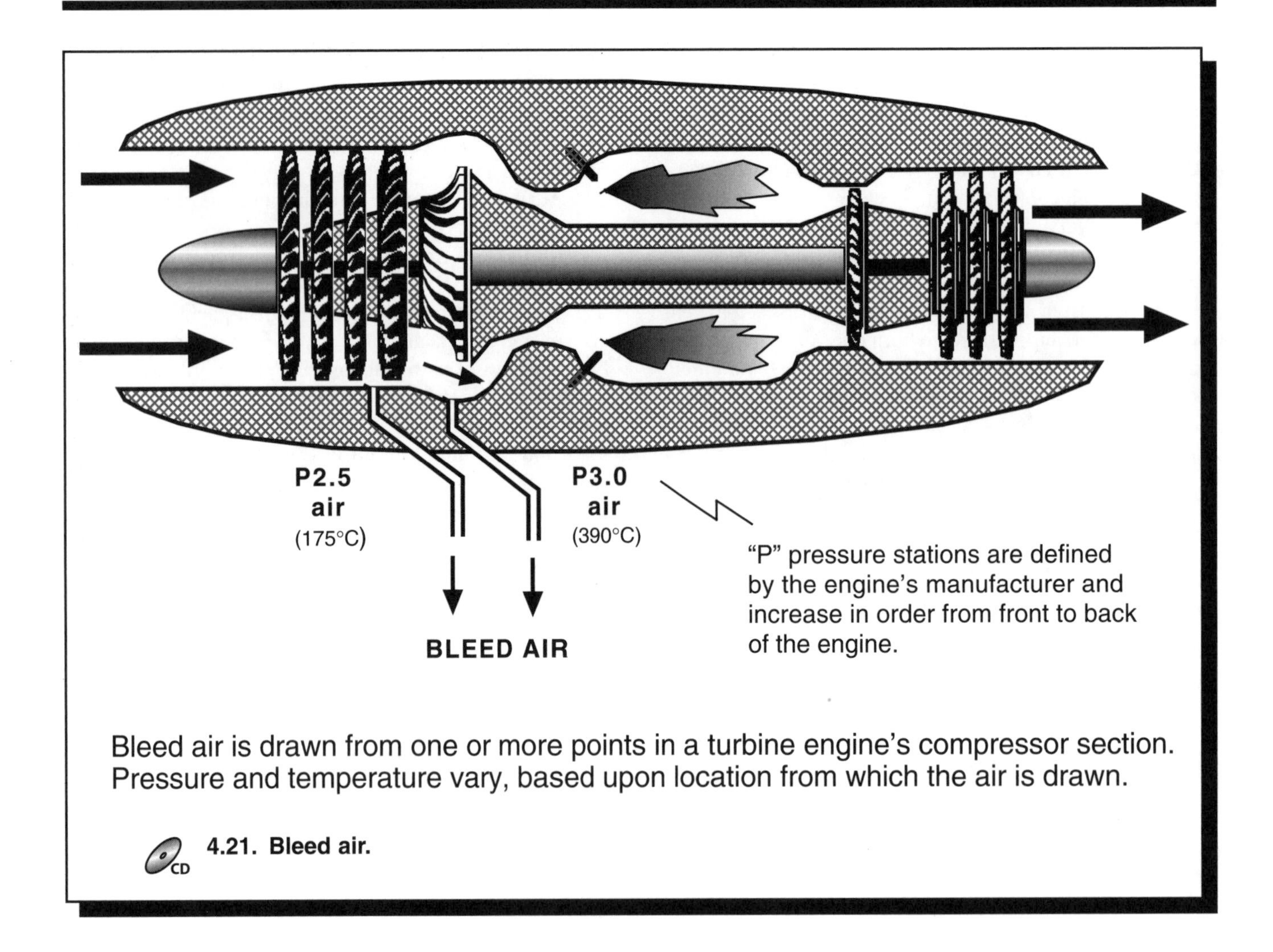

4.21. Bleed air.

piston airplane pneumatic (vacuum) systems you may be familiar with. As such, low-pressure air in turbine aircraft drives air gyros and pneumatic deicing systems, if installed. Low-pressure air may also be used for mechanical control functions such as valve actuation (pneumatic and/or hydraulic), control of pressurization outflow valves, and inflation of door seals to sustain pressurization (Fig. 4.22).

Bleed Hazards and Protections

Since high-pressure bleed air is drawn directly from the engines, bleed usage on many turbine aircraft must be managed by phase of flight. For example, use of bleed air impacts the power produced by the engine supplying it, since air and pressure are being bled away from the engine prior to the combustion chamber. Therefore, in most turbine aircraft, high-pressure bleed air use is restricted during takeoff and go-around situations. Low-pressure air remains operative on takeoff to power flight instruments, but other bleed-powered items may be turned off for departure, which are then reactivated as part of the after-takeoff checklist, if appropriate.

Sometimes weather conditions at takeoff do require use of high-draw bleed items, such as engine inlet anti-ice systems. For these situations, most turbine aircraft have correction factors in their takeoff power and performance charts, often restricting takeoff weights and/or speeds.

Use of bleed air for thermal wing anti-ice is virtually never permitted for takeoff. Not only do the bleeds sap lots of engine power, but on the ground, the 600°–800°F bleed temperatures can weaken the aluminum wing leading edges. (See "Ice and Rain Protection" in Chapter 6.)

Bleed pneumatic systems do have some other hazards associated with them. Due to the very high temperatures involved, fire can result from unattended bleed air leaks within the aircraft. Therefore, bleed systems are normally monitored in the cockpit. Temperature probes are often used to monitor passages carrying bleed air lines. Other systems run pressurized EVA (ethylvinylacetate) tubing along bleed lines. This material melts and signals a sensor when a leak

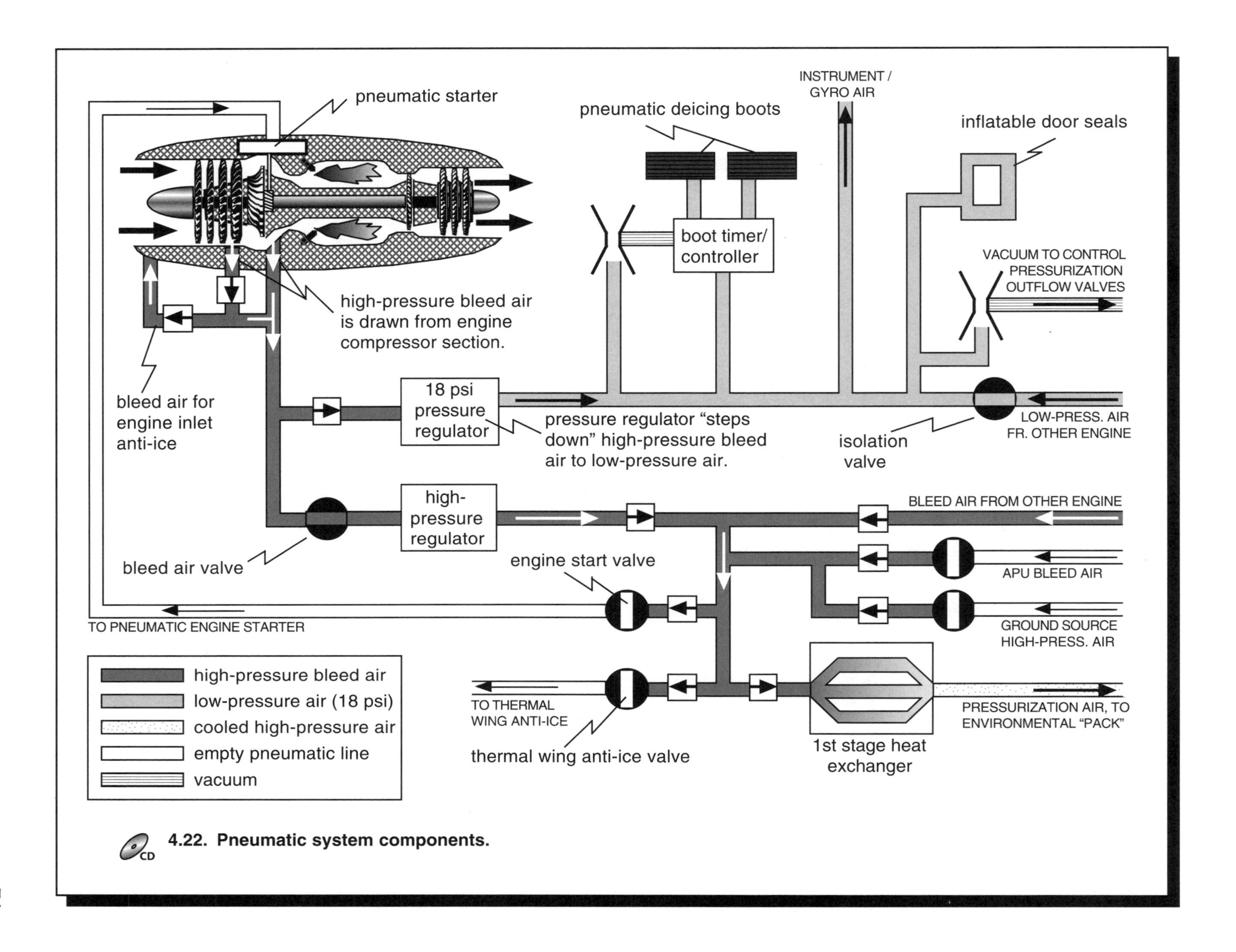

4.22. Pneumatic system components.

occurs. In the event of a high-temperature leak, the associated bleed source must be shut off, followed by execution of the appropriate "abnormal" checklist. Some larger aircraft also have fire extinguishing capability for engine bleed lines in critical areas. (See also "Fire Protection Systems" in Chapter 6.)

Pneumatic engine starting systems offer some interesting hazards of their own. Pneumatic system pressure must be monitored throughout the start cycle (described earlier) because, if the start valve fails to close as the engine starts, bleed pressure from the now operating engine can blow back through the starter at high pressure. There have been several cases where a sticking start valve caused the pneumatic starter to be literally blown off the airplane, through the engine cowl, and quite some distance from the aircraft. Therefore, engine start must be aborted immediately if the start valve fails to close when it should.

Like other aircraft systems, pneumatic systems have multiple power sources and protective devices to minimize the effects of system failures. Redundancy is accomplished through use of bleed sources from all engines and sometimes from the auxiliary power unit (APU), when installed.

Check valves, like the ones in our reference water system, are used to restrict pneumatic flow in certain areas to one direction. *Isolation valves* (automatic or manual) allow separation of interconnected bleed systems in the event of leaks or other failures. Normally, if one bleed is shut off at its source, most systems can be operated on the remaining bleed air source(s), except under very high-draw situations. However, if a serious leak occurs somewhere in the pneumatic system, requiring use of an isolation valve, some bleed-powered items will likely become inoperative.

Let's say, for example, that a leak develops on the left side low-pressure pneumatic system on your deice boot-equipped aircraft. First, you'll have to turn off the left bleed valve in order to prevent bleed air leakage from sapping engine and system power. You'll also have to isolate that side of the low-pressure system, which means probable loss of some air-powered flight instruments and perhaps some pressurization control functions. Finally, before using your pneumatic deice boots, you'll need to consider how they are affected. (Most modern aircraft have boots allocated so that deicing capability may be lost on some control surfaces, but not asymmetrically.) Obviously with this sort of problem you'll want to get on the ground promptly in icing conditions.

These are the kinds of scenarios you'll have to memorize in ground school for your particular aircraft type.

Auxiliary Power Units

Ever wonder about that little exhaust port in the tail of many turbine aircraft that looks like it's from a tiny jet engine? Well, that's exactly what it is. An *APU* (*auxiliary power unit*) is a small turbine engine installed to provide

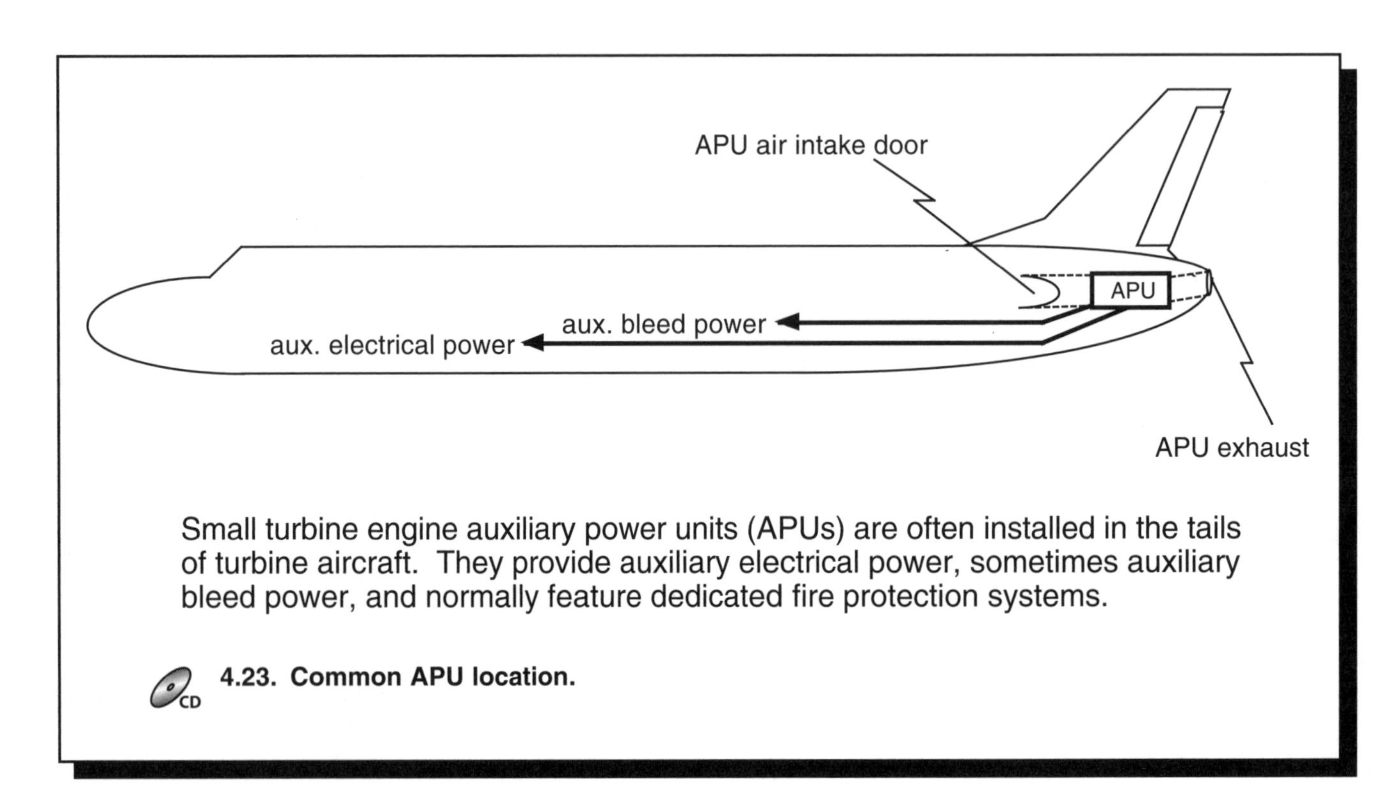

Small turbine engine auxiliary power units (APUs) are often installed in the tails of turbine aircraft. They provide auxiliary electrical power, sometimes auxiliary bleed power, and normally feature dedicated fire protection systems.

4.23. Common APU location.

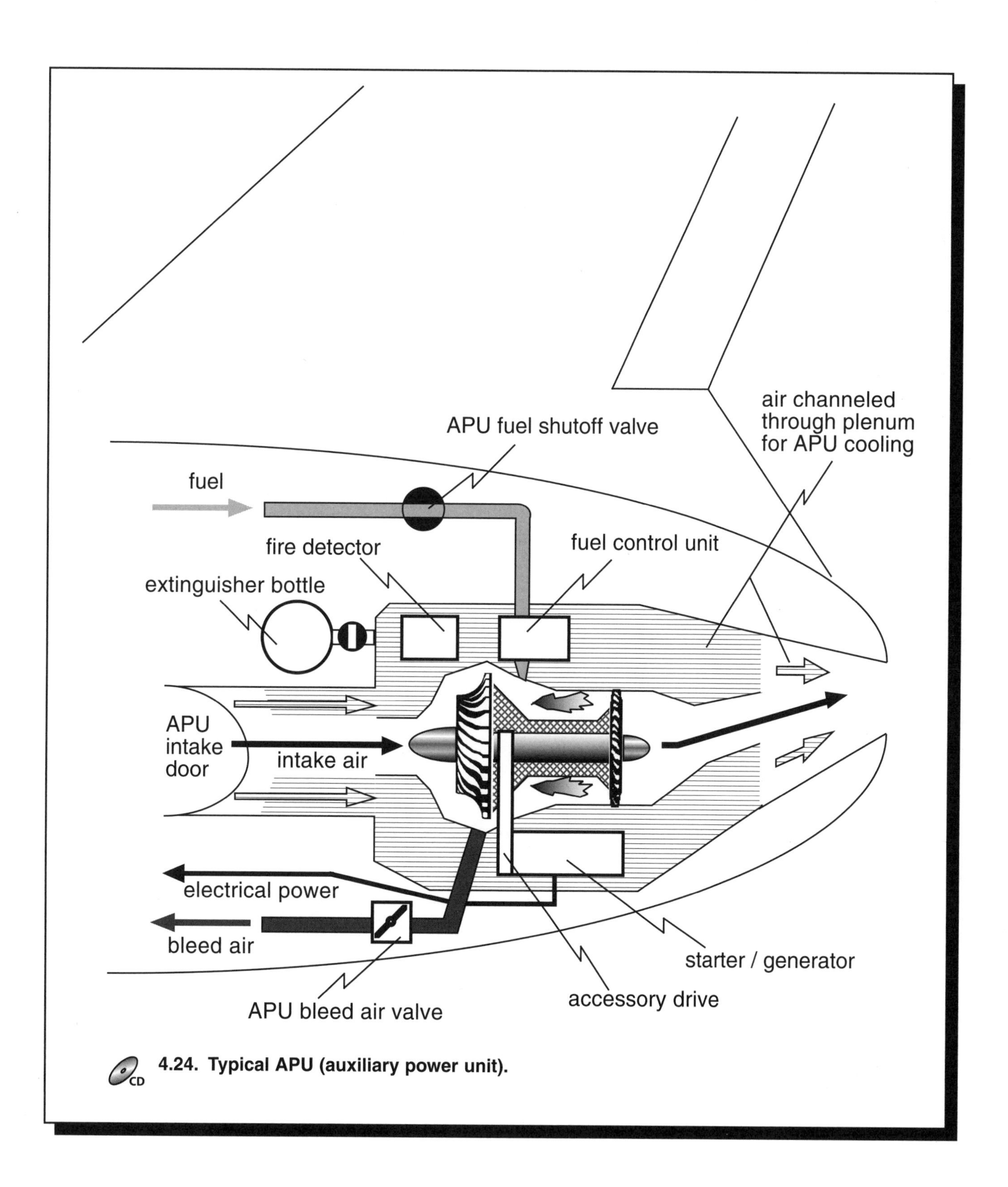

4.24. Typical APU (auxiliary power unit).

supplementary aircraft power. APUs are often found in the tails of aircraft ranging from larger turboprops to jets (or in the wing root or fuselage on tri-jet aircraft). APUs serve a number of useful purposes. (See Figure 4.23.)

APUs are installed with dedicated generators to provide auxiliary electrical power, in addition to that provided by aircraft engine-driven generators. This is valuable for running aircraft systems on the ground without powering up the engines, especially at facilities where no ground electrical power is available. Applications include powering environmental systems such as air conditioning, supporting maintenance of major electrical systems, and providing power for crew functions such as preflight, cabin cleanup, and galley (kitchen) operation. APUs on many (but not all) aircraft may be operated in flight, providing backup power for engine generators.

Also, APUs on larger aircraft are plumbed to provide an auxiliary bleed air source. As you remember, the large jet engines used on airline aircraft must be started using pneumatic power. Unless a ground pneumatic source is available, there is no way to start a large turbine engine without an operating APU (unless another engine is already running, of course). For this purpose, the APU's small turbine is started electrically. Once up and running, APU bleed air is routed through start valves to pneumatic starters on the main engines. These, in turn, spin up the engine compressors for starting. (See Figure 4.24.)

APUs on many aircraft can be used to provide backup pneumatic power for pressurization in flight and to back up environmental systems on the ground and in the air.

CHAPTER 5

Major Aircraft Systems

Flight Controls

Most smaller corporate and commuter turboprops use the same rudimentary flight control systems found in light aircraft, incorporating simple cable- or pushrod-actuated control surfaces and electric trailing edge flaps.

Control systems become much more complex, however, as aircraft get larger and operate under greater speed ranges. One reason is that most turbine-powered aircraft must resolve the aerodynamics of high-speed cruise with the slow flight and high lift required for safe takeoffs and landings.

High-speed flight is best achieved with small, thin, low-camber swept wings of high wing loading, since these characteristics minimize drag. Safe takeoffs and landings, on the other hand, are best accomplished using thick, high-camber, high-lift wings with low wing loading (and therefore high drag).

So, in order to create practical high-speed aircraft, in essence, two different airplanes are required: one that can go fast and one that can get everybody off the ground in less than 10 miles of runway.

Control Surfaces

Flaps and Leading Edge Devices

The combination of high-speed cruise and acceptable takeoff, approach, and landing performance is achieved through extensive use of *flap* and *leading edge device* (*LED*) systems to convert the wing of an aircraft from one configuration to the other. These devices are extended only when needed and then can be tucked away in the wing structure for high-speed flight (Fig. 5.1).

As on the piston aircraft that you may be familiar with, flaps on turbine aircraft are high-lift devices. They increase a wing's lift by increasing the camber of the wing, and with certain types of flaps they increase the surface area of the wing as well (e.g., Fowler flaps). Turbine aircraft, almost without exception, use slotted flaps, Fowler flaps, or a combination of the two.

Slotted flaps, when extended, form a space or slot between the leading edge of the flap and the inside of the wing (see Fig. 5.2). This gap allows fast-moving, high-pressure air to pass through the slot and over the upper portion of the lowered flap, preventing air separation and increasing the lift produced by the wing when the flap is lowered. This arrangement is so effective that many aircraft employ double- or even triple-slotted flaps.

Fowler flaps, when extended, move on tracks aft of the trailing edge of the wing structure instead of hinging downward (Fig. 5.2). The benefit of Fowler flaps is that they change the wing's camber and also increase its surface area. During initial flap extension settings, Fowler flaps increase the wing's lift because both camber and surface area are effectively increased. Drag, on the other hand, increases only slightly. Further flap extension (midrange) drives the flaps downward, increasing both lift and drag. The last few flap settings (fully extended) increase drag without adding much lift, thereby providing better approach performance by steepening the descent without increasing approach speed.

Leading edge devices (LEDs) are another type of control surface used on large turbine aircraft to increase lift produced by the wing. By effectively lowering the angle of attack for the leading edge of the wing, LEDs increase lift by keeping airflow attached over the top of the wing at high angles of attack. The most common leading edge devices are slats and leading edge flaps (Fig. 5.2).

Slats are secondary airfoils mounted on the leading edge of the wing, which, when extended, increase the wing's camber. The flap lever extends the slats hydraulically (and automatically) as flaps are extended, typically during the

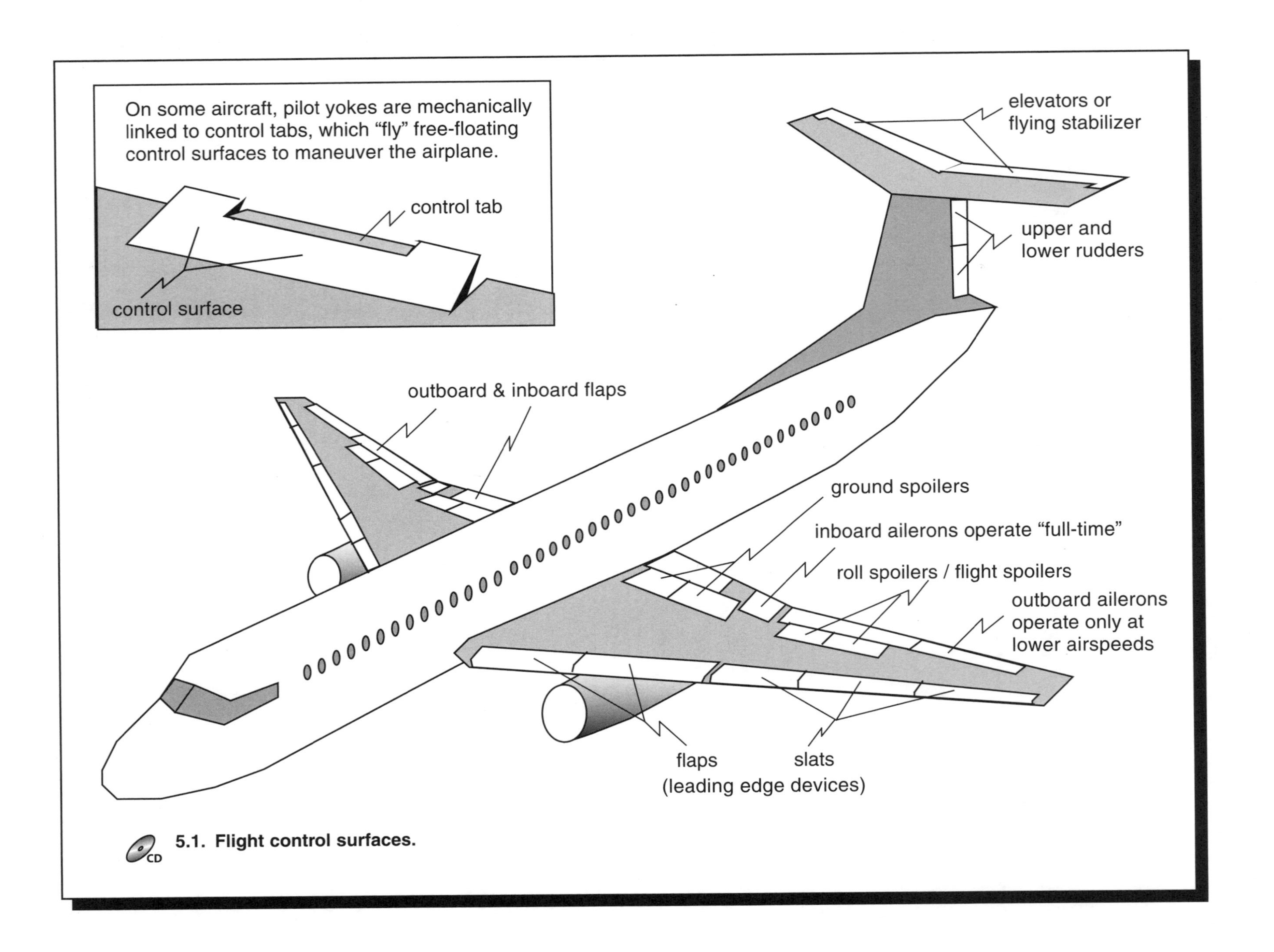

5.1. Flight control surfaces.

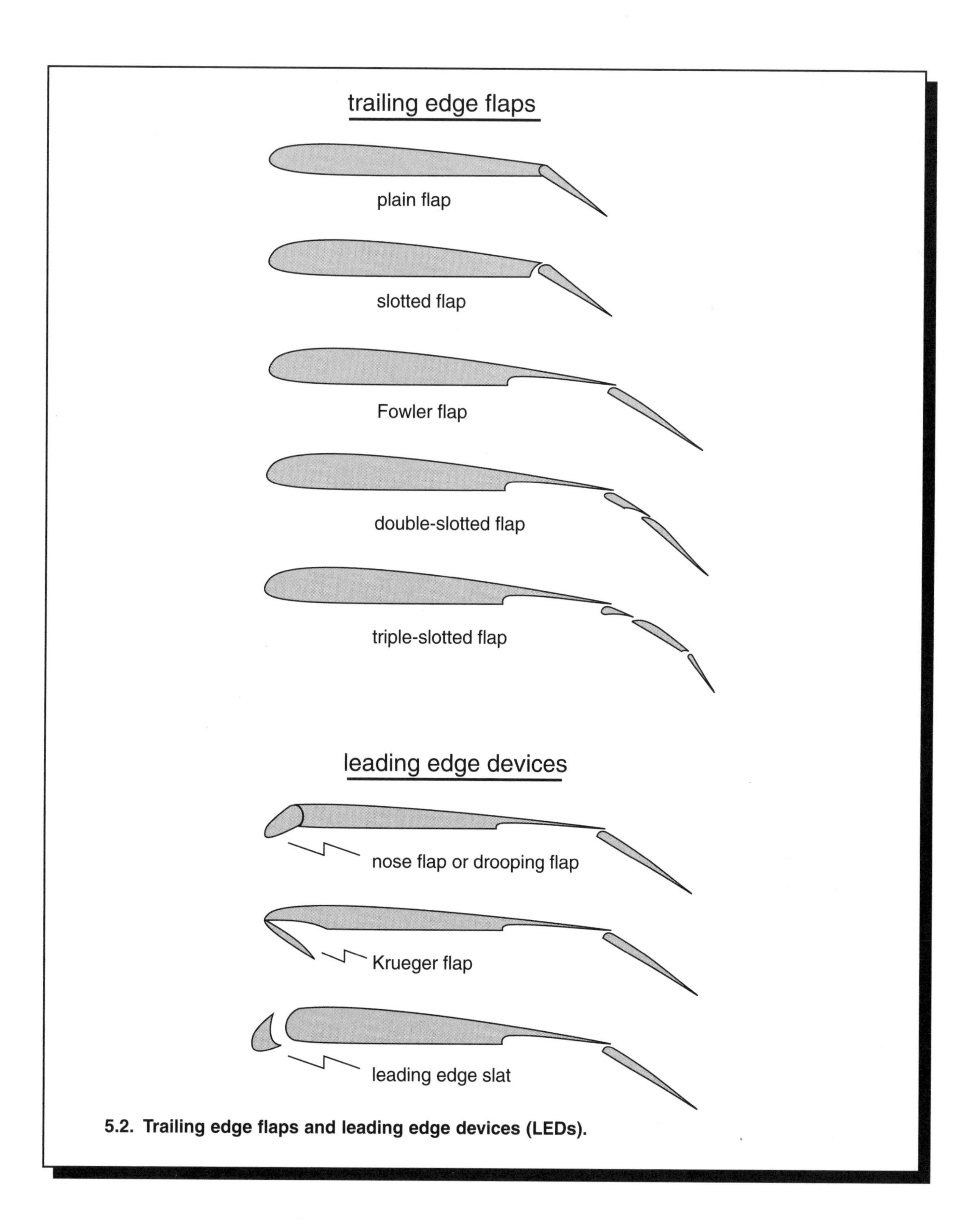

5.2. Trailing edge flaps and leading edge devices (LEDs).

first few flap extension settings. As slats extend, they form a slotted leading edge, allowing high-pressure air from below the wing to flow over the upper portion of the wing and delaying air separation as trailing edge flaps are extended.

Leading edge flaps include nose flaps, sometimes called *drooping flaps* because of their appearance when extended, and *Krueger flaps*. The effect of each type of flap is to increase camber and attach airflow over the top of the wing. Nose flaps are hinged so as to effectively droop the leading edge of the wing downward. Krueger flaps hinge forward from under the leading edge of the wing, thereby acting as air dams to force airflow over the top of the wing (see Fig. 5.2).

Ailerons

In order to make wing size (and drag) small for optimum cruise speeds, high-lift devices (LEDs and flaps) must extend across as much of the wing's span as possible for adequate low-speed effectiveness. This arrangement leads to some interesting design challenges when it comes to roll control.

The huge speed range of such aircraft results in vastly differing roll control needs in the same airplane. Although small ailerons are sufficient for high-speed cruise, they're generally inadequate for effective low-speed roll response. Conversely, ailerons large enough for effective low-speed performance may cause overcontrol at high airspeeds.

The common solution to this problem is installation of multiple ailerons on each wing, separately activated as a function of airspeed. The Boeing 767, for example, has two ailerons per wing; the inboards operate continuously but outboard ailerons operate only below 240 knots to assist roll control at slow speeds.

A related problem is that aileron size on high-speed turbine aircraft is severely limited by those big spanwise flap requirements. There's only so much room on the wing for control surfaces, and often it's simply not enough for adequate ailerons. As a result, many turbine aircraft are equipped with roll spoilers to supplement the ailerons in imparting roll.

Roll Spoilers

Roll spoilers are flat panels mounted on the upper wing surfaces and are used to assist the ailerons in roll control. In a turn, they work by deploying, one-at-a-time, up into the slipstream on the down wing, disturbing lift and thereby aiding the down wing aileron in effecting the turn. (The down wing aileron is least effective in a turn, as you remember from basic aerodynamics.)

Roll spoilers are interconnected with the ailerons so as to perform in harmony with them, meaning that pilot yoke movements automatically operate both ailerons and spoilers as a single control movement. As with ailerons, roll spoiler deployment on many turbine aircraft is regulated as a function of airspeed. For example, on the Dash-8 two roll spoilers deploy on each wing with aileron usage below 140 knots, but only one operates per wing above that airspeed (Fig. 5.3).

Roll spoiler operation also varies in other ways by specific aircraft model. For example, on the Lockheed L-1011 roll spoilers operate only when flaps are extended, while on Boeing 737s they function anytime the control wheel is displaced more than 10 degrees.

On a few aircraft types, such as the Mitsubishi MU-2, large spanwise flap requirements, on very small wings, dictate roll control strictly through the use of spoilers. (There are no ailerons on an MU-2.) Since spoilers effect roll by destroying lift (to a much greater degree than ailerons), crosswind techniques for such aircraft must be modified for safe operation under marginal takeoff and landing situations.

Ground Spoilers and Lift Dump Mechanisms

Most large aircraft also require the use of *ground spoilers* for landing. With all of those high-lift devices in action on final approach, it becomes desirable to dump as much lift as possible upon touchdown in order to control landing distance and improve braking. Ground spoilers usually deploy automatically under some combination of power lever position and weight on landing gear (see Fig. 5.4). (This, again, is the sort of information you'll learn in ground school about the systems of your particular aircraft.)

Some aircraft use "lift dump" mechanisms for this purpose instead of ground spoilers. In these aircraft, when the weight-on-wheels switch senses the plane is on the ground, flaps are quickly and automatically extended from landing position to nearly 90 degrees. As the slotted flaps on such systems extend, they effectively dump the lift of the wing and create a large amount of drag.

Flight Spoilers and Speed Brakes

Finally, many aircraft have *flight spoiler* systems that can be deployed in the air (see Fig. 5.4). Modern jets are so aerodynamically "clean" that even with power reduced to minimum it is often difficult to attain desired descent rates. This problem is compounded when rapid descents at low airspeeds are requested by air traffic control (ATC). Flight spoilers disturb overwing lift in order to increase descent rate. Activation is usually via a lever on the power quadrant.

Speed brakes are installed on some aircraft. They're related to flight spoilers in that they also are extended in flight. However, speed brakes are designed to increase drag,

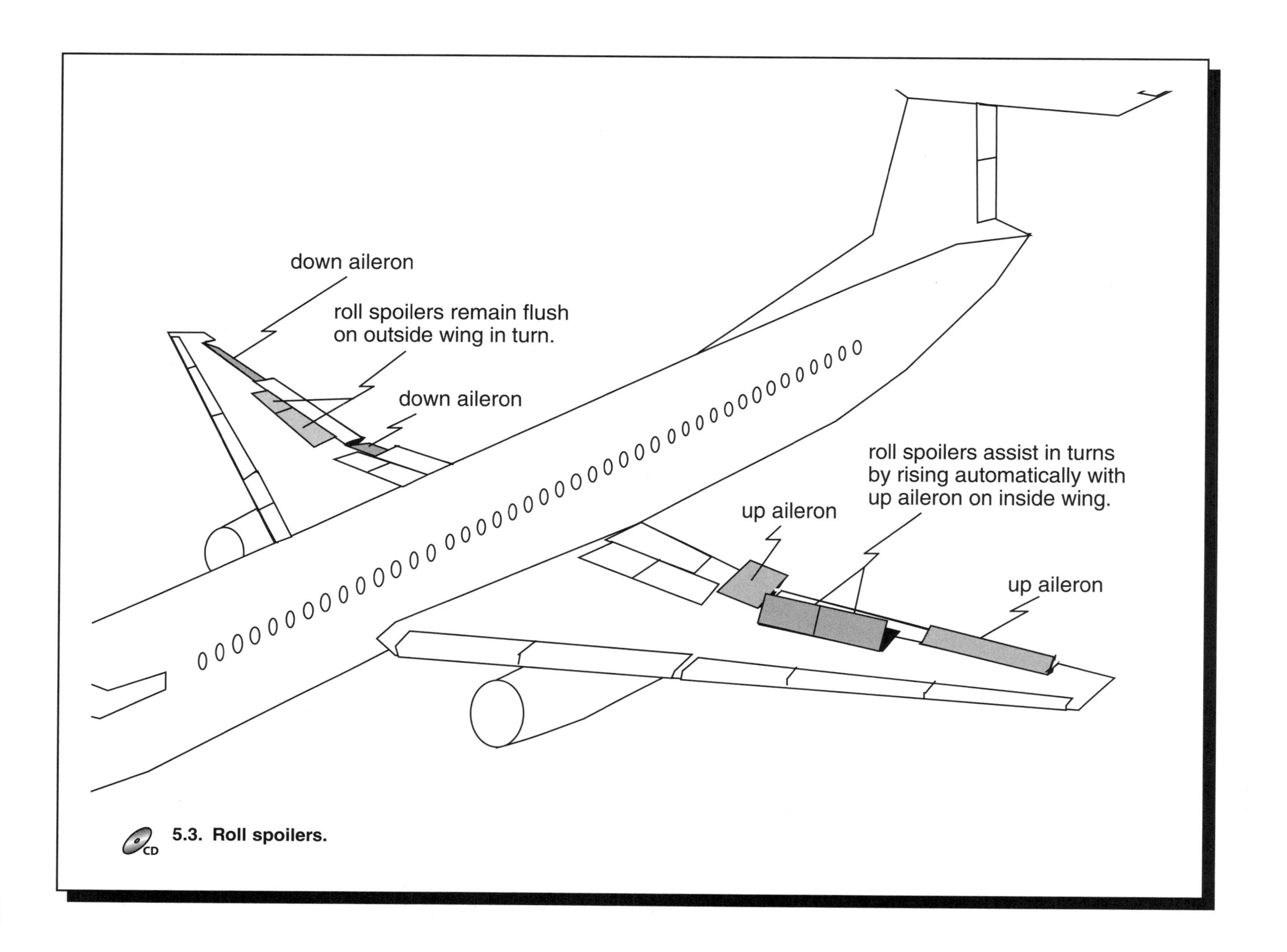

5.3. Roll spoilers.

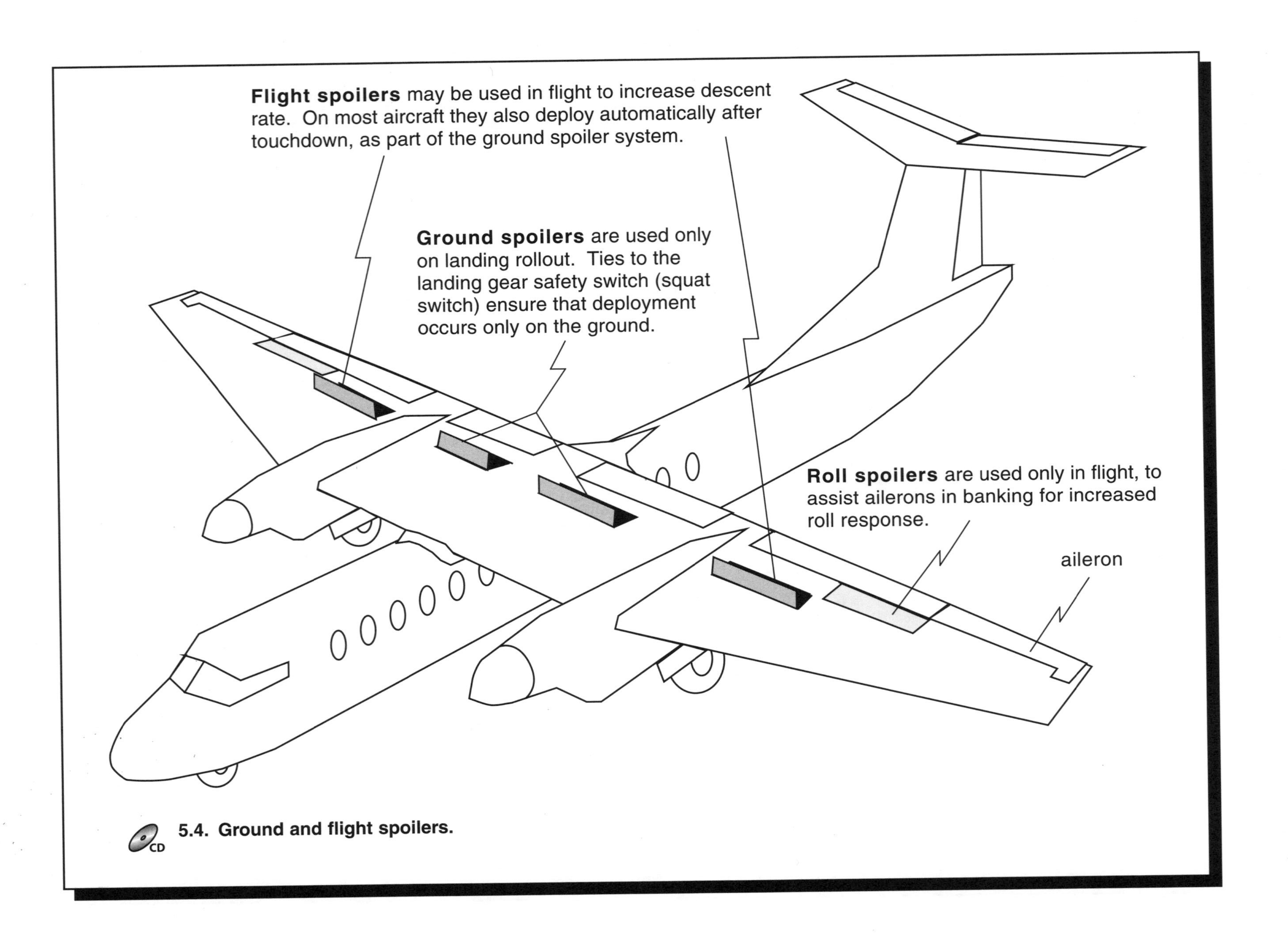

5.4. Ground and flight spoilers.

rather than to spoil lift, and are accordingly often installed on the fuselage or tail cone. In comparison with flight spoilers, speed brakes slow an airplane, rather than increase its sink rate. (Although for a given airspeed, they in effect do increase sink rate.)

Any combination of flight, ground, and roll spoilers may be installed on a given type of aircraft. Spoilers are usually hydraulically powered and may have dual modes of function for any one panel. (For example, roll spoilers may deploy on both sides upon landing in order to function as ground spoilers.)

Control Tabs and Power-Assisted Controls

A further complicating factor on large aircraft is control forces. The elevators on a 757, for example, are heavy and a heck of a long way back from the cockpit. Therefore, medium to large aircraft often require power-assisted controls so the pilots can handle them. This is usually accomplished through hydraulic boost and is carefully designed to give the pilot normal control feel. These power-assisted control systems, of course, are double or triple redundant. *Manual reversion,* meaning operating the controls on such aircraft using only human strength, is at least as tough as it sounds. You'll get to try it one day, but hopefully only in the simulator!

Another way to deal with large aircraft control forces does not require powered control boost. *Control tabs* are used on such aircraft as the McDonnell Douglas DC-9/MD-80 series and their successor, the Boeing 717 (see inset, Figure 5.1). On these airplanes, ailerons and elevators are not directly controlled by the pilots. Rather, pilot control yokes are mechanically connected to control tabs located on each aileron and elevator. Each yoke input moves the control tabs, which in turn "fly" each control surface to its desired position. Control tabs look and operate very much like trim tabs. If you've ever taken any flack from your instructor for flying with the trim instead of the yoke, you're already expert on the operating principles of control tabs!

Flight Control System Redundancy

Aircraft certified under FAR Part 25 ("Airworthiness Standards: Transport Category Airplanes") must also meet another interesting requirement: redundancy or separation of primary flight controls in order to overcome system jamming.

In some airplanes, for example, roll spoilers (usually hydraulic) are directly connected to one pilot's flight controls, while ailerons (usually mechanical, possibly with hydraulic assist) are attached to the control yoke of the other. As long as both yokes are connected to one another, all controls act in harmony. But in the event of a jammed or inoperable system, a clutch connecting the two systems is released or overcome, allowing one pilot to fly via the operating system.

In pitch, redundancy may also be accomplished by splitting the elevator controls, or it may be achieved through the use of alternate power sources for the elevators. In mechanical systems, separate lines from each control yoke are again connected via a clutch, which may be overpowered or disconnected in a jamming situation. On the Dash-8, for example, there's a manually released clutch between the yokes. When the controls are separated, each pilot controls only his or her own half of the elevator. You may have noticed, when taxiing behind a DC-9 or MD-80, that one elevator sometimes droops well below the other. That's because the elevators aren't mechanically linked together in movement; until there's enough airspeed for control tab effectiveness, each elevator seeks its own position, based on airflow over the control surfaces. A stiff breeze can lift the upwind elevator on the taxiing aircraft.

Flight Control Surface Position Indicating Systems

Depending on what you've been flying until now, you might be amazed to learn that on larger aircraft it's often impossible for pilots to see any portion of the wing from the cockpit, let alone the aircraft's tail. It is therefore impossible to look out the window and confirm movement of primary flight controls or spoilers when moving control yokes or rudder pedals. To deal with this problem, some large turbine aircraft come equipped with a *flight control surface position indicating system,* which gives the flight crew a continuous visual display of primary control surface and spoiler positions. As you might imagine, this display is especially valuable when performing pre-takeoff control checks (see Fig. 5.5)!

Fly-by-Wire Control Systems

Until relatively recently, the primary flight control surfaces on civilian aircraft were always mechanically linked to pilot yokes. Controls might be hydraulically or electrically boosted, or separately powered, but always there were cables or pushrods to allow for emergency manual control of the aircraft. That's all rapidly changing, however.

Fly-by-wire flight control systems are regarded by many as the technology of the future. On fly-by-wire aircraft, all control inputs are converted into electrical impulses via transducers in the yoke or joystick. These impulses direct hydraulically or electrically powered control surfaces through a computer. There are no mechanical linkages between yoke and control surfaces, only the electrical wiring that conveys the signals. (Hence the name, "fly-by-wire.") (See Figure 5.6.)

Redundancy in fly-by-wire systems is achieved in several different ways: Multiple routings are used for control

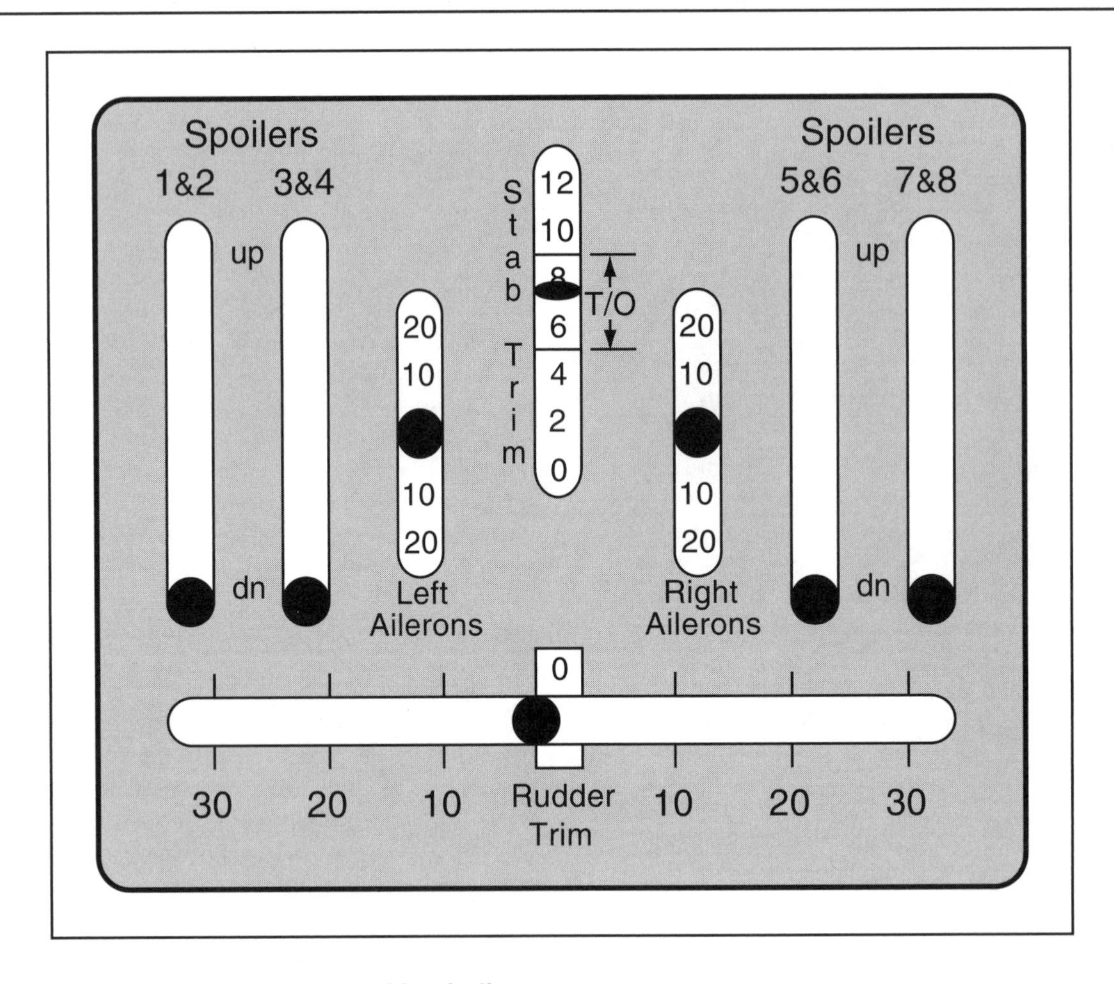

5.5. Flight control surface position indicator.

wiring in order to provide alternate electrical paths; Computers compare control inputs with control surface responses; Finally, the flight control computers themselves are redundant.

Advantages of fly-by-wire over conventional systems are many. Control responses can be varied automatically to accommodate different flight situations. Reductions in aircraft weight and mechanical maintenance can be significant. Cockpit and instrument panel space is reclaimed by removing the massive linkages of mechanical control yokes. In some fly-by-wire aircraft (like the A-320 and other newer Airbus aircraft), yokes have been replaced by "sidesticks," joysticks mounted at the pilots' sides (see Fig. 5.7). In others, such as the Boeing 777, conventional control yokes drive fly-by-wire control systems.

While many pilots are leery of flying aircraft without mechanical control linkages, fly-by-wire systems have been used by the military for years, and the benefits are such that pilot acceptance continues to grow. It's important that all of us pilots get used to the idea of alternate control systems, since manufacturer interest in them is very high right now. Another recent development, for example, is fly-by-light technology, where pilots are connected to their precious control surfaces via optical interfaces.

Pressurization

Pressurization is one of those aircraft systems that is very simple in concept but surprisingly complicated in execution. The principle, of course, is to seal up the airplane's cabin into a *pressure vessel.* Air is then pumped in to maintain internal pressure as close as possible to that at sea level.

The complexity, however, begins with the aircraft fuselage as a pressure container. The pressure vessel does not

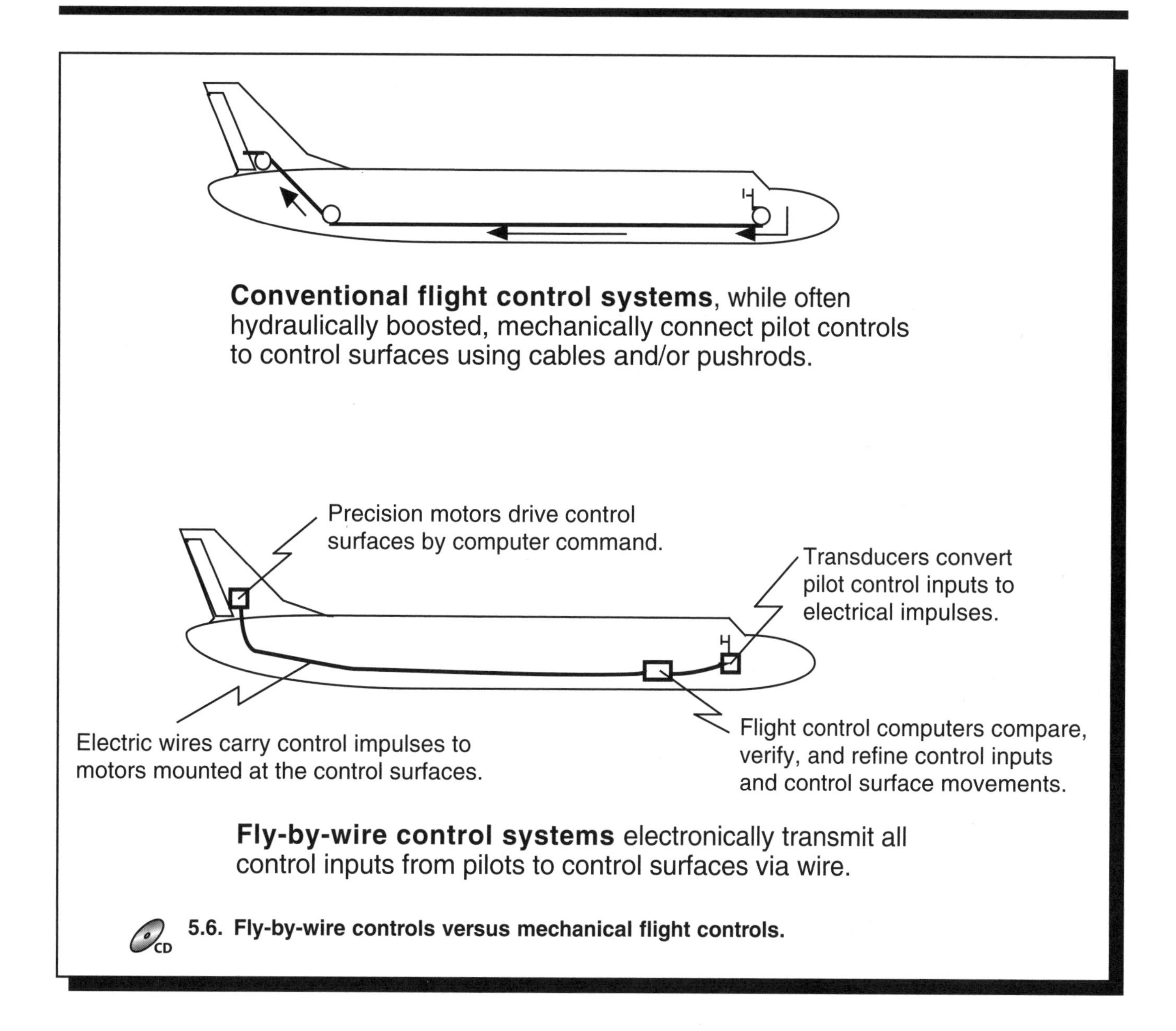

5.6. Fly-by-wire controls versus mechanical flight controls.

occupy the entire fuselage but rather uses pressure bulkheads, plus the outer skin, to contain the passenger cabin and some or all cargo areas. Control cables, wiring, and plumbing must pass through the pressure vessel, with further perforation by exits, windows, and emergency exits. To make matters worse, the aircraft fuselage changes dimensionally with every pressurization cycle. Obviously, sophisticated engineering and maintenance is required for such an aircraft. (See Figure 5.8.)

In turbine aircraft, a steady supply of engine bleed air is used to pressurize the cabin. Cabin pressure is then controlled by modulating the exhaust of cabin air via *outflow valves*. Outflow valves are manipulated via a pressurization controller operated by the pilot.

The main measure of a pressurization system's efficiency is known as its *maximum differential* (or *max diff*). This is simply the maximum ratio of cabin pressure to outside air pressure that the pressurization system and vessel can sustain. Max diff varies significantly by aircraft type. This is due to many factors, including pressure vessel design, engine bleed air capacity, and aircraft weight and power considerations. For many pressurized aircraft, certified maximum operating altitude is determined not by the airplane's service ceiling but by the ability of the pressurization to meet supplemental oxygen requirements of the FARs. Maximum operating altitudes, in such cases, are defined by the greatest altitude the aircraft can attain and still maintain legal cabin altitudes (10,000 feet under FAR Parts 135 and 121, and 12,500 feet for Part 91).

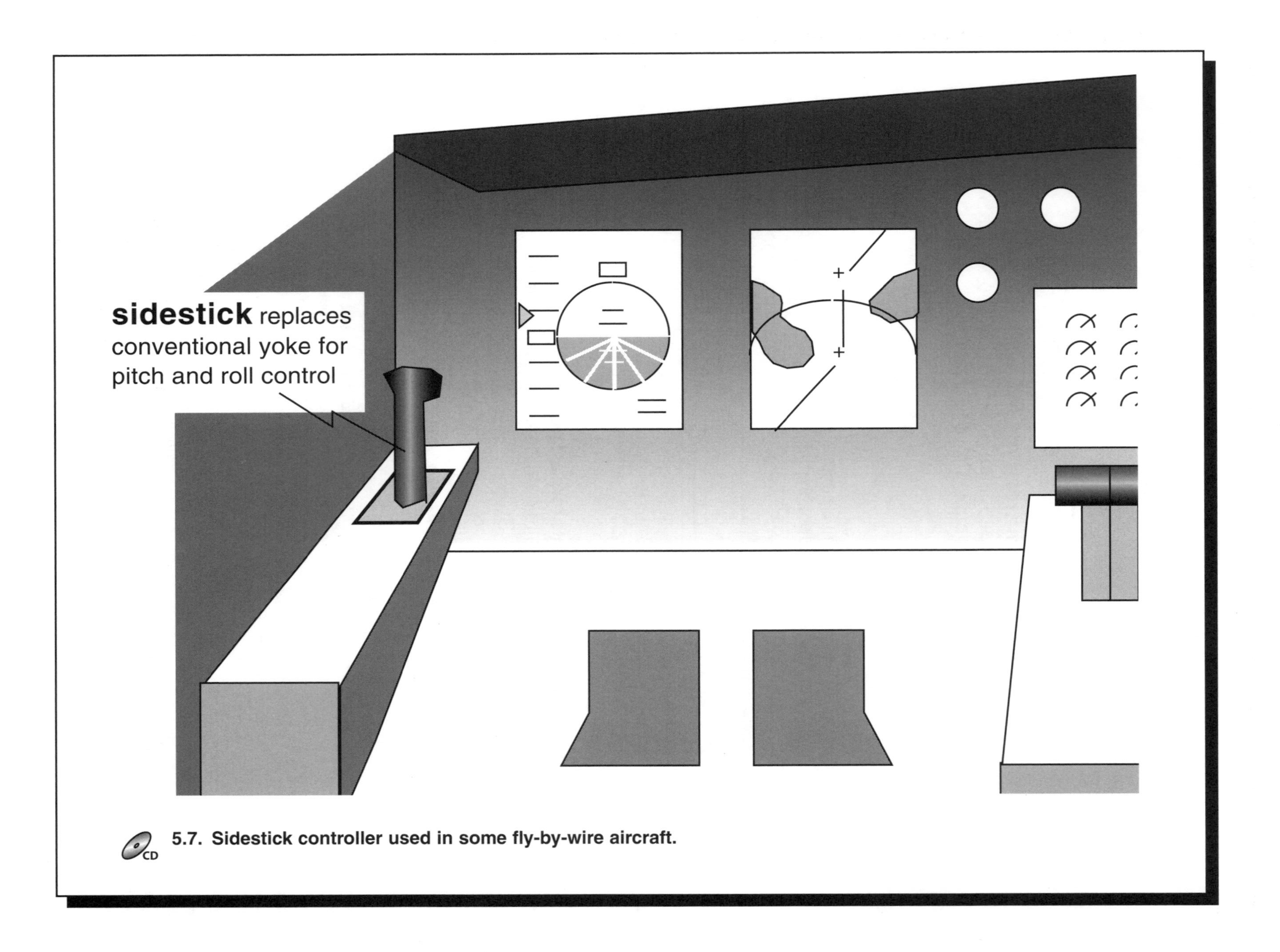

5.7. Sidestick controller used in some fly-by-wire aircraft.

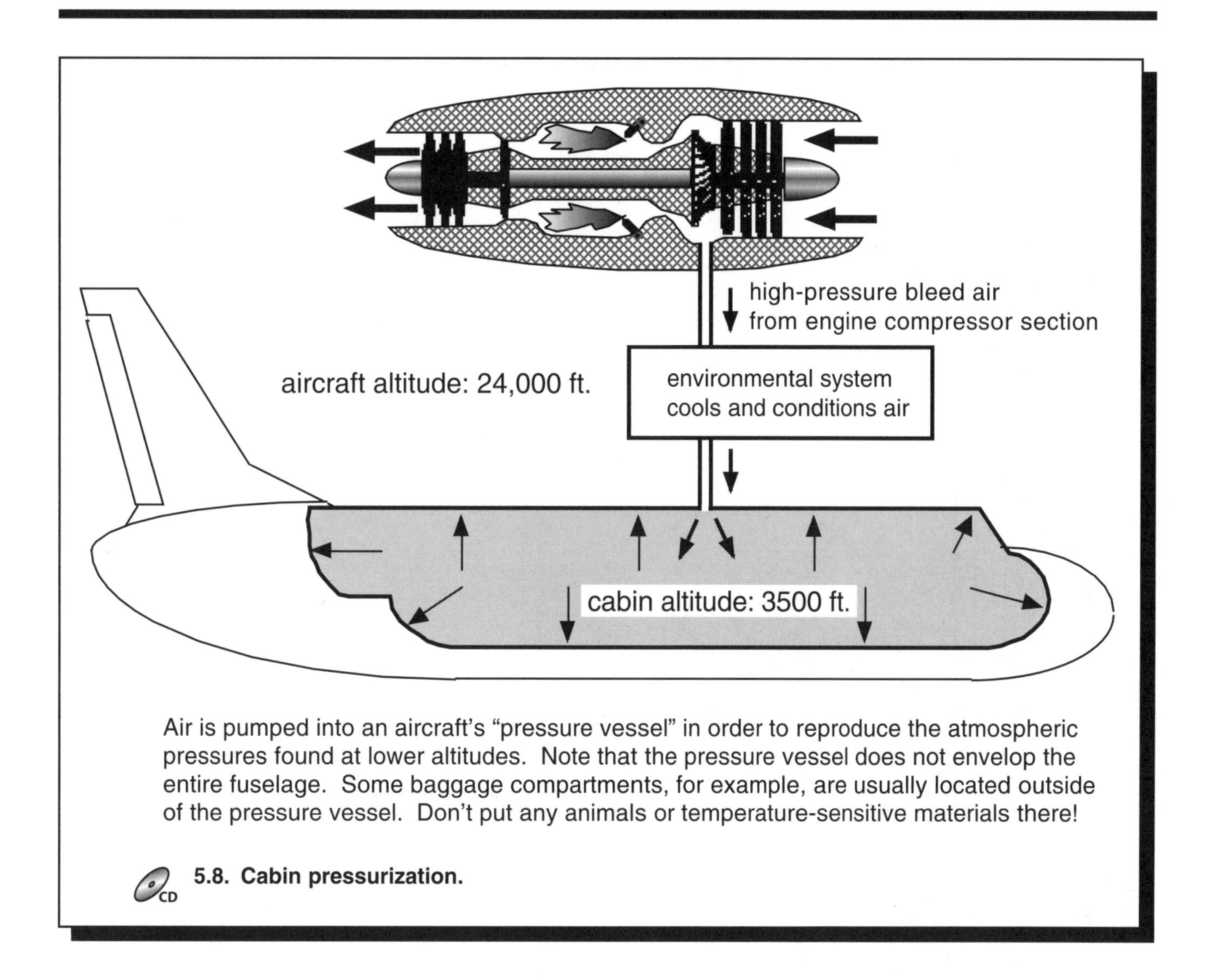

Air is pumped into an aircraft's "pressure vessel" in order to reproduce the atmospheric pressures found at lower altitudes. Note that the pressure vessel does not envelop the entire fuselage. Some baggage compartments, for example, are usually located outside of the pressure vessel. Don't put any animals or temperature-sensitive materials there!

CD **5.8. Cabin pressurization.**

Pressurization Indicators and Controls

Pilots control pressurization by setting one or more variables in the pressurization controller. Keep in mind that there are two types of aircraft performance involved in pressurization: the airplane's climb/descent performance and that of the cabin. Pressurization systems of all types are monitored in the cockpit via *cabin altitude, cabin rate of climb,* and *pressure differential indicators* (Fig. 5.9).

Pressurization controllers vary in design from simple to complex. On old airplanes, pilots manually controlled the outflow valves to meet charted values for maximum differential. On those systems, pilots had to manually calculate and adjust cabin rate of climb for every climb and descent. This is not particularly difficult, but it requires lots of attention compared with newer systems. It's worth taking a quick look at the calculations, since when a newer system fails the copilot (probably you!) often gets stuck regulating pressurization the old way. Besides, it helps in understanding how pressurization controllers work (see Fig. 5.10).

Let's say that an aircraft is flying at 15,000 feet with its cabin pressurized to 5000 feet. The pilots are cleared for descent to a sea-level elevation airport. Calculating that they're ten minutes out, they decide on an aircraft descent rate of 1500 fpm. It's easy to see that the cabin rate of descent must be proportional to that of the aircraft, if both are to reach sea level at the same time. Therefore the cabin, which must descend 5000 feet during that same ten minutes, will need a selected descent rate of 500 fpm.

On newer pressurization systems, the pilot enters some combination of aircraft cruising altitude, departure field elevation, and/or landing field elevation. The controller does all the climb and descent calculations but still requires some start and endpoints for each climb or descent.

Let's consider a flight, say, from Los Angeles to Denver. Prior to takeoff, the pilot sets projected cruising altitude

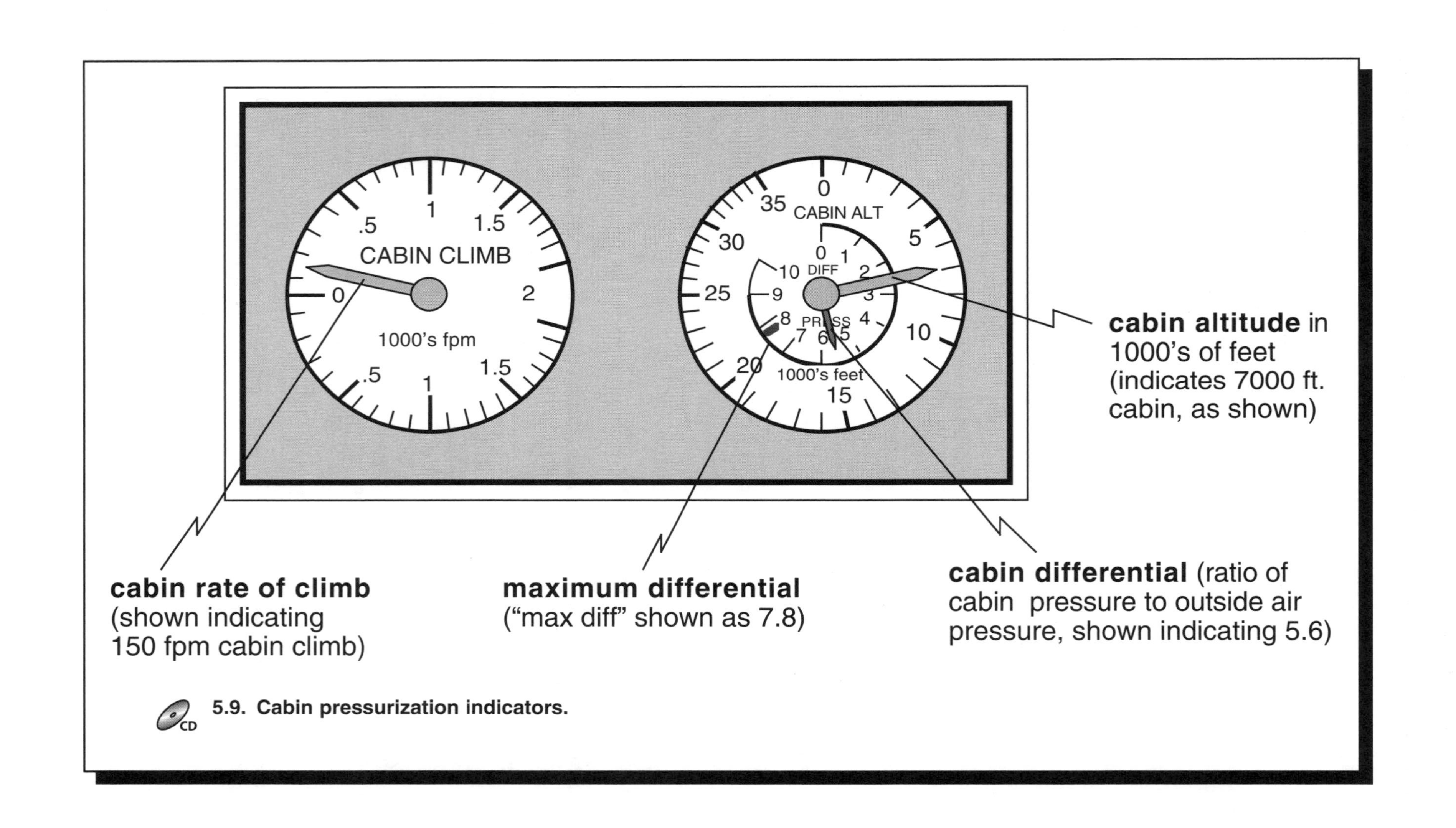

5.9. Cabin pressurization indicators.

into the pressurization controller. The controller must know the target aircraft altitude in order to smoothly "climb the cabin" from sea level up to its normal value for that cruising altitude. Once there, of course, cabin altitude remains constant.

Upon initiating descent, the pilot must reset the controller to the destination airport elevation. This is how the controller "knows" what the landing pressurization value must be, so it can proportionally descend the cabin to landing field elevation. (Some controllers automatically schedule cabin descent when sensing aircraft departure from selected cruising altitude. If the aircraft for some reason has leveled off at an altitude lower than planned, the cabin descent sequence may not begin as expected.)

If the aircraft in our example were returning to Los Angeles, the cabin pressurization performance could simply be reversed for descent; the controller would be set a little above sea level. However, the aircraft's Denver destination has an elevation of 5280 feet, so the pressurization must be scheduled to descend to that altitude, not to sea level. (Note that some aircraft can maintain cabin altitudes well below that 5280 feet during the cruise portion of this trip and could actually require climbing of the cabin altitude prior to landing.) What if you forget to reset for landing in this situation? We'll get to that in a moment.

The latest pressurization controllers do almost all of this work for you. Most include a computer-controlled "auto" mode requiring little pilot input. These systems use information from an *air data computer* and cabin sensors to automatically control all aspects of the pressurization cycle. A "standby" mode allows semiautomatic control of cabin pressure if the auto mode computer is not operating correctly. "Manual" controls bypass the electronic pressurization controller entirely, sometimes using vacuum or low-pressure air to control the outflow valves directly.

Pressurization System Safety Features

Regardless of pressurization controller design, it is important that the aircraft land unpressurized or at a low predetermined pressurization value. This is because pressurization stresses the fuselage significantly, and a hard landing while pressurized could cause structural failure. (Some large aircraft, such as the B-737, are actually structurally strengthened by pressurization. These are designed to operate slightly pressurized through the takeoff and landing phases. They are depressurized once on the ground.) After landing, opening and closing passenger doors while pressurized can be nearly impossible, not to mention the effects on the ears of the passengers!

Safety devices are installed in the pressurization system, in order to address some of these issues. *Positive pressure relief valves* ("safety valves") vent excess pressure overboard any time maximum differential is exceeded. This is to prevent overpressurization of the aircraft in the event of malfunctioning controller or outflow valves (Fig. 5.10).

Dump valves allow pilots to manually vent cabin pressurization in an emergency. These may be used in the event of a pressurization malfunction or to "dump" the cabin in the event of smoke or other cabin air contamination.

Negative pressure relief valves ensure that cabin pressure never falls below ambient pressure. They come into play when the pilots forget, upon initiating descent, to set the controller for landing. Say that the airplane is cruising at 25,000 feet, with a cabin altitude of 4000 feet. (We'll assume a sea-level destination.) When properly set for descent, the controller gradually descends the cabin at a few hundred feet per minute, while the plane itself may be descending at several thousand feet per minute. However, if the controller is not reset for landing, the cabin will stay at 4000 feet until the airplane reaches the matching 4000-foot pressure altitude. From that point on, the negative pressure relief valves will vent the cabin so as to prevent cabin altitude from being lower than the actual altitude. This situation, known as "catching the cabin," is not particularly serious, except that the cabin's descent rate now matches that of the airplane. In a fast airplane, high descent rates are required to get down in the same distance as a slower plane. (Descent rates of 3000 fpm are common in turbine aircraft.) Catching the cabin, therefore, makes it difficult to keep both ATC and the passengers' ears happy. (Even if the passengers don't complain, the captain will!)

Note that while outflow, dump, and pressure relief valve functions are used in every pressurization system, some of them may be combined into multipurpose valve units.

Pressurization systems are electrically tied to *squat switches* on the landing gear (Fig. 5.10). These ensure that the cabin is depressurized (or pressurized to a prescribed value) on the ground, prior to takeoff, and after landing. Remember our aircraft traveling from Los Angeles to Denver? If the pilot (on most aircraft) forgets to reset the cabin controller to Denver's 5000-foot field elevation, the cabin will descend toward sea level on approach. Upon touchdown, the squat switch will activate the dump valve, and the cabin will depressurize within seconds from sea-level pressure to ambient at 5000 feet! Enthusiastic passenger reaction prevents most pilots from making this mistake more than once.

Loss of Cabin Pressure in Flight

Loss of pressurization under certain situations can be a very serious emergency. Pilots must memorize the emergency procedures for getting their own oxygen masks on and activated, for getting passengers' masks activated, and for initiating an emergency descent of the airplane immediately.

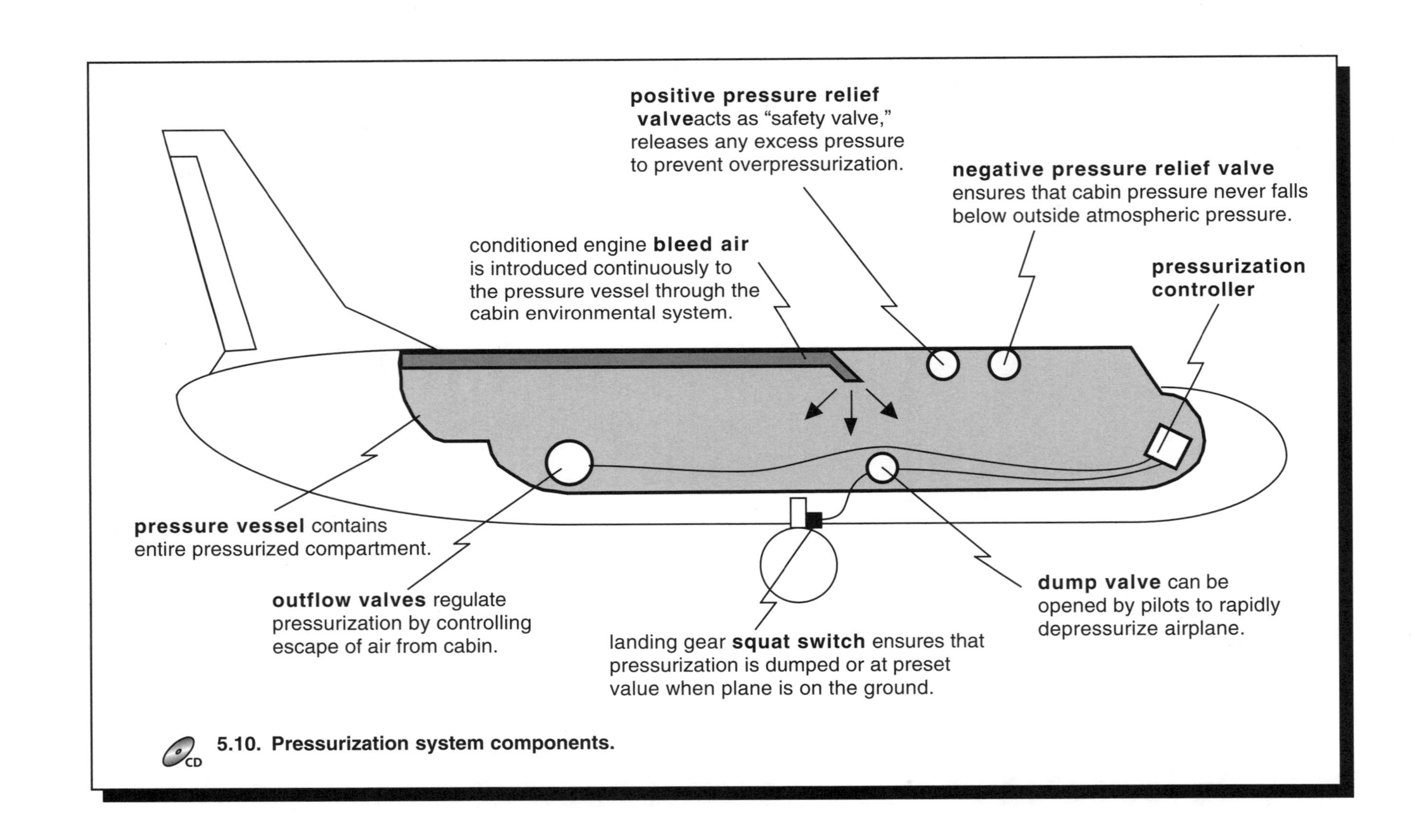

5.10. Pressurization system components.

It's critical to recognize the signs of lost pressurization as soon as possible. If the aircraft is operating at high altitudes and depressurizes, there is the problem of keeping everyone conscious long enough even to get the masks on. The "time of useful consciousness" rapidly shortens with altitude, so immediate action is required by the pilots in the event of rapid depressurization. To give you an idea of just how serious this problem is, check out the table below: The time of useful consciousness at 25,000 feet is 3–5 minutes, at 35,000 feet it is less than one minute, and at 40,000 feet only 15–20 seconds! (For this reason, the FAA requires quick-donning oxygen masks in the cockpit when operating above 25,000 feet. If one pilot leaves the cockpit for any reason, the other must wear his or her mask until the first returns.)

Altitude	Time of useful consciousness
FL 500 and above	9 to 12 seconds
FL 430	9 to 12 seconds
FL 400	15 to 20 seconds
FL 350	30 to 60 seconds
FL 300	1 to 2 minutes
FL 280	2.5 to 3 minutes
FL 250	3 to 5 minutes
FL 220	10 minutes
FL 180	20 to 30 minutes

Source: *Aerospace Physiology/Human Factors,* T37/T-38, U.S. Air Force Air Training Command, 1986. Note that these times are for healthy individuals and do not take into account such factors as fatigue, smoking, or physical fitness.

Perhaps the most obviously serious situation occurs in the event of an "explosive" or *rapid decompression* (*RD*). Although rare, people can and have been blown out of the aircraft in cases of a pressure vessel failure.

If decompression is rapid, a fog of condensing water vapor may pass through the cabin. This is due to the sudden change in air pressure, which causes instant condensation of excess moisture in the air.

Sometimes, however, cabin decompression can be insidiously slow, in which case pilots must be sensitive to indications of depressurization. A cabin altitude alert should occur in the cockpit as cabin pressure climbs through 10,000 feet or 12,500 feet, depending on type of operation.

Another key indicator, under such circumstances, is how the pilots feel under the effects of hypoxia (oxygen deprivation). In an effort to increase pilot understanding of this issue, the FAA now requires physiological training of pilots operating pressurized aircraft. (Altitude chamber training is a great way to learn about your own reactions to oxygen deprivation. Don't miss the opportunity!)

There are other potentially debilitating effects of high altitude on the human body, in addition to hypoxia. Gases increase tremendously in volume at high altitudes. They double in volume from sea level to about 17,000 feet and expand by a factor of ten between sea level and 43,000 feet. Think of the effects of decompression on your ears and on your circulatory system, or consider the impact of a bean burrito lunch on your abdomen.

Military pilots wear pressure suits to protect themselves from such problems at high altitudes, yet many civilian turbine pilots find themselves aviating in shirtsleeves at FL 430. That's why it's imperative to learn everything you can about high-altitude physiology before operating pressurized aircraft.

Emergency Descent Maneuvers

Many pilots transitioning for the first time from piston-engine aircraft to turbine airplanes are used to cruising around at altitudes that don't require the use of pressurization or even supplemental oxygen. Turbine aircraft, on the other hand, fly at altitudes where pressurization is a necessity and the loss of pressurization can mean complete incapacitation in as little as 15–60 seconds at higher altitudes. For this reason, the FAA requires flight crews to be trained to bring the airplane down smoothly to a safe altitude—typically 10,000 MSL, where pressurization is not required—in a minimum amount of time.

Emergency descent maneuvers are generally divided into two different types: high-speed descent and low-speed descent. *High-speed emergency descents* are used when the flight crew must descend from high altitudes without cabin pressure, but the aircraft is otherwise structurally sound. Descent is usually made at Mmo/Vmo with throttles at idle, gear up, and speed brakes (if installed) deployed.

Low-speed emergency descents are used when a known structural failure has occurred (like a cracked window or missing door). In these cases, the aircraft customarily descends with throttles at idle and at an airspeed as slow as possible, so as to minimize structural loads that might further damage the airframe. Landing gear is extended to further increase descent rate, and sometimes flaps are extended a notch or two. Low-speed descents are also used when pressurization is lost in areas of severe turbulence, where higher airspeeds could damage the airframe.

As mentioned earlier, most turbine aircraft are equipped with a cabin pressure warning alarm to alert pilots if the cabin altitude rises above limits. When such an alarm activates, the flight crew immediately dons their oxygen masks and completes the appropriate checklist, including emergency descent if necessary. Some aircraft automatically deploy passenger oxygen masks with the cabin pressurization alarm.

In order to fully appreciate the importance of properly and promptly executing emergency descent maneuvers when required, it's worth noting that while pilots must be provided with two hours or more of supplemental oxygen supply, passengers must be provided only a limited supply, which is designed to accommodate them briefly until the airplane descends to a breathable altitude. (At less extreme cruising altitudes not all passengers must even be accommodated.)

Therefore, emergency descent maneuvers serve not only to quickly escape altitudes where pressure suits are required for survival but also to get your precious passengers down to an altitude where they can breathe adequate oxygen before the aircraft's supply runs out. Thought provoking, isn't it? Pressurization failure is perhaps the most underestimated and unanticipated potential emergency faced by pilots.

Cockpit Oxygen Breathing Systems

With that in mind, one type of emergency equipment with which you must become familiar is *cockpit oxygen breathing systems.* Pilots use these systems when supplemental oxygen is needed, as in cases of pressurization failure. Cockpit oxygen breathing systems come in three common types*: continuous-flow/diluter systems, diluter-demand systems,* and *pressure-demand systems.*

Continuous-flow oxygen systems are found on some turboprops and most general aviation aircraft. Oxygen flow is controlled by a simple on/off valve. When oxygen is required, the flight crew turns the valve ON and oxygen is dispensed from an oxygen storage tank through an oxygen pressure regulator and into pilots' oxygen masks. The oxygen tank is typically located in or near the cockpit and may include a pressure gauge visible to the flight crew.

Depending on the system's design, 100 percent oxygen may be delivered to the flight crew, or in most cases oxygen is mixed with either ambient air from the cockpit or with air from an attached rebreather bag.

A visual flow indicator is installed in the oxygen line leading to the mask to indicate proper oxygen flow. Most systems use green to indicate oxygen flow and clear or black to indicate no oxygen flow.

Diluter-demand oxygen systems are the most commonly found type on turbine aircraft. Diluter-demand oxygen regulators dilute the oxygen supplied to the mask with predetermined levels of air from the cockpit when the aircraft is flying at lower altitudes. At sea level the mask supplies virtually no supplemental oxygen unless the pilot manually selects 100 percent oxygen (as when there is smoke in the cockpit). As cabin altitude increases, so does the percentage of oxygen supplied to the mask, with the amount of cockpit air mixed with oxygen diminishing until about FL 350 (35,000 feet), above which the pilots receive 100 percent oxygen.

An oxygen bottle located in or near the cockpit supplies all pilot positions; an oxygen pressure gauge may be installed on an overhead panel, regulator panel, flight engineer panel, or mounted atop the bottle itself. An oxygen regulator control panel, located within easy reach of each pilot station, generally includes on/off switch, oxygen selector for diluted or pure oxygen, and test mask switch (Fig. 5.11).

Some diluter-demand systems include an emergency switch for delivering a continuous flow of 100 percent oxygen under positive pressure to the mask. This is designed for use primarily during loss of cabin pressure at high altitudes, above FL 350, where positive pressure is required to force oxygen into the lungs and bloodstream.

Pressure-demand oxygen systems are found on a number of high-performance aircraft that regularly fly at altitudes above FL 350. Below this altitude, the oxygen system operates like a diluter-demand system. Above FL 350, the pressure-demand regulator automatically furnishes pilot oxygen masks continuously with 100 percent oxygen under positive pressure in the event of lost cabin pressure.

Passenger Oxygen Systems

Passenger oxygen systems are designed to provide passengers with supplemental oxygen during emergency situations, such as smoke in the cabin or loss of cabin pressure. Two types are commonly used on turbine aircraft: *gaseous oxygen systems* (oxygen supplied from a tank) and *solid chemical generator systems,* which generate oxygen through chemical reaction. Depending on the manufacturer, either passenger oxygen system can be activated manually or automatically when cabin altitude exceeds a predetermined altitude (e.g., 13,000 feet). Most systems include a cockpit-mounted indicator light informing the flight crew when passenger masks have been deployed.

From a passenger standpoint, both systems operate similarly. When the passenger oxygen system is activated and pressurized, masks are exposed, or dropped, at each passenger seat. Oxygen flow to each mask does not begin until the passenger tugs it, opening a valve to initiate flow. Passenger oxygen masks may be stowed in overhead panels above passengers' seats or in a wall compartment next to each seat. Oxygen masks are also stowed in every lavatory and flight attendant station.

Solid chemical generators are typically found only on larger turbine aircraft having one hundred or more passenger seats. The advantage of chemical oxygen generating systems is that they eliminate the need to route oxygen lines throughout the cabin in order to supply every passenger seat. Instead, each seat row may be equipped with a separate self-

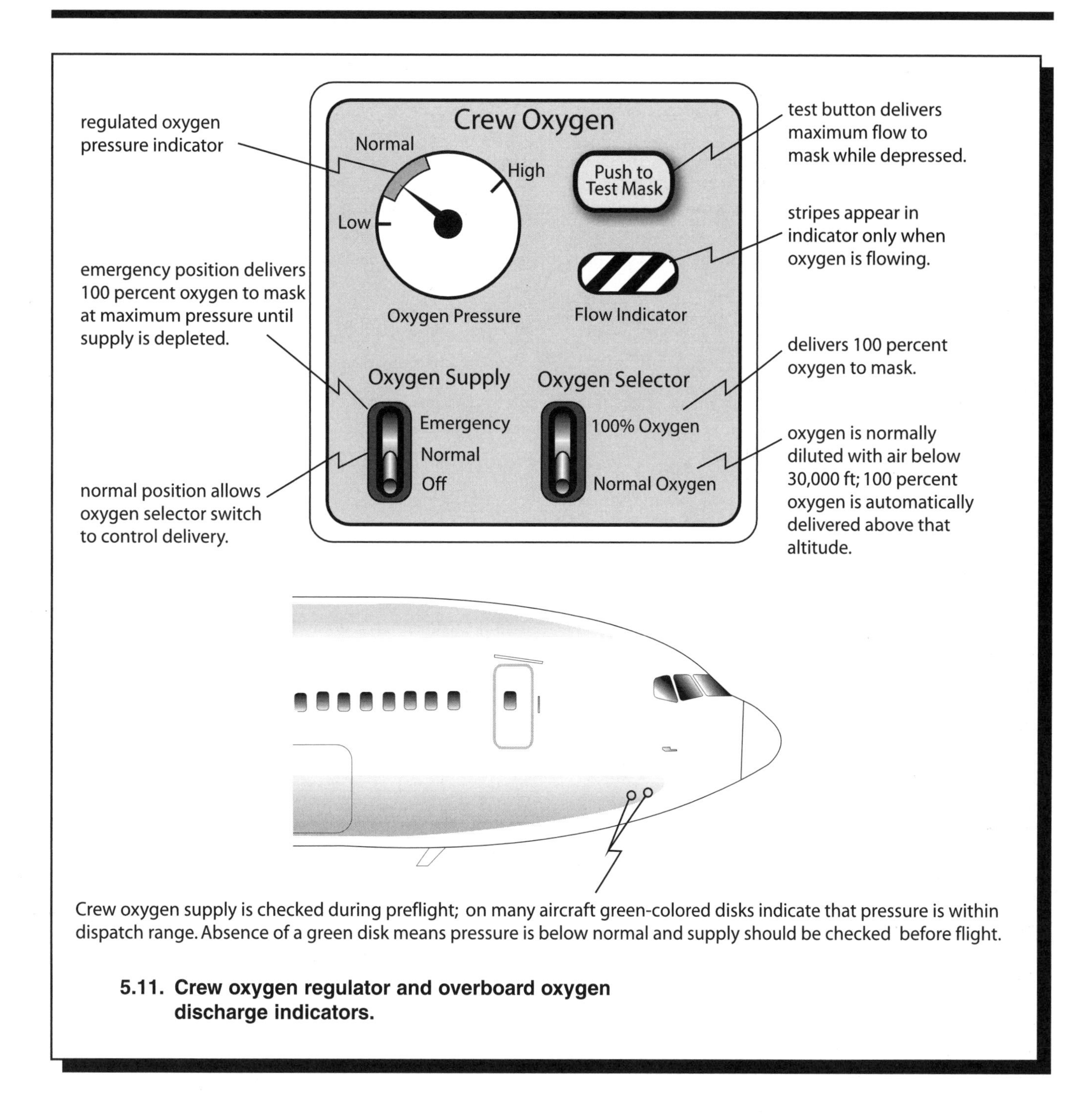

5.11. Crew oxygen regulator and overboard oxygen discharge indicators.

contained generating unit. There are several important disadvantages to chemical generator systems, the first being that once generators are activated there is no way to turn off the flow of oxygen. Units continue to supply oxygen until all of the chemicals are expended. Another disadvantage is that the chemical reaction required to produce oxygen generates heat, posing the risk of passengers burning themselves if the generating unit is touched.

Environmental Systems

Environmental systems on large turbine-powered aircraft are complex, compared with the light aircraft systems that you may have experienced in the past. An airline aircraft is so large that balancing temperatures throughout the vehicle can be challenging. Besides, these aircraft operate through such a huge range of temperatures that their systems must be quite versatile. A given airplane might sit for two hours on a 105°F ramp, then experience temperatures of –40°F in flight.

In piston aircraft, cabin heat is generated through combustion heaters or engine exhaust heat. Turbine environmental systems, however, get their heat from engine bleed air. Modified bleed air is introduced into the aircraft for pressurization and heating, and in many cases, for cooling.

Proper operation of environmental systems is rather critical in pressurized turbine aircraft. Since pressurization air is high-pressure bleed air (read "hot"), all environmental air must be cooled, to some degree, before it's put into the cabin. You may be used to thinking of air conditioning as a luxury in aircraft, but in many turbine aircraft out-of-service environmental cooling equipment means "no-go."

A combination of heat exchangers, air cycle machines, and/or vapor cycle machines are used to modify and control cabin temperatures in turbine airplanes.

Heat Exchangers

Heat exchangers are simple, passive devices that transfer heat between two different fluids (Fig. 5.12). Your car radiator is an excellent example of a heat exchanger. Air passing through the radiator from the grill absorbs heat from the engine coolant pumped through the radiator core. In aircraft, heat exchangers are used to absorb and remove heat in a variety of applications within the environmental system and elsewhere.

Air and Vapor Cycle Machines

Since bleed air coming from the engines is already hot, the environmental system's challenge is to cool it. The two devices commonly used for this purpose, *air cycle machines* (*ACMs*) and *vapor cycle machines* (*VCMs* or "Freon units"), are related in that each works on similar physical principles. When a gas is compressed, it gets hot. When expanded, gas cools, meaning that it transfers heat to the surrounding air. The amount of heating or cooling is proportional to the change in volume of the gas (see Fig. 5.13).

If you start with a liter of gas at a given temperature and compress it to a smaller volume, the compressed gas will be hotter than it was originally. Now if you remove some of the heat from that compressed gas by blowing some cool air past it (say, through a heat exchanger) and then expand it back to its original volume again, it will be cooler than it was to begin with. This is the basic operating principle of both air and vapor cycle machines.

Air Cycle Machine (ACM)

In air cycle machines, high-pressure bleed air from the engines is first passed through a compressor, further squeezing the already hot gas. It is then routed through a heat exchanger or two to remove heat. The now cooler but still highly compressed air then passes through an *expansion turbine* into a larger chamber. The combined effects of driving the turbine and expanding into a larger chamber dramatically cools the air (usually down close to freezing; water traps are critical in the system to prevent freeze-up).

The expansion turbine is connected by shaft to the ACM's compressor, so expanding air works to compress upstream bleed air similar to the way a turbine engine or a piston engine turbocharger works. This cycle may be repeated several times, with the end result that system air temperature is cooled far below ambient temperature (Fig. 5.14).

Vapor Cycle Machine (VCM)

A vapor cycle machine, when installed in your car or home, is otherwise known as an air conditioner. There are several key differences between an ACM and a VCM. One is in what material is compressed and expanded to do the cooling. While air cycle machines use air for this purpose, vapor cycle machines use refrigerants specially selected for cooling capacity. (Refrigerants have higher thermal capacities than air, so they transfer more heat on each cycle.)

The most important difference is that VCMs take advantage of another physical property that greatly adds to their efficiency. As you may remember from high school physics, a great deal of energy is absorbed when a substance changes phase from liquid to gas. Refrigerants (such as Freon) are designed to undergo phase changes with every cycle of temperature, compression, and expansion. Refrigerant gas is compressed in a VCM's compressor. It is then run through a special heat exchanger, known as a *condenser*, where heat is removed. As the gas cools under pressure, it condenses into a liquid (hence the name "condenser"). The liquified refrigerant continues on its journey to another heat exchanger, the *evaporator*, which interacts with cabin air. As the name implies, the refrigerant is allowed to drop in pressure in the evaporator. As it evaporates (another phase change), the refrigerant absorbs a tremendous amount of heat from the passing cabin air. The cooled air is returned to the cabin, while for the refrigerant it's off to the compressor again to start a new cycle. (See Figure 5.15.)

Why two different types of machines? Air cycle machines are ideally suited for turbine aircraft due to the supply of (already) compressed bleed air, reasonably simple systems, and no need for special coolants. On the other hand, ACMs require significant volumes of bleed air, and turbine components make ACMs relatively expensive. Large

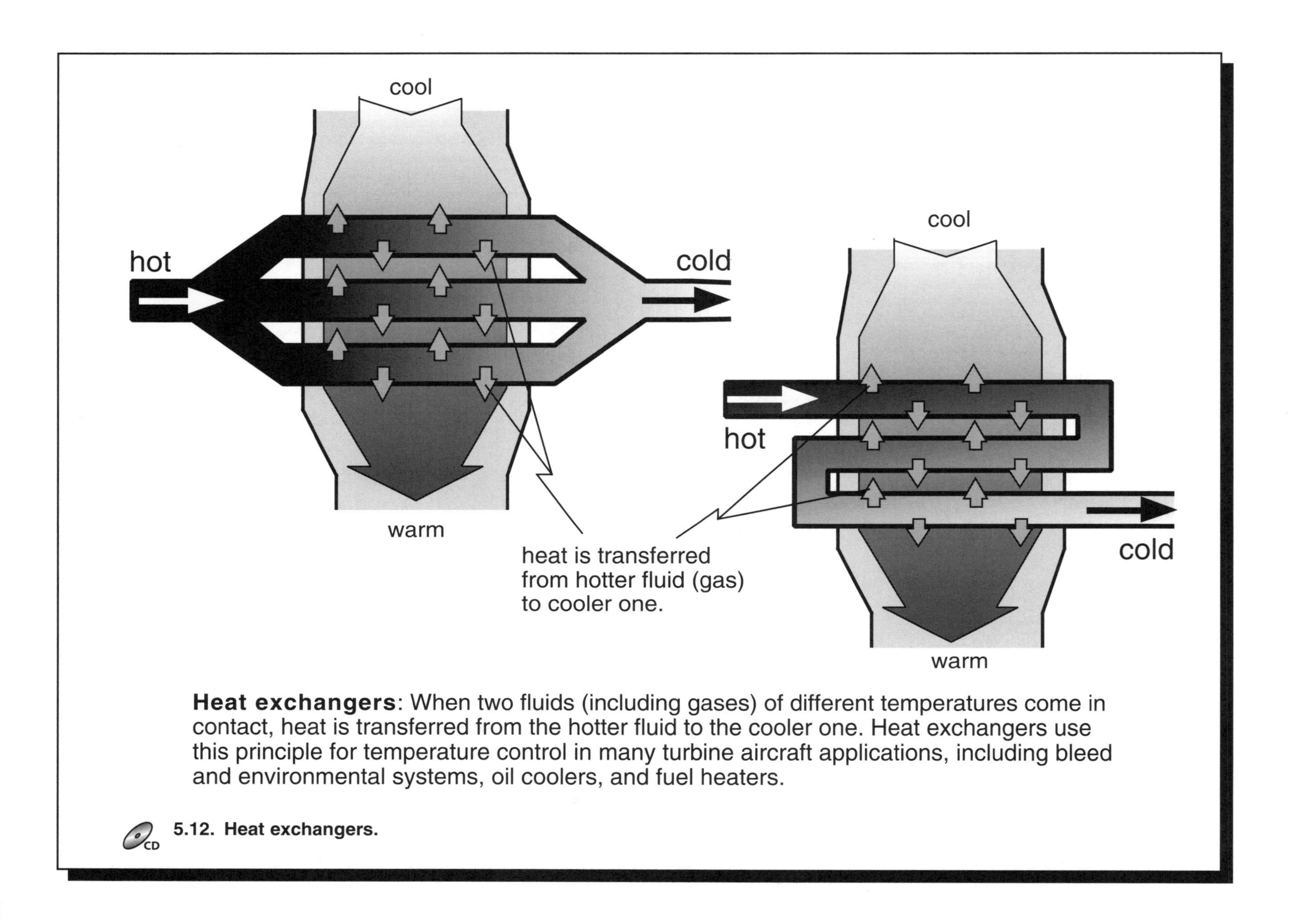

Heat exchangers: When two fluids (including gases) of different temperatures come in contact, heat is transferred from the hotter fluid to the cooler one. Heat exchangers use this principle for temperature control in many turbine aircraft applications, including bleed and environmental systems, oil coolers, and fuel heaters.

5.12. Heat exchangers.

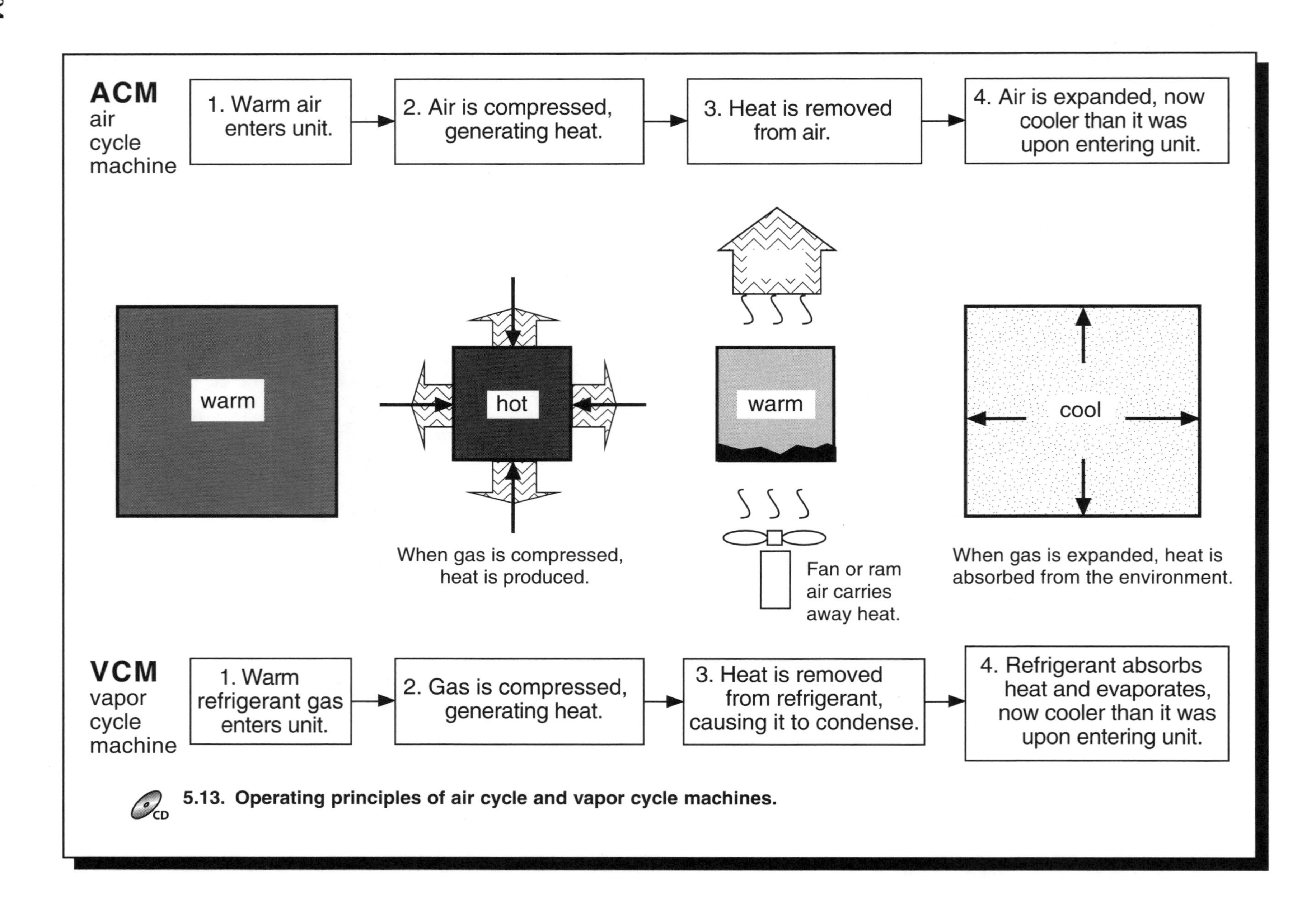

5.13. Operating principles of air cycle and vapor cycle machines.

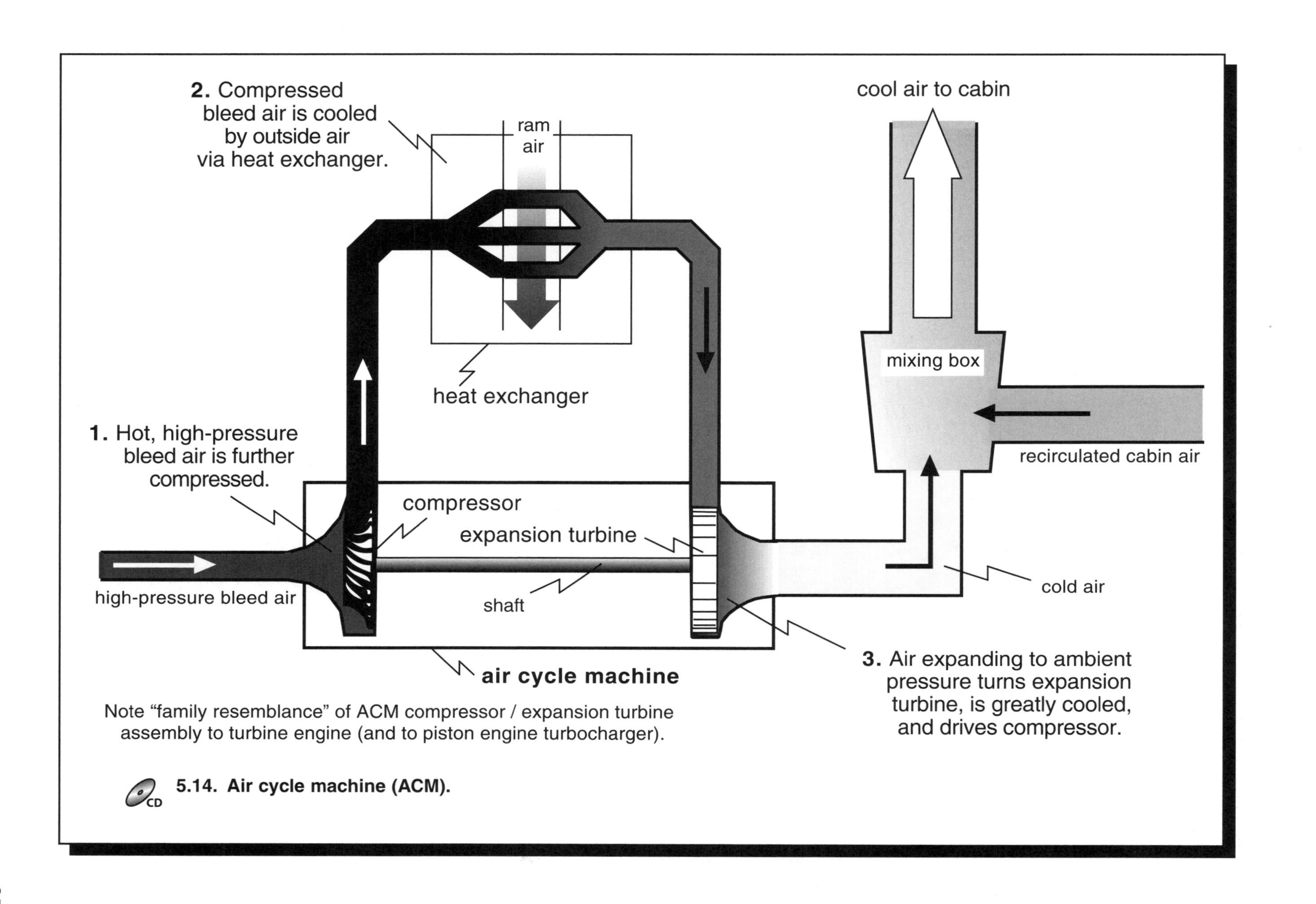

Note "family resemblance" of ACM compressor / expansion turbine assembly to turbine engine (and to piston engine turbocharger).

5.14. Air cycle machine (ACM).

aircraft always have ACMs installed because of their economy of use, hefty pressurized air (bleed) sources, and the need to process large volumes of air.

VCMs, on the other hand, are efficient, significantly less expensive, and are well suited to aircraft with limited engine bleed capacity. A vapor cycle machine does, however, require a separate mechanical compressor, which adds complexity and weight.

Small turboprops and corporate jets, in many cases, have vapor cycle systems installed. This is particularly true of older corporate aircraft, which tended to have less available power and, therefore, less bleed capacity. Since newer turbine aircraft generally have more powerful engines, and ACM technology has developed rapidly for smaller aircraft, ACMs are more common in newer models.

One other advantage of a VCM is that it can be set up to provide cooling on the ground, without an operating engine, APU, or external high-pressure air source. While many VCM compressors are engine-driven, they can also be set up to operate from electric motors. That way, a crew sitting on a hot ramp can plug in ground power and cool down the passenger cabin before start-up. Given the different efficiencies and benefits of ACMs and VCMs, many aircraft have both systems installed.

Aircraft Environmental System

To control cabin temperature the cooled air from ACM or VCM is simply mixed with hot bleed air. Two or more *temperature mixing valves* are used for this purpose. In large aircraft, the whole environmental heating/cooling system is bundled together, including ACM, bleed heat source, VCM (if installed), and mixing valves (see Fig. 5.16). This package is commonly referred to as a "PACK." Normally two are installed for capacity and redundancy.

In most modern aircraft, comfort is controlled via automatic temperature control systems tied to cabin air temperature sensors. Manual control of the temperature mixing valves is usually available to back up the automatic systems.

Fuel Systems

The fuel system in any aircraft is designed to store fuel for flight, then deliver it to the engines in the proper amounts and at the correct pressures. In turbine aircraft, this simple assignment becomes rather complex. Large (often huge) amounts of fuel must be carried and distributed, many tanks and engines may be installed, and allowances must be made for the characteristics of jet fuel and turbine engines themselves.

Most civilian turbine fuel systems are designed for long-range operations plus IFR reserve capacity. Fuel must be manageable during engine-out flight operations and in some cases must be shifted to keep the aircraft within CG (center of gravity) range. Therefore, most turbine fuel systems support fuel transfer between tanks and from one side of the aircraft to the other. As with every turbine aircraft system, redundancy is added for safety.

Fuel Tanks

The increased complexity of a large turbine aircraft fuel system is immediately apparent when considering the fuel tanks. Depending on aircraft type, fuel tanks come in all sizes and shapes. They may be located in the wings, the fuselage, or even the tail.

Fuselage tanks are normally constructed of formed aluminum. Most wing tanks, however, are made by sealing part of the wing structure so that fuel can be held inside. These *wet wings* have largely replaced the rubber bladder tanks found in many older aircraft. (Bladder tanks were prone to deterioration over time and to water and fuel entrapment due to distortion of the tanks.) External wing tanks, while not found on airliners, are fairly common on corporate aircraft. Tip tanks are found on some Learjets and Cheyenne turboprops, while the Lockheed Jetstar was equipped with underwing tanks.

Engines on turbine aircraft may draw from each tank separately, from interconnected fuel tanks, or from a main tank fed by some combination of reserve or auxiliary tanks.

In order to ensure a steady fuel supply to the engines in all flight attitudes, fuel tanks are often divided or interconnected. Large turbine aircraft have long wings. You can imagine that unporting of the fuel supply to the engines could easily occur without proper system design. ("Unporting" refers to fuel interruption, caused by the fuel in a tank sloshing away from the intake that draws it for the engine.)

One common solution to this problem is the use of *collector bays.* A collector bay is simply a separated segment of a larger fuel tank. It is located near the engine and directly supplies it with fuel. Each collector bay is in turn fed by the regular selected tanks. In the event of a momentary fuel supply fluctuation from the regular tank, the collector bay takes up the slack to provide continuous flow. Flapper valves (spring-loaded one-way doors) are sometimes installed between the collector bay and the rest of the fuel tank in order to prevent reverse flow back to the main tanks. Some aircraft are fitted with *header tanks,* small separate tanks that serve the same function as collector bays. Other specialized fuel tanks are sometimes installed in aircraft for various purposes. *Surge tanks,* for example, are often located near the wing tips of large aircraft to manage fuel movement, overflow, and venting.

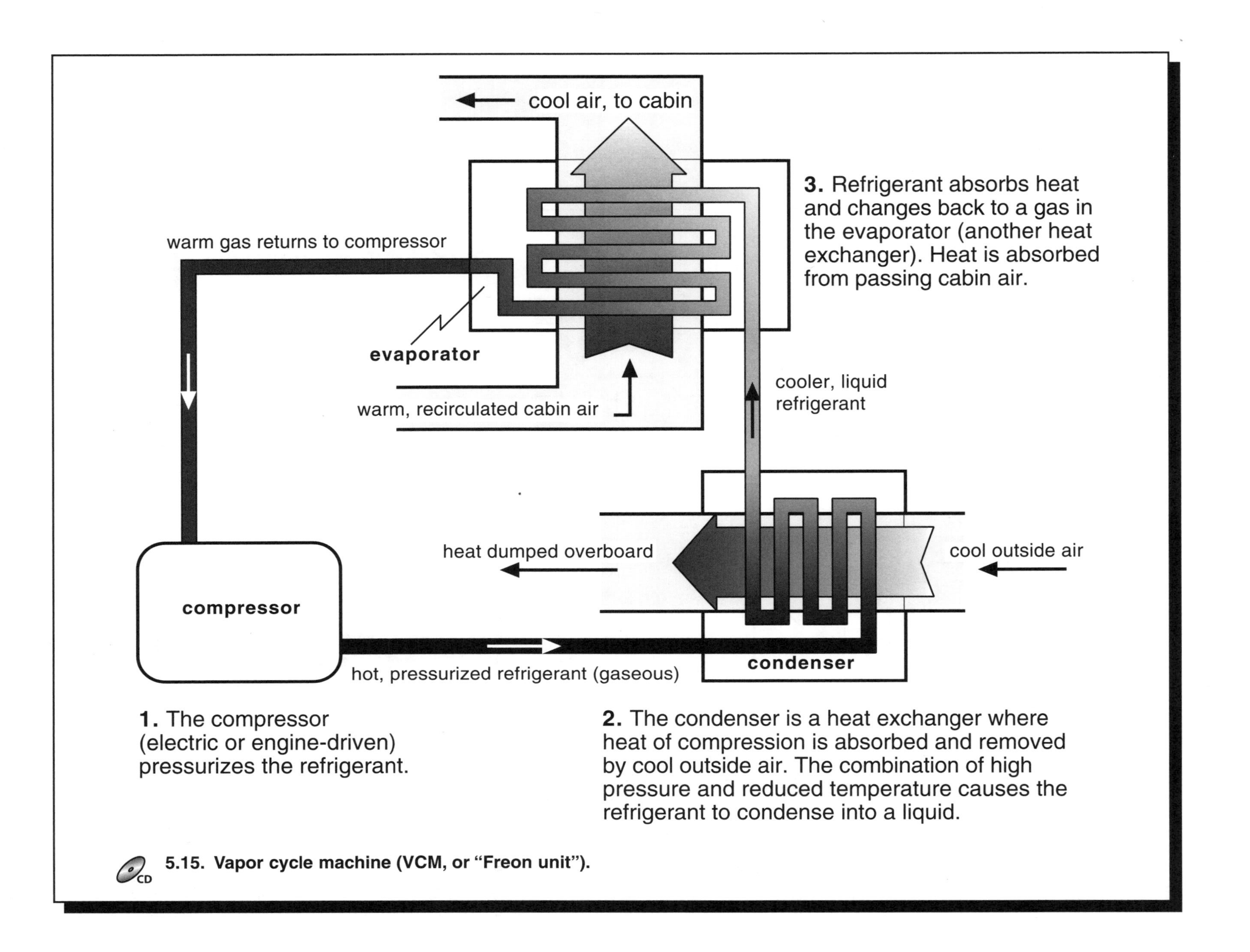

5.15. Vapor cycle machine (VCM, or "Freon unit").

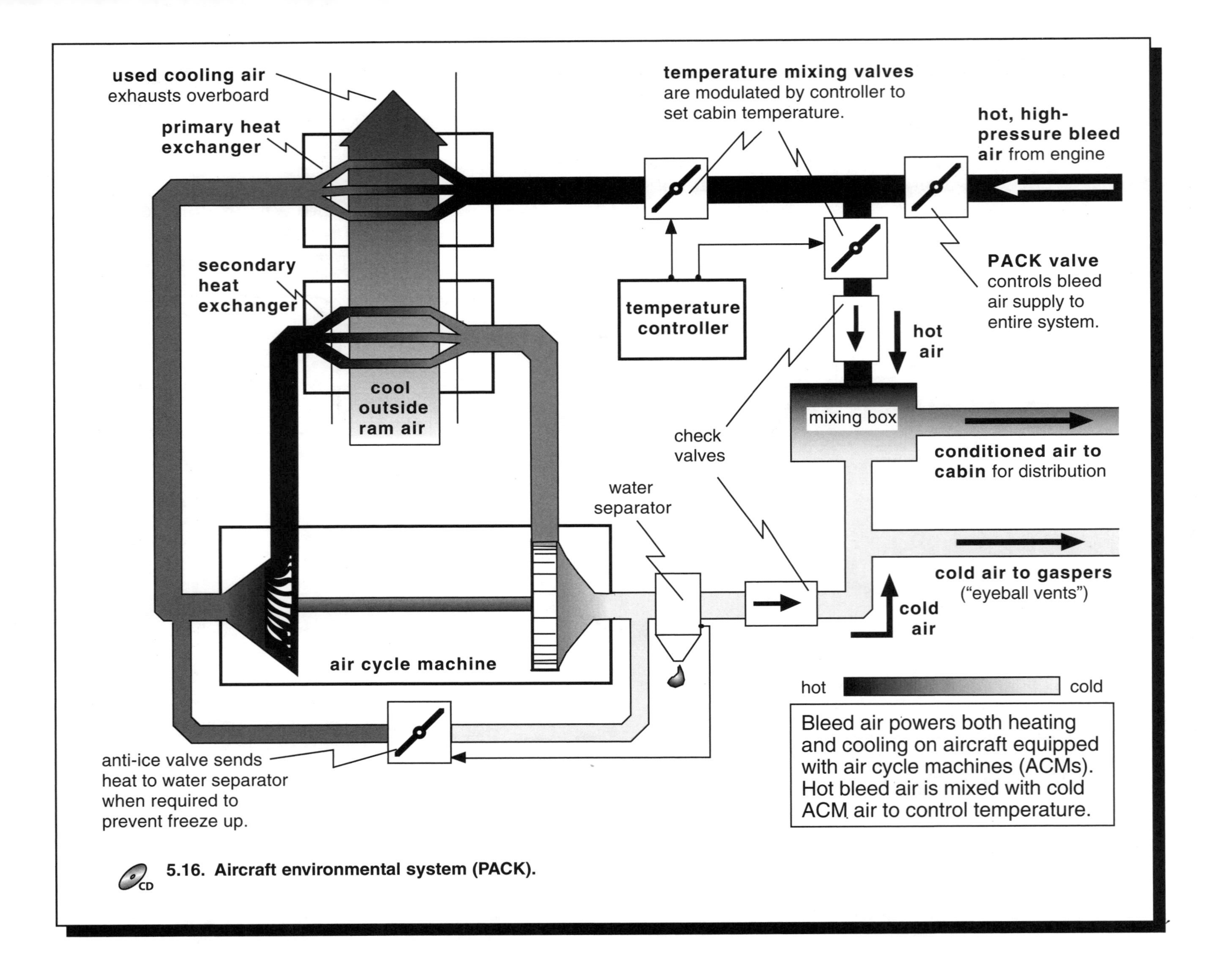

5.16. Aircraft environmental system (PACK).

Fuel Pumps

A variety of pumps are installed in turbine fuel systems to move fuel from tanks to engines and to move fuel from tank to tank. The most common categories of fuel pumps are high-pressure pumps, low-pressure pumps, auxiliary pumps, and jet pumps (Fig. 5.17).

An engine-driven high-pressure fuel pump supplies fuel to each engine at pressures in the neighborhood of 900–1000 psi. Use of high pressure for this purpose is important for several reasons. First, the amount of fuel required on takeoff may be two or three times greater than that needed at normal cruise. Therefore, both pressure and flow capacity must be available to rapidly provide large increases in fuel delivery. (Excess fuel delivered by the pump, at any given time, is routed back to the tanks via return line.) High pressure also supports proper fuel delivery into the engine. Fuel is sprayed into the engine's combustor at a pattern and pressure to optimize combustion, center the flame in the chamber, and keep the fuel nozzles from overheating.

The *low-pressure pump* is usually also engine driven. It draws fuel from the tanks and supplies it to the high-pressure pump, often through a fuel filter and heater.

Most turbine fuel systems are also equipped with electric *auxiliary fuel pumps.* (These are otherwise known as *aux pumps, standby pumps, fuel transfer pumps,* or *boost pumps.*) Aux pumps have various purposes depending on specific aircraft fuel system design. They are often used to transfer fuel from one tank to another and to provide emergency backup for the engine-driven low-pressure pump. In large aircraft, boost pumps operate continuously in tanks being used, since suction from low-pressure engine pumps is not sufficient to draw fuel at altitude.

In most systems, aux pumps earn glory for crossfeeding fuel in the event of engine failure. *Crossfeeding* refers to fuel transfer from the tanks on one side of an aircraft to the engine(s) or tanks on the other. It's an important safety feature in the event of engine failure, allowing the operating engine to utilize the dead engine's fuel supply. Balancing fuel weight from side to side can also be critical under such circumstances.

Auxiliary pumps of most varieties are operated from a fuel control panel located in the cockpit. (Auxiliary fuel control panels may be located out on the wing for use by fuelers on larger aircraft equipped with single-point refueling systems.)

Ground school will cover the purpose of each fuel pump in the aircraft you'll be flying, as well as how to monitor operation from the cockpit. Be aware that most fuel pumps cool and lubricate themselves with the fuel they pump. So be careful. You may burn up an aux pump one day by operating it from an empty tank!

It's also interesting to note that turbine engines are quite versatile in the fuels they'll burn. Many smaller engines, for example, are approved for limited use of gasoline. Due to its lesser lubricating qualities, however, gasoline usage in fuel pumps is often limited to only a few hours or the pumps must be rebuilt.

One other type of pumping action is commonly found in turbine aircraft fuel systems. *Motive flow* refers to the use of small venturi-type ports, which are used to draw fuel into collection lines. It is desirable to collect fuel from several places in each tank in order to prevent unporting and to back up any clogged pickups. Rather than installing lots of mechanically driven pumps all over the tank, these venturi devices, or *jet pumps,* draw fuel into the lines by creating low-pressure areas in the passing fuel moving through the lines. Therefore, jet pumps cannot pump fuel by themselves. Rather, they act as localized secondary pumps (sometimes called scavenge pumps), effectively powered by the main pump on the line. (Usually it's the low-pressure engine-driven pump or an electric auxiliary pump.)

Fuel Control Unit

A *fuel control unit* (*FCU*) is a precise hydromechanical or computerized electronic device that delivers fuel to the engine. The FCU collects such inputs as power or thrust lever position, air pressure, and engine temperatures. It then meters the proper amount of fuel from the high-pressure fuel pump to the gas turbine engine's combustion chamber. Different types of electronic fuel control units are sometimes known as "FADECs" (full-authority digital engine controls) or as "ECUs" (electronic control units). For redundancy, some aircraft have mechanical FCUs backing up ECUs in the same fuel system.

Fuel Valves

Various valves are used to manage fuel flow. Most valves simply open to allow fuel to flow or close to stop fuel flow. *Check valves,* like the ones in our reference water system, allow fuel to flow in one direction at various points within the system but not in the other direction.

Fuel selector valves are used by the pilots to assign fuel supply to each engine from selected tanks. *Crossfeed valves* direct fuel from the tanks on one side of an aircraft to engine(s) or tanks on the other. (In some aircraft, crossfeed is authorized only for emergency use in the event of engine failure. In other vehicles, crossfeed is approved for routine use to balance lateral fuel loads.)

Fuel dump valves are installed in many aircraft where maximum allowable takeoff weight (MTOW) is significantly greater than maximum allowable landing weight (MLW). If a heavily loaded aircraft must return for landing

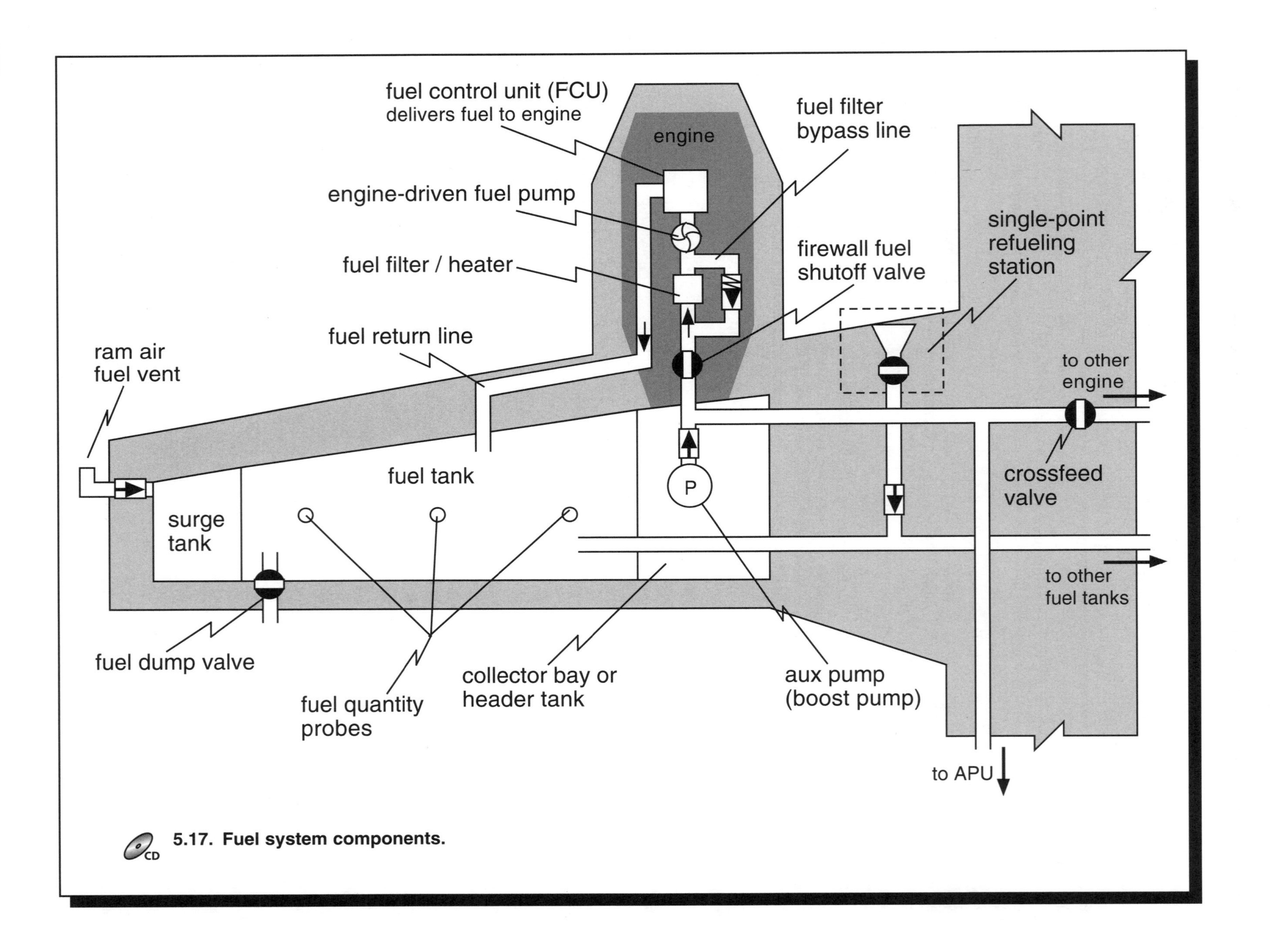

5.17. Fuel system components.

immediately after takeoff, fuel must be dumped overboard in order to bring the aircraft's weight below its MLW for a safe landing. (That fuel would normally have been consumed on the planned trip.) Fuel dump valves are used for this purpose. (See also Chapter 11, Weight and Balance.)

All turbine-powered aircraft have some type of emergency fuel shutoff valve. *Firewall fuel shutoff valves* may be mechanically or electrically operated, depending on manufacturer. These valves are used to eliminate fuel supply to the engine compartment during emergency engine shutdown only. The "fire handles" (or "T-handles"), which operate firewall fuel shutoff valves, are among the most visible controls in any cockpit. (See also "Fire Protection Systems" in Chapter 6.)

Fuel Heaters

Jet fuel has the unfortunate characteristic of absorbing water. When a turbine aircraft is flying at high altitudes, low outside air temperatures can cause this absorbed water to crystalize in the fuel system. As you might imagine, ice crystals clog the fuel filters and can accumulate to the point of causing engine flameout. (*Flameout* means that combustion ceases in the engine: "the fire goes out.") *Fuel heaters* are used to warm the fuel and prevent this from happening. There are two common ways of warming the fuel: fuel-oil and fuel-air heat exchangers. Fuel-oil heat exchangers transfer heat from engine oil to warm the fuel. Fuel-air heat exchangers use hot bleed air for the same purpose.

Fuel Quantity Measurement Systems

Fuel quantity in large turbine aircraft is generally measured using a different system than the floats with which you may be familiar. This is true for two reasons. One is that the fuel moves around quite a bit in very large tanks, even with baffles, making it hard to measure. The other has to do with the nature of jet fuel (specially formulated kerosene). Jet fuel changes volume measurably as a function of temperature. Therefore, the amount of energy in a gallon of jet fuel also varies with temperature. To get around this problem, turbine airplane fuel consumption and quantities are measured in pounds, not gallons. (The volume of fuel in a tank may change, but its weight won't.)

Capacitance fuel quantity indicator systems are used to measure fuel mass (and therefore weight). These systems consist of a series of long metallic probes stretched from top to bottom across the fuel tanks. By measuring electrical capacitance across the probes, the fuel indicating system determines the amount of fuel in one or more tanks. This information is sent to the fuel quantity gauges in the cockpit, which, of course, read out in pounds of fuel. This type of system is easily calibrated to compensate for fuel movement within the tanks.

Just to make your life as a jet jockey (or turboprop driver) more interesting, the fuel trucks on the ground usually measure fuel delivery in gallons because that's how their metering systems work. So when refueling, turbine pilots calculate their fuel needs for the next leg in pounds, but they often must order that fuel load *in gallons.* When accuracy is critical, a fuel conversion chart may be used. However, most of the time, fuel conversions are estimated at around 6.7 pounds per U.S. gallon. (See Appendix 1, Handy Rules of Thumb for Turbine Pilots, for an easy estimating method.) Fuel is calculated and ordered by company dispatch for many scheduled carriers, so those pilots often escape the computations.

Fuel Quantity Measuring Sticks

Sometimes it's desirable to physically measure the exact amount of fuel in a given tank. (This may be for calibrating the aircraft's fuel quantity measuring system or to permit aircraft dispatch when a fuel gauge is out of order.) Obviously, it would be no fun to measure tens or hundreds of thousands of pounds of fuel in a large aircraft using handheld dipsticks or five-gallon buckets. Therefore, many manufacturers build in relatively convenient mechanical fuel measurement systems consisting of *fuel quantity measuring sticks* (Fig. 5.18).

These handy devices are used to measure fuel quantity from beneath the wings and are located in one or more hollow tubes that run vertically through each fuel tank. Under normal circumstances, the measuring sticks are locked in place in each fuel tank. To manually measure fuel quantity, each stick is unlocked and lowered until its top is level with the fuel surface. On older "dripsticks" the fuel level was identified when fuel began dripping out of the bottom of the hollow stick.

Newer magnetic models are located in sealed tubes. A magnetic float rides along each tube's outer surface on top of the fuel. When the "Magnastick" is unlocked, it slides freely out of the wing's bottom until another magnet atop the stick aligns itself with the magnetic float. Both dripsticks and Magnasticks are calibrated so that fuel quantity can be read directly, based upon how far each stick projects out of the wing.

Fuel Vents

Vents are crucial to proper operation of any fuel system. As fuel is drawn from each tank, outside air must enter to replace it, or fuel flow will stop due to development of a vacuum. At the same time, fuel vents must also provide for relief of fuel tank overpressure due to thermal expansion,

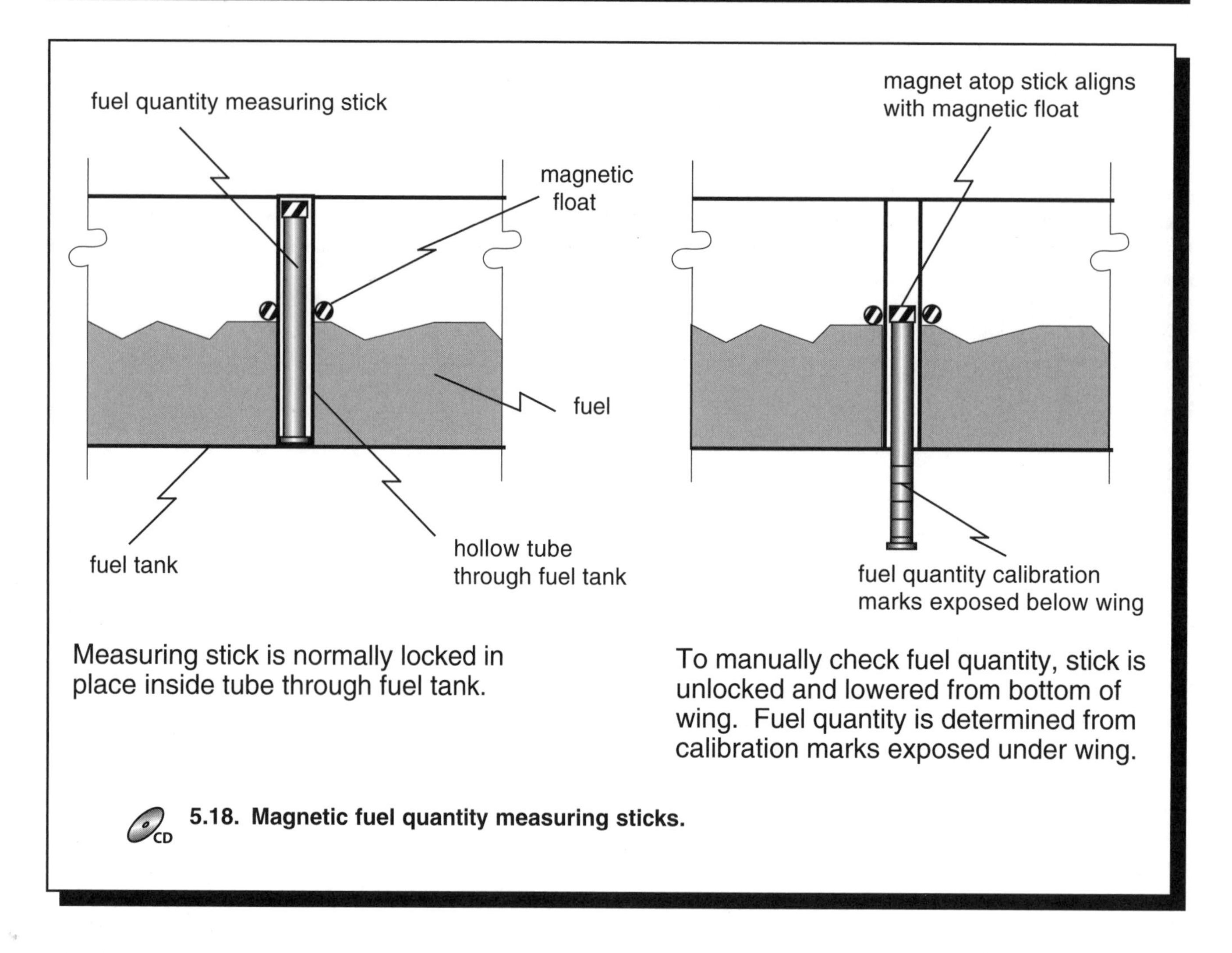

5.18. Magnetic fuel quantity measuring sticks.

but they must be designed to prevent fuel from venting overboard under normal conditions.

The delicate balance of fuel tank pressures is further complicated by the need for specialized vent designs in different locations. *Heated vents,* for example, are used for protection from structural ice, while *flame arrestor vents* protect fuel tanks from ignition by hot exhaust. *Ram air vents* scoop ram air in flight to pressurize the tanks and enhance fuel flow. Flush *NACA vents* are installed in some locations due to low-drag characteristics and resistance to icing. (NACA, the National Advisory Committee for Aeronautics, was established in 1917 and replaced by NASA, the National Air and Space Administration, in 1958. NACA's many contributions to aviation include drag reduction innovations such as the NACA vent along with the NACA airfoil classification system.) It's common to find at least two different vent types out at the wing tip of a turbine airplane.

Fuel Management

Proper *fuel management* is very important in turbine aircraft due to the large fuel flows and weights involved. (See also Chapter 10, Performance.) Add to that the fuel system peculiarities of each aircraft type. In many airplanes, tanks must be "burned off" in a specific order during flight, or excess fuel returning from the engine high-pressure pump may be indirectly pumped overboard. Also, from the standpoint of weight and balance, it is entirely possible in many aircraft to create an unbalanced situation through poor fuel management. (For more on this, see Chapter 11, Weight and Balance.) These complexities require the turbine pilot to know his or her fuel system, to constantly monitor fuel levels, and to keep a balanced fuel load.

CHAPTER 6

Dedicated Aircraft Systems

Ice and Rain Protection

As you're undoubtedly aware, flight under certain environmental conditions requires specialized aircraft systems. Instrument flying, for example, was not practicable under many circumstances until the development of deice and anti-ice systems for aircraft. Even today, with all of the advanced systems and knowledge available, accidents occur all too regularly due to ice accumulation on aircraft. The types and causes of icing, and strategies for safely dealing with ice, are critical knowledge for any all-weather pilot. No matter how well equipped your aircraft, be sure to receive proper training before considering operation under icing conditions.

Several different types of icing are of concern to pilots operating turbine aircraft. These include structural icing, engine inlet icing, and fuel icing. Let's start by examining *structural icing,* meaning accumulation of ice on the external structures of the aircraft, including wings, tail surfaces, and fuselage. For our purposes, we'll consider structural icing in two categories: ground and in-flight.

In-Flight Structural Icing

The vast majority of installed aircraft ice protection systems are designed to address in-flight structural icing. The protected flight surfaces commonly include leading edges of wings, tail surfaces, propellers, and in some cases struts or auxiliary stabilizer surfaces.

Two categories of structural ice protection equipment may be installed for this purpose. *Deice* (or "deicing") systems are designed to remove ice that has already accumulated. *Anti-ice* (or "anti-icing") systems are designed to prevent icing before it occurs

The terms "deice" and "anti-ice" sound so similar that they're often mistakenly used interchangeably in casual conversation and are sometimes mislabeled even in technical publications. However, proper use of each type of system is different and critical to flight safety. For example, since anti-icing systems are preventative, they must often be turned on at the earliest threat of icing conditions in order to work effectively. Deicing systems, on the other hand, sometimes carry stipulations that they not be activated until a certain minimum amount of ice accretion (accumulation) has already occurred. Clearly, it's very important to understand the fundamental difference between these two entirely different systems.

Note that under extreme conditions even the most effective ice protection systems can be overwhelmed. The *Aeronautical Information Manual* (*AIM*) defines "severe icing" pilot weather reports (PIREPs) to mean that the rate of accumulation is so great that it cannot be reduced or controlled by installed ice protection systems. Immediate flight diversion is the only option. Clearly, along with proper use of ice protection equipment, a big part of any good pilot's strategy when encountering significant icing is to minimize exposure by getting out of those conditions as quickly as possible. You can see why proper training is imperative before flying in conditions conducive to icing.

Pneumatic Leading Edge Deice Boots

Many smaller turbine aircraft, especially turboprops, incorporate *pneumatic leading edge deice boots* on wing

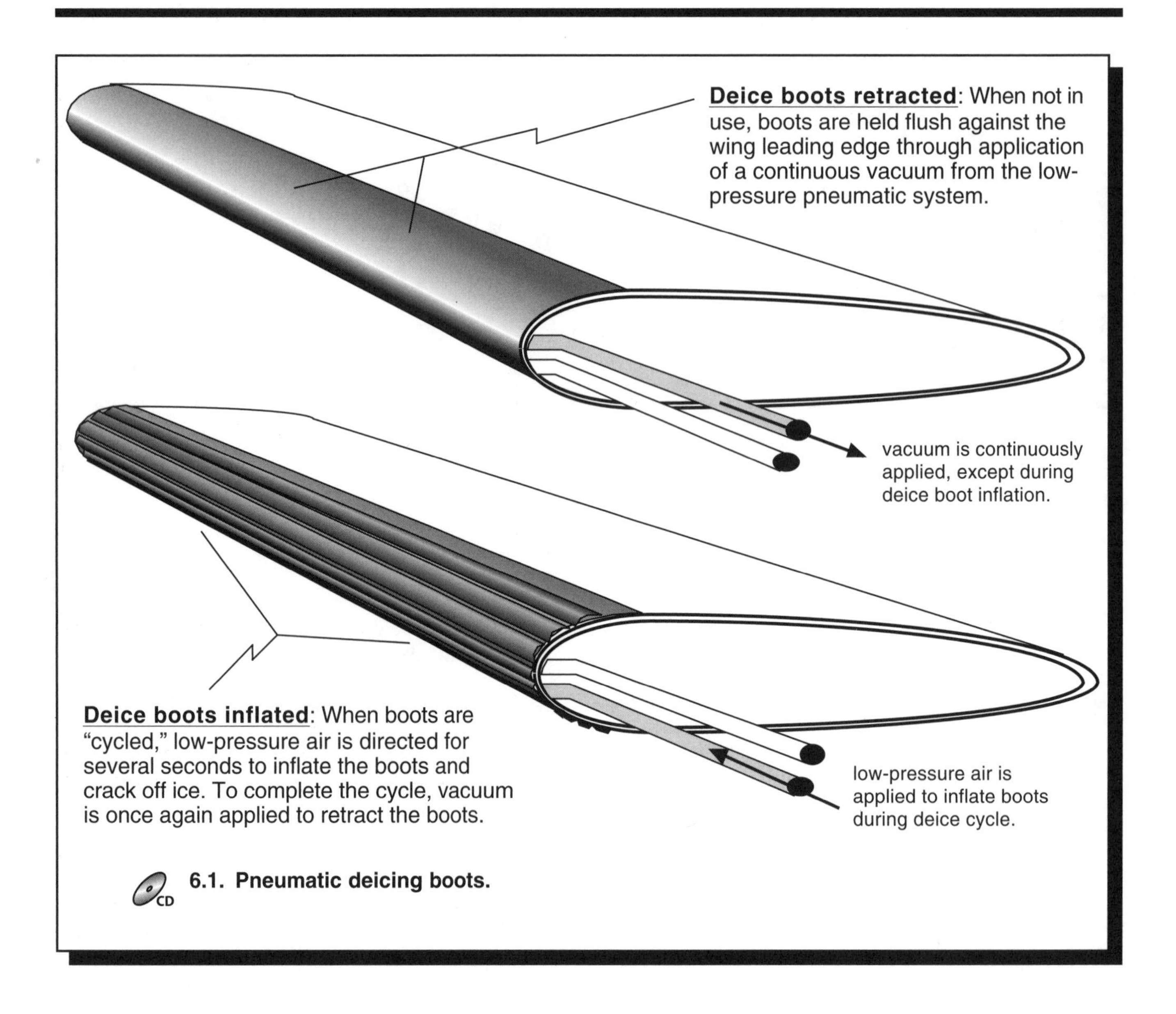

6.1. Pneumatic deicing boots.

and tail surfaces (like the ones used on many piston aircraft). These boots are powered by low-pressure air and its associated vacuum (Fig. 6.1). Pneumatic boots are a deicing system, since ice is normally allowed to accumulate before the system is used. Operation of these systems traditionally calls for allowing buildup of some minimum thickness of ice (usually 1/2 to 1 inch) on the surface. Then the boots are "cycled": first inflated to expand and crack off the ice, then drawn back down to the wing by vacuum. (A small amount of vacuum is applied at all times to keep the boot retracted to the wing contour, except at those moments when inflation is necessary.)

It has long been considered important not to cycle pneumatic boots before ice has accumulated to the minimum stated in the aircraft's operating manual. This is because thin layers of ice can theoretically "stretch," leaving an airspace between ice and boot or holding the boot in a partially extended configuration. If this happens, the boot may not operate effectively later, when more ice has accumulated and the system is really needed. However, aspects of this theory have recently come under investigation, so be sure to consult the latest operating procedures for pneumatic deice boots on aircraft you fly.

Most pneumatic deice systems incorporate a two-mode controller. One mode allows manual cycling of the boots by the pilot. An additional auto mode provides automatic cycling of the boots at regular time intervals.

Bleed Air Thermal Leading Edge Anti-Ice Systems

Large jets are commonly equipped with *thermal leading edge anti-ice systems,* where hot, high-pressure bleed air

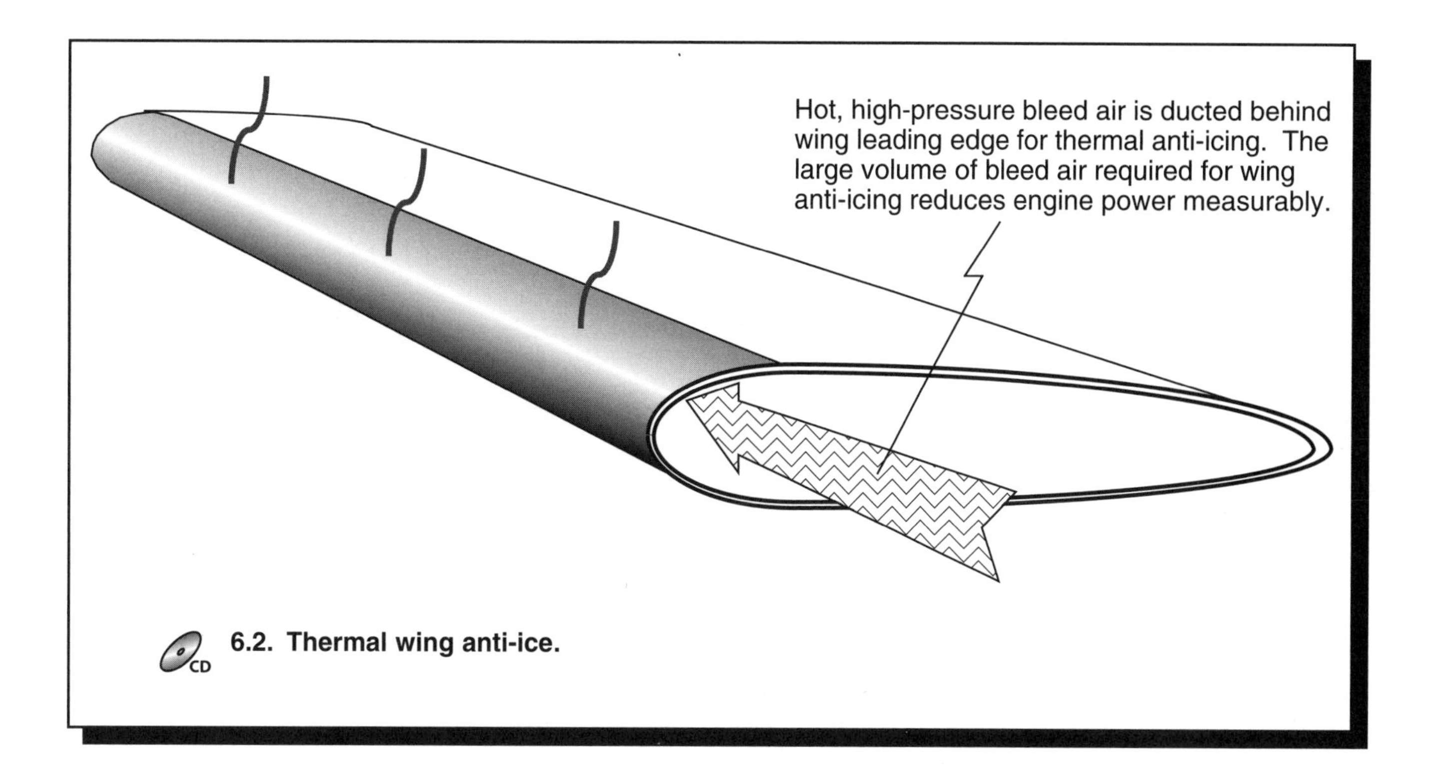

6.2. Thermal wing anti-ice.

is routed under the leading edge surfaces to prevent the formation of ice. These are normally designed as anti-ice systems to be turned on preventatively when icing conditions exist (usually visible moisture and outside air temperatures below +10°C). While effective, bleed air anti-ice systems sap a good deal of engine power, due to the drawing off of air from the engine compressor sections (Fig. 6.2).

Another problem is that bleed air is so hot (often 600°–800°F) that use on the ground could lead to deformation of aluminum leading edge components. Therefore, bleed-powered wing anti-ice is normally restricted to in-flight use only. (These systems are often tied electrically to a landing gear squat switch in order to prevent inadvertent operation on the ground.) When icing conditions are anticipated on takeoff for aircraft having these systems, ground-applied anti-icing chemicals must protect the leading edges until the plane is airborne.

Liquid Ice Protection Systems

A few aircraft, such as the Raytheon Hawker jet series and the Cessna Citation S-II, incorporate *TKS liquid ice protection systems,* sometimes known as *weeping wing.* On these anti-ice systems, an ethylene glycol-based deicing fluid is forced through tiny pores in a stainless steel or titanium wing leading edge, where it coats the surface with fluid (Fig. 6.3).

Advantages of this system are that it generates less aerodynamic drag than pneumatic boots, and it does not draw engine bleed air like thermal systems. Therefore, TKS can be used on takeoff without penalty. Since protective fluid runs back from the leading edges, ice does not readily develop behind the leading edge (which it sometimes does on other systems).

Disadvantages include the need to carry and replenish a supply of deicing fluid. The system can be somewhat heavy on older large-capacity systems.

A diminishing number of older turboprop aircraft use alcohol prop anti-icing systems, where anti-ice fluid is centrifugally spread over propeller leading edge boots from outlets mounted near the prop hubs. Alcohol windshield anti-icing systems are also found on some older models.

Electrically Heated Anti-Icing Systems

Electrically heated deicing boots are generally too energy demanding for use on large surfaces like wing leading edges, but they are commonly used for propeller anti-icing on turboprops and piston aircraft. These “hot prop” systems draw too much electrical current for all boots to operate at the same time. Therefore, a prop anti-ice controller cycles power to each prop boot in a timed, symmetrical manner. Normally, pilots must memorize the prop boot electrical interval and sequence for their aircraft so as to monitor system operation on a prop ammeter (see Fig. 6.4).

Miscellaneous other structures are normally protected by electrically heated anti-ice systems. These include electrically heated windshields, sensors (stall warning, air data

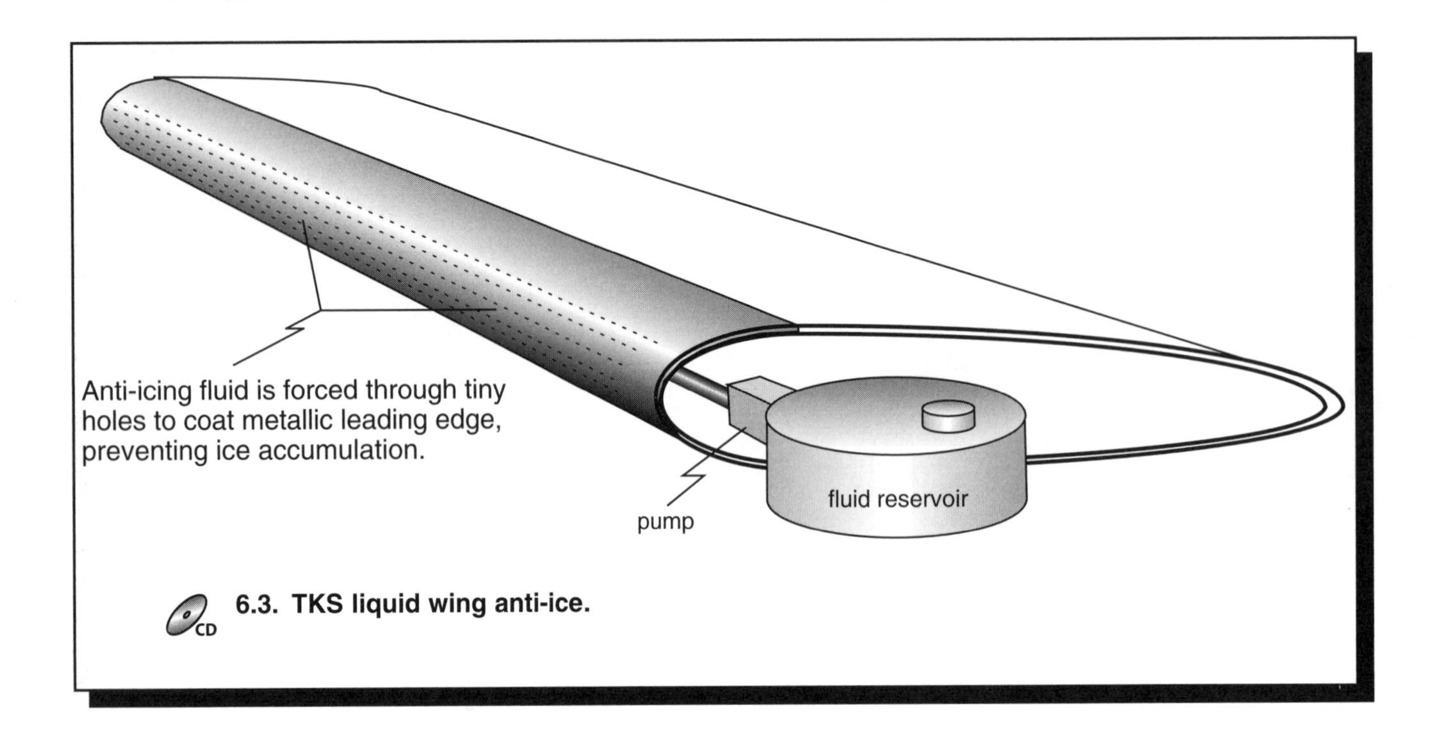

6.3. TKS liquid wing anti-ice.

computer, angle-of-attack, pitot, static, and other probes), and various types of vents (fuel system, among others). Incidentally, on many airplanes electric windshield heat serves a second purpose in addition to anti-ice—birdstrike protection. Windshields are less brittle when heated and, therefore, less subject to failure when struck in flight. Therefore, in many turbine aircraft windshield heat is always on in flight, and some airplanes are speed-limited when windshield heat is inoperative.

New Technology Anti-Ice Systems

In-flight icing is such a serious problem that researchers are always looking for better ways of doing it, with associated objectives of minimizing drag, weight, and power consumption. From the standpoint of new technology, a good deal of research has been conducted on vibration deicing systems, where the wing leading edges are distorted slightly at high frequencies to crack off ice. Electromagnetic deicing is another promising new system that uses far less electrical power than traditional electrically heated surfaces.

Airframe Ice Detectors

We talked previously about how in today's large turbine aircraft pilots can rarely see much of the plane's exterior from the cockpit. This lack of visibility also raises problems in identifying and treating airframe ice when it occurs.

Accordingly, ice detector probes are increasingly being installed to detect and warn flight crews of ice accumulation on the airframe. Ice detectors may be found on all types of turbine aircraft from the smallest turboprop to the largest airliners. Probes are mounted in key locations where icing would usually first occur, often on the forward portion of an aircraft's fuselage.

The most common type of ice detector incorporates a vibrating element in the probe housing, which vibrates continuously when electrical power is applied. When ice accumulates on the airframe, it also forms on the ice detector probe, immobilizing the vibrating element (i.e., it's frozen solid). The ice detector control unit senses stopped vibration at the probe and sends a warning signal to the cockpit. Depending on aircraft type, this warning signal comes in the form of a bell or chime and an ice warning light.

Once the warning signal is sent to the cockpit, the ice detector probe deices itself with an internal heating element, effectively resetting itself. The vibrating element will then again vibrate continuously until ice builds up and another warning signal is sent to the cockpit. This process continues until the aircraft exits icing conditions.

As you can imagine, ice detectors are very useful, particularly at night when it's especially difficult for the flight crew to see ice accumulating on the aircraft.

Engine Icing

On piston aircraft, induction system icing (carburetor and/or intake) is the main engine icing problem. Turbine engines, on the other hand, are most threatened by ingestion of solid ice, due to the potential for mechanical damage. (Imagine the effect of a fist-sized chunk of ice penetrating a 37,000 rpm compressor wheel.) Therefore, most turbine engines utilize anti-ice (rather than deice) systems for protection.

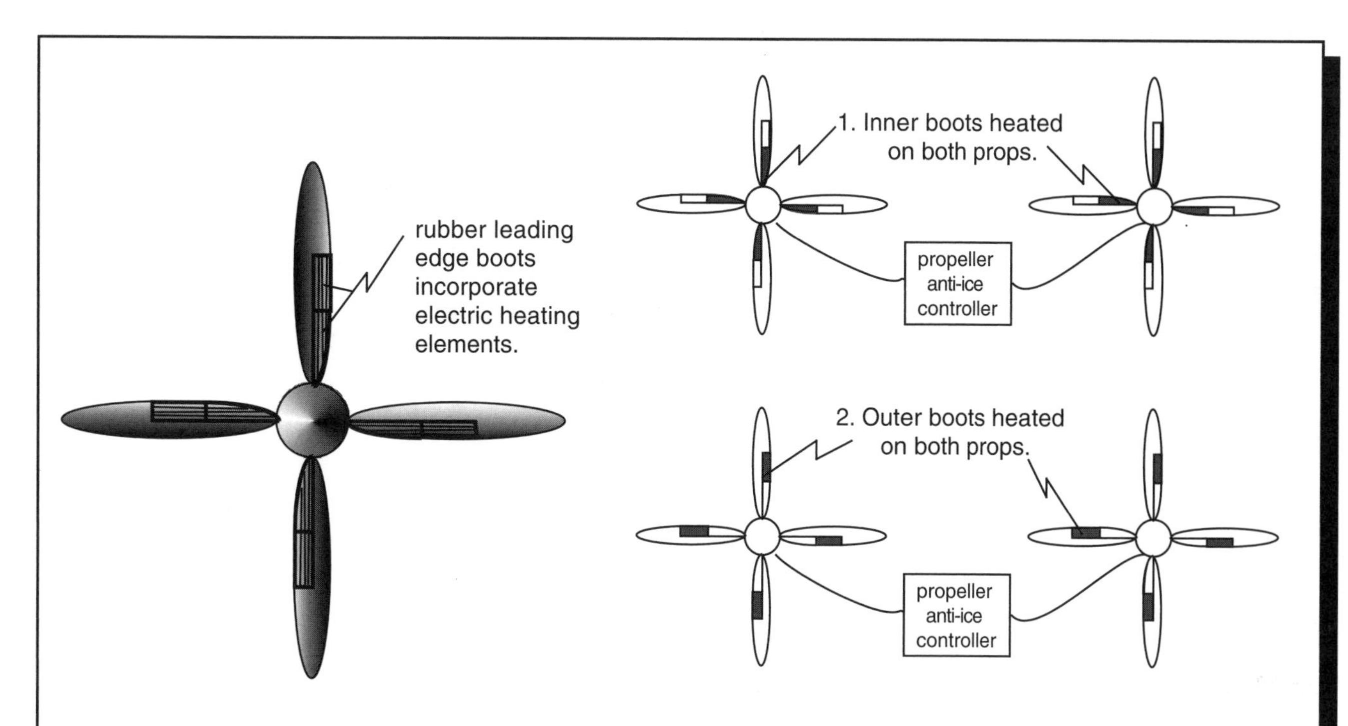

“Hot prop” anti-ice systems incorporate electric heating elements in rubber leading edge propeller boots. Due to high amperage requirements of electrical heating, all boots cannot be heated at the same time. A propeller anti-ice controller (or “timer”) sequentially cycles power to symmetrical sets of boots. An anti-ice electrical loadmeter (ammeter) is often associated with the system. Reduced or no current draw for one heating cycle indicates failed boot heating segments.

6.4. Electrically heated propeller anti-ice system (“hot props”).

Engine Inlet Anti-Ice

Engine intake lips and compressor domes may be protected using any number of systems: high-pressure bleed air, ducted exhaust heat, or electric heating elements. (A few turboprops even use pneumatic boots to protect their intake lips.) The threat of damage is so significant on some aircraft that *engine inlet anti-ice systems* are kept permanently on while engines are running.

Bleed-powered engine inlet anti-ice systems, as you might expect, sap power from the engines, potentially reducing takeoff power measurably. Therefore on many jets it can be used on takeoff only under restrictions such as reduced takeoff weight. Some newer aircraft, however, such as the Boeing 757 and 767, have enough excess engine power available that they are not restricted in this manner.

Inertial Separators

Turboprops have an additional engine ice liability, in that the cyclic operating nature of hot prop anti-ice systems causes chunks of ice to be released under icing conditions. In order to prevent ingestion of this ice into the engines (and to prevent FOD, or *foreign object damage,* on the ground), some turboprop engines are equipped with *inertial separators* (or *ice vanes*) in their induction systems. Ice vanes are “doors” located at sharp corners in the induction system (Fig. 6.5). The idea is that foreign objects (such as chunks of ice) “can’t make the corner” into the engine and are thrown out. Most ice vane systems are mechanically opened by the pilot under potential icing conditions. A few aircraft (such as the DHC-8) have spring-loaded automatic versions that are pushed open only when FOD contamination occurs.

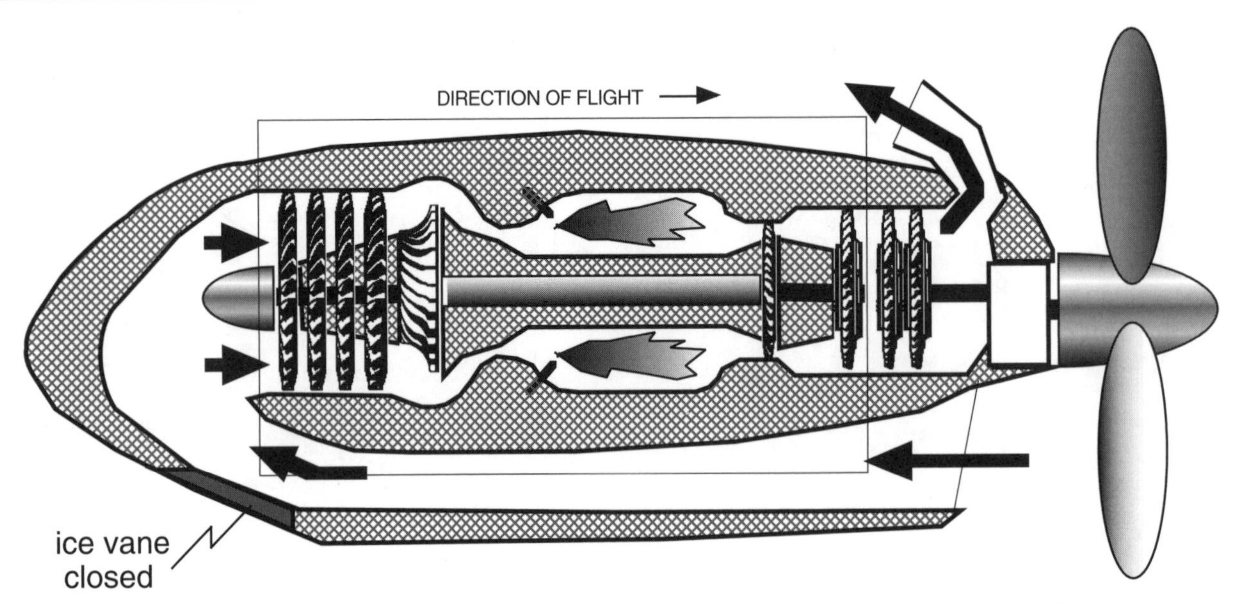

Under normal (non-icing) conditions, ice vanes are closed, providing maximum ram air effect to engine, and therefore optimum power.

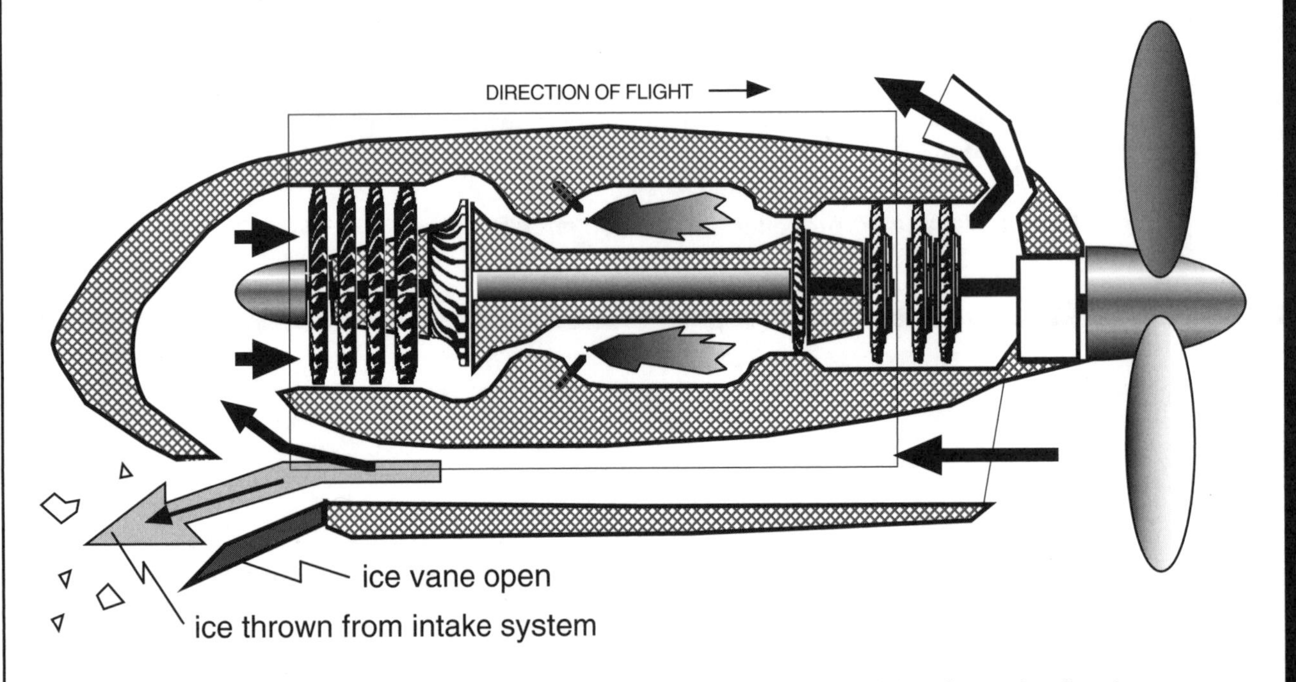

Under icing conditions, ice vanes are opened to expel ice from the intake system before it can enter the engine. However, much of the intake ram air effect is lost, so engine power is reduced while ice vanes are activated.

6.5. Inertial separator engine anti-ice system ("ice vanes" shown on a reverse-flow, free-turbine turboprop).

The downside of most ice vane systems is that when they're selected induction air is dumped out of the system along with any ice, thereby reducing engine power. For this reason, inertial separators are usually activated only in icing conditions or sometimes on the ground.

Fuel System Icing

It may surprise former piston pilots to learn that jet fuel readily absorbs water. Therefore, when fuel temperature falls too low, entrapped water crystallizes into ice, which can then accumulate to clog filters and other restrictions in the fuel system.

Several different approaches are used to combat this problem. Chemical anti-ice additives may be added to the fuel. (Additives also serve to prevent growth of microorganisms in the fuel that could clog the system.)

Fuel supply lines are often heated by bleed air or waste heat from other sources (such as engine oil return lines). Other fuel system restrictions, such as fuel filters, may be further equipped with pilot-selectable electric or bleed air heat. (Fuel temperatures in these cases are monitored from the cockpit via fuel temperature gauges or annunciator lights.) Other key fuel system components, such as fuel tank vents and overflow ducts, may also be electrically heated to sustain proper operation by protecting against both fuel system icing and structural icing that might clog the vents.

The Role of the Pilot

You can see that as simple as ice protection may sound upon first mention, it actually presents some of the most challenging management tasks faced by a pilot. Along with the icing identification and avoidance issues mentioned earlier, pilots are also challenged in properly coordinating use of so many different systems. On most aircraft, each of these different anti-icing and deicing systems is separately controlled, perhaps with use of some systems called for at different times than others. It's easy to forget to activate one ice protection system or to fail to notice when another is not operating properly. Obviously, this could prove deadly. So could asymmetrical operation of a system, in the event of a malfunction. Ice protection systems are not particularly difficult to use, but they require careful and intelligent pilot monitoring for safe operation. Then there is the added challenge of removing ice and snow from the aircraft on the ground prior to takeoff, and preventing its subsequent accumulation.

Ground Icing

There are few integral aircraft systems for dealing with icing on the ground. Yet the accumulation of frost, snow, and other precipitation under freezing conditions requires a great deal of attention. Most aircraft are deiced by ground crews when ground icing occurs or is anticipated. The key issues are determining when the aircraft requires deicing prior to takeoff and how long the effects of deicing will last under adverse weather conditions. The FAA now requires specific examinations by crew members for ground icing under many operations.

Ground deicing is accomplished by using ground equipment to spray a glycol-based liquid on the airframe. Deicing fluids are categorized into three types, Type I, Type II, and Type IV (See Table below). Type I is a low-viscosity (thin) glycol-based fluid used to remove ice and snow from the airframe. It is used primary to remove ice and snow already accumulated on the wing and provides only a small amount of preventative "anti-ice" protection.

Type II fluid is a high-viscosity (thick) fluid that may also be used to deice the airframe. However, since Type II fluid is thicker, it clings to the airframe for many minutes, providing an extended anti-ice capability. Type II fluid is designed to shear off on the takeoff roll at airspeeds between 80 and 100 knots. Therefore it is not to be used on aircraft with a rotation speed of less than 100 knots. (Some aircraft are not certified for use of Type II fluids.)

Type IV fluid is a "new and improved" deicing fluid that shares many characteristics of Type II fluid. Its main benefit is anti-icing effectiveness far better than Type II. Type IV fluid has some distinct visual characteristics (when applied to the wings, the extra thickness may cause the fluid to appear wavy or bumpy).

Deicing Fluid Characteristics

Type I	Type II	Type IV
Orange or pink color	Amber color	Pale green color
Thin fluid	Thicker fluid than Type I	Much thicker than Type II fluid
Limited anti-icing effects	Provides anti-icing benefits	Provides better anti-icing effectiveness than Type II fluid
Used primarily for ice and snow removal	Used for ice and snow removal and to provide some anti-ice protection ≈ after application	Used for deice and anti-ice
Drains off aircraft after application	Shears off on takeoff roll	Shears off on takeoff roll

We mentioned earlier that one of the key issues concerning ground deicing and anti-icing was determining how long the effects of deicing will last under adverse weather conditions. Aircraft operators use FAA-approved criteria for determining how long the anti-icing fluid will prevent formation of frozen contaminates on treated surfaces of aircraft during taxi for takeoff. This time period is called "holdover" time. Specific holdover times, which begin at first fluid application, are determined from approved Ground De/Anti-Ice Guideline charts. These charts take into consideration environmental factors such as outside air temperature (OAT), types of precipitation, and of course what type of fluid was applied. Using a holdover chart, flight crews are better able to judge just how long the effects of de/anti-icing will last.

A few aircraft do have built-in ground anti-icing systems in use or under development. For example, at least one popular transport aircraft has in the past had problems with ice formation on upper wing surfaces, even at above-freezing temperatures, due to the presence of cold fuel in the wet wings. ("Wet wings" refers to integral wing fuel tanks. See "Fuel Systems" in Chapter 5.) Fuel lines were rerouted in that design to address the problem, and electrically heated thermal panels were developed for ground heating of its upper wing surfaces.

Rain Protection

In addition to icing liabilities, turbine engines are also vulnerable to *flameout* (engine failure due to smothered combustion) if large amounts of water are ingested in flight. In precipitation this problem is addressed by selecting "igniters on" and in some cases operating at higher power settings.

Many turbine aircraft also have windshield rain protection systems, for purposes of preserving visibility on takeoff and landing. Along with heavy-duty windshield wipers, many larger jets use *rain repellent chemicals,* which may be released onto the windshield from the cockpit. Rain repellent chemicals reduce adherence of water to the windshield by changing its viscosity and slipperiness. The result is reduced formation of water droplets on the windshield and, therefore, enhanced visibility. Neither wipers nor rain repellent fluid should ever be used on a dry windshield!

Landing Gear Systems

While there are a few fixed gear models around (such as the DHC-6 Twin Otter and the single-engine Cessna Caravan), retractable landing gear is found on most turbine aircraft. Some older airplanes used electrically powered, gear-driven landing gear retraction systems. These proved to be heavy and electrically demanding, with high maintenance requirements due to the large forces required to extend and retract landing gear. Hydraulic power has proven to be ideal for this purpose and is found on virtually all of today's retractable landing gear systems.

Light turbine aircraft, such as the smaller King Airs, often use the same types of electrically powered hydraulic power packs found in general aviation piston aircraft. On these systems, an electric motor drives a dedicated hydraulic pump, based on pilot selection of the gear handle. Hydraulic lines transmit the generated hydraulic pressure to power landing gear operation.

Most large aircraft have generalized, multipurpose hydraulic systems, of which one major duty is powering landing gear operation. Gear doors on these systems are usually also hydraulically powered through the use of piston-type hydraulic actuators. Door operation may be separately controlled or may be made part of the gear retraction cycle through the use of hydraulic sequencing valves in the landing gear hydraulic circuit. On some aircraft, gear doors are linked mechanically to landing gear operation, using some combination of pushrods, cables, or pins.

All aircraft having retractable gear are required to have pilot warning systems in order to reduce the likelihood of unintentional gear-up landings. In most cases, gear warning horns are designed to sound when the landing gear is retracted, and some combination of flap position, low power setting, and perhaps airspeed suggests imminent landing. You'll be required to memorize the gear warning conditions for aircraft you fly.

Other safety features are built into landing gear systems to confirm proper position and operation of the gear. Sensors known as *landing gear position switches* are located at every landing gear strut to confirm gear position as up and locked or down and locked, as appropriate. These are usually mechanically actuated electric switches, all tied to a central warning system that monitors and compares gear positions. Cockpit warnings are triggered if any gear position doesn't match that selected by the pilots.

Landing Gear Squat Switch

One particularly noteworthy type of landing gear position switch is known as a *landing gear squat switch* (sometimes called a "weight-on-wheels switch" or "landing gear safety switch"). These electric switches are mechanically actuated by the weight of the airplane compressing the gear struts, thereby positively indicating whether or not the plane is on the ground.

Why does the squat switch rate its own section? These switches are very important. First of all, landing gear safety systems use squat switch position to prevent inadvertent gear retraction on the ground. If the gear struts are compressed, the system will not allow gear retraction, regardless of gear selector handle position in the cockpit.

This alone would be a valuable safety feature, but the lowly squat switch sends its simple message, "The airplane is . . ." or "The airplane is not on the ground," to all sorts of interesting places.

As with the gear selector, many aircraft systems have built-in safeguards to prevent accidental activation at times when it could be dangerous. These automatic safeguards prevent selected systems from working unless certain conditions are met. We've noted that if the squat switch says the plane is on the ground the landing gear will not retract. This information is also valuable for safe operation of many other systems.

Ground spoilers, for example, should never be activated in the air; their effects in flight could be fatal. One condition for their deployment, on many aircraft, is an on-the-ground message from the squat switch. The engine thrust reversers of some aircraft are tied to the squat switch for the same reason; reverser actuation aloft would be (and has been) disastrous for certain aircraft.

On many airplanes, the cabin pressurization dump valve is activated by the squat switch when the airplane touches down, thereby depressurizing the plane upon landing. Pressurization begins again after takeoff, based on switch position. In other vehicles, the switch signals the system to change pressurization criteria to meet differing flight or ground parameters.

The APUs of some aircraft are not certified for operation in the air. The squat switch turns them off upon takeoff if shutoff is missed in the cockpit. Bleed-powered wing anti-ice systems, on the other hand, generally cannot be used on the ground. Guess what closes the wing leading edge anti-ice valves at touchdown if the pilots forget?

In ground school, you'll have to learn all of the squat switch tie-ins for the aircraft you fly. This is important for many reasons, not the least of which is understanding what else *doesn't* work if the squat switch fails in one position or the other. You'll want to check all visible landing gear switches during preflight inspection, but pay extra close attention to the landing gear squat switch(es) tied in to all those safety systems.

Brakes

On light airplanes brakes are conceptually very simple. A pilot presses the toe pedals, moving pistons in the brake master cylinders. The pistons push hydraulic fluid through hydraulic lines to slave cylinders out at the wheels. Pistons in the slave cylinders convert hydraulic pressure back to mechanical movement, which squeezes brake caliper pads against brake disks ("rotors") mounted on the wheels, slowing the airplane. These systems are found on light turbine, as well as light piston aircraft.

Even something as conceptually simple as brakes becomes complex, however, when the object is to stop several hundred thousand pounds of airplane. Brake mass and contact area have to be very large in order to stop a jet airliner in a reasonable distance. For the "heavies," this requires multidisk rotor-and-stator brakes mounted on many wheels. Power assist is required to multiply the strength of the pilots in doing the job.

In addition, the energy of the airplane's rolling mass is converted to heat (and lots of it) as the vehicle is brought to a halt. All that heat causes some interesting problems that must be addressed by large aircraft braking systems and operating procedures. First of all, as the brakes heat up during rollout, brake effectiveness decreases. This phenomenon is known as *brake fade.*

With this in mind, brake systems on large aircraft are designed to deliver at least one maximum effort stop from landing speed or V_1 without overheating. (This assumes that thrust reversers are not used.) If a second use of the brakes is required any time soon afterward, overheating is a distinct possibility, resulting in reduced brake performance. In fact, it is entirely possible to overheat the brakes to the point of blowing out the tires or even causing a fire. (Many aircraft have fire extinguisher systems installed in their wheel wells to suppress fires resulting from hot brakes and other causes.)

What does this mean to the pilots? On an aborted takeoff, for example, the nonflying pilot notes airspeed when braking is initiated. Once the airplane has slowed, he or she refers to a *brake cooling chart* to determine when braking may be safely used again. The answer could very well mean a takeoff delay, but you wouldn't want less than optimum braking available in the event of another aborted takeoff. In some cases, it may not even be allowable to taxi the airplane off the runway until some period of cooling has taken place.

Brake Antiskid and Anti-Ice Systems

Obviously, it's no fun to slide around on a wet or icy runway in a fast airplane. To prevent this from happening, *brake antiskid systems* are installed on many turbine-powered aircraft. Brake antiskid systems use computer-controlled hydraulics to maximize braking effectiveness under such conditions. Use of antiskid by the pilots is straightforward: apply steady maximum braking at the pedals. The system works by monitoring rotational speed of each wheel and comparing it with the expected value based on a dry runway. If actual wheel speed is less than 85 percent or so of normal, braking is momentarily released to let the wheel speed up. By pulsing the brakes in this manner, antiskid systems maintain the best wheel speed for optimum braking.

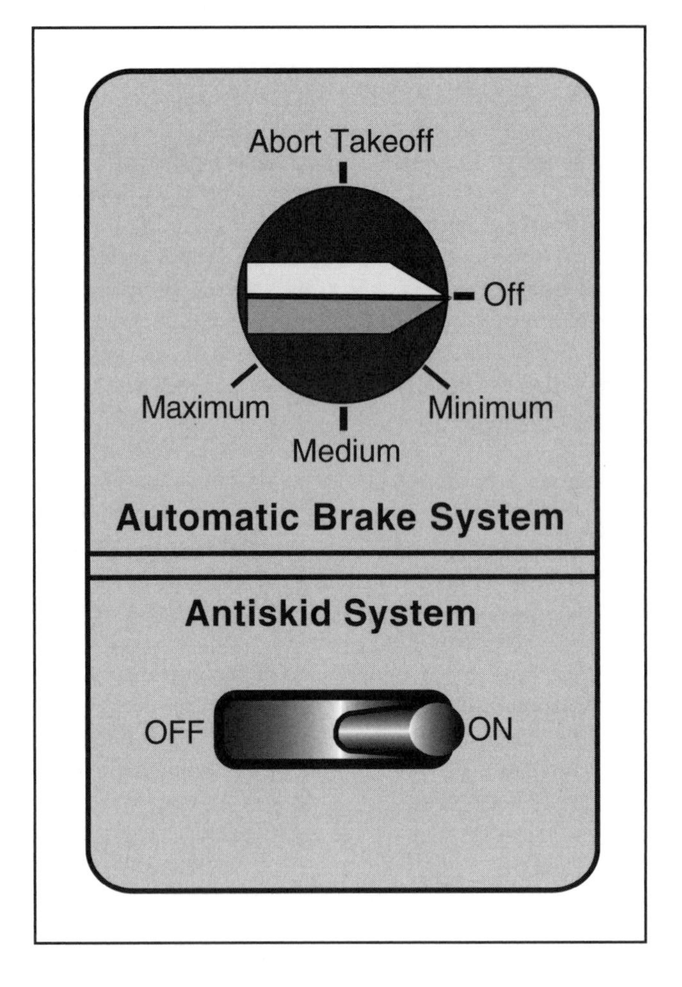

6.6. Automatic brake.

A few aircraft models are also available with *brake anti-ice systems,* which use bleed air heat to prevent the brakes from icing up. This kind of problem can occur when runway slush sticks to cold landing gear and brake surfaces. It's also a problem when the aircraft takes off from a wet or slushy runway, then climbs to below-freezing temperatures at altitude. By landing time, everything can be pretty well frozen if brake heat is not available.

Automatic Brake System

Automatic braking systems are installed in some aircraft to deliver a preselected rate of deceleration during landing roll-out. Various levels of deceleration can be selected by the flight crew prior to landing (e.g., minimum, medium, maximum). Typically, automatic brakes are designed to apply brake pressure when thrust levers are retarded to idle position and the main landing gear antiskid system computer has sensed wheel spin-up. The automatic brake system will then bring the aircraft to a complete stop, unless the pilot either selects the system to off with the automatic brake switch or applies brake pressure manually, which disables the auto brake operation.

Automatic brake systems offer another important safety function on departure—immediate, maximum braking in the event of an aborted takeoff. For this purpose, the automatic brakes are generally selected prior to takeoff. The aborted takeoff mode is then armed by some combination of throttle position and main landing gear wheel speed on the takeoff roll (for example, wheel speed greater than 80 knots and throttles advanced past 50 percent of their travel). Once the system is armed and takeoff roll initiated, maximum braking will be applied if the throttles are retarded to idle. The amount of braking force corresponds to the pilot applying maximum manual braking (Fig. 6.6).

Nosewheel Steering

If you've been flying light aircraft much, you're used to the concept that ground steering is mechanically controlled by the rudder pedals. However, on more sophisticated aircraft, hydraulic nosewheel steering systems are normally used for a variety of reasons. First of all, the distances and forces involved make it impractical to run mechanical linkages all the way from the rudder pedals down to the retractable nose gear on a jetliner. Then there's the fact that steering sensitivity needs to be different for parking than for rolling down the runway at 120 knots.

Hydraulic steering systems suit these purposes well. Hydraulic power lines, as you remember, are easy to route, and systems can be designed to offer two or more steering sensitivities for different circumstances. The optional hydraulic nosewheel steering on the Beech 1900, for example, offers "taxi" and "parking" modes for ground maneuvering. Once that aircraft is lined up on the runway, nosewheel power steering is selected "off," and differential braking is used to assist in steering the castering nosewheel before rudder effectiveness is achieved. Other vehicles provide power steering over a broad range for ground maneuvering (± almost 90° in the case of Boeings), with limited rudder pedal steering (±10°) through the takeoff roll.

Another point is worth noting to save you embarrassment on your first flight as copilot on an aircraft with hydraulic steering. Ground maneuvering, with most powered nosewheel steering systems, is done exclusively by the captain. Somewhere around his or her left knee is a steering wheel or tiller with which the deed is done. On your flying legs, the captain will taxi the plane to and onto the runway. Once the plane's lined up and ready to go, the captain will inform you that "it's your aircraft." After landing, as the plane slows, the captain again takes control of the plane for the journey back to the gate. Ah, the privileges of command!

Annunciator and Warning Systems

While cockpit layout obviously varies tremendously by aircraft, there's been a trend toward standardization over the years. Each manufacturer has been working to standardize its own cockpit designs (and systems designs, for that matter), in order to reduce pilot training costs across various models.

Airbus Industries, for example, has worked to make all of its fly-by-wire models virtually identical, not only in cockpit layout but also in control response and feel. The twin-engine A-320 and four-engine A-340 are tremendously different in size and capacity but are designed to feel similar in the cockpit.

The same is true with the Boeing 757 and 767, vastly different aircraft that share a common type rating. (Boeing, for years, has had "logic" in the orientation of cockpit toggle switches in all its aircraft. Toggle any switch toward the windshield for "on," away from the windshield for "off.")

Annunciator or Advisory Panels

One cockpit system is already well standardized through most of the turbine fleet. Every aircraft, from King Airs through the "big guys," is equipped with a centrally located pilot *advisory panel* (or *annunciator panel*), which depicts everything from system status to emergency warnings (Fig. 6.7). Advisory panels are designed to alert pilots to system problems that require further investigation or action. They are normally divided and color coded into three levels of information.

Warning Lights

Warning lights are always red. These are triggered by emergencies that require immediate crew action in order to maintain safe flight. Fire emergencies are the single most important class of emergencies associated with warning lights. Warning indicators are always tied to a master warning light and sometimes also to a "fire bell" or horn. This master warning light is located at eye level directly in front of the pilots, and it illuminates or flashes so it can't be missed. The idea is for pilots that see it to then check the annunciator panel to determine exactly what the problem is and to follow up with instrumentation, as necessary, to verify the problem. Normally, master warning lights may be pressed, after illumination, to dim or extinguish them after the problem has been addressed (see Fig. 6.8).

Caution Lights

Caution lights are always yellow and generally refer attention to serious system malfunctions that require crew action as soon as time permits but do not require immediate emergency action. Caution lights on a turbine aircraft cover nonemergency system malfunctions such as generator and electrical bus failures, overheating of various components, and fuel system anomalies. The key point about caution lights is that, while they reflect a nonemergency situation when first activated, prompt crew action is normally required to prevent development of a more serious situation.

"Master caution lights" are also installed in some aircraft (see Fig. 6.8). They illuminate whenever a caution annunciator is activated, thereby directing the pilots to look into the problem. Master caution lights may be extinguished by pressing them after the problem has been addressed.

Warning Lights (red) refer pilots to emergency or "immediate action required" situations.

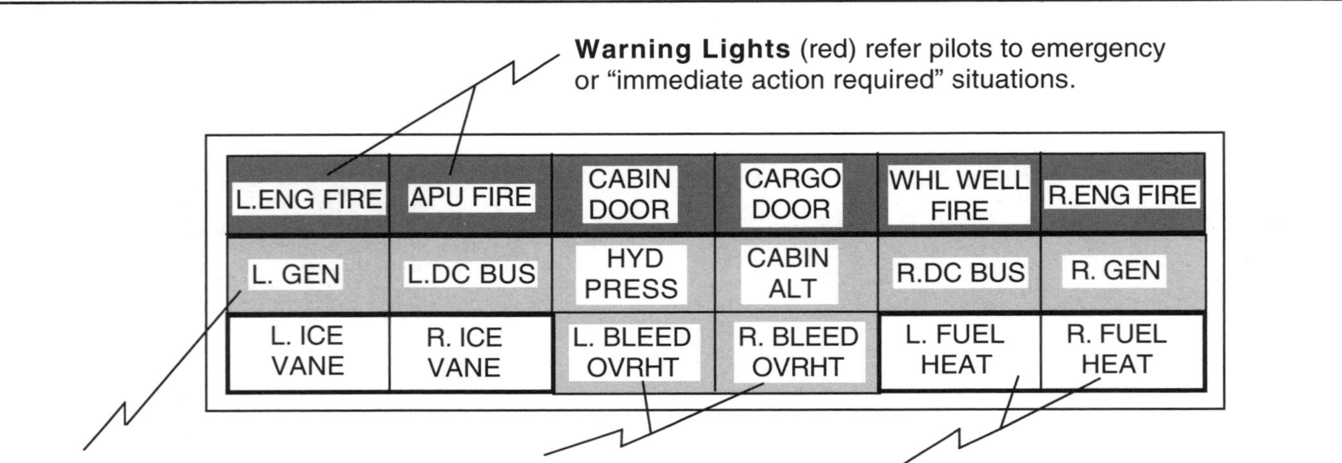

Caution Lights (yellow or amber) indicate problems that should be addressed as soon as time permits. An emergency could occur if these problems are left unaddressed for an extended period.

Status Lights (white, green, or blue) indicate status or operation of components that could affect the flight, if not properly selected for conditions.

Annunciators alert or inform pilots of conditions that are important to safety of the flight. This particular panel directs pilots to specific components or problems that require attention.

Fuel Panel	Overhead Panel	Environmental
Electrical	Hydraulic	Pneumatic

Some annunciator panels simply refer pilot attention to caution lights or other indicators activated on specific panels where they may not be immediately seen.

6.7. Annunciator or advisory panels.

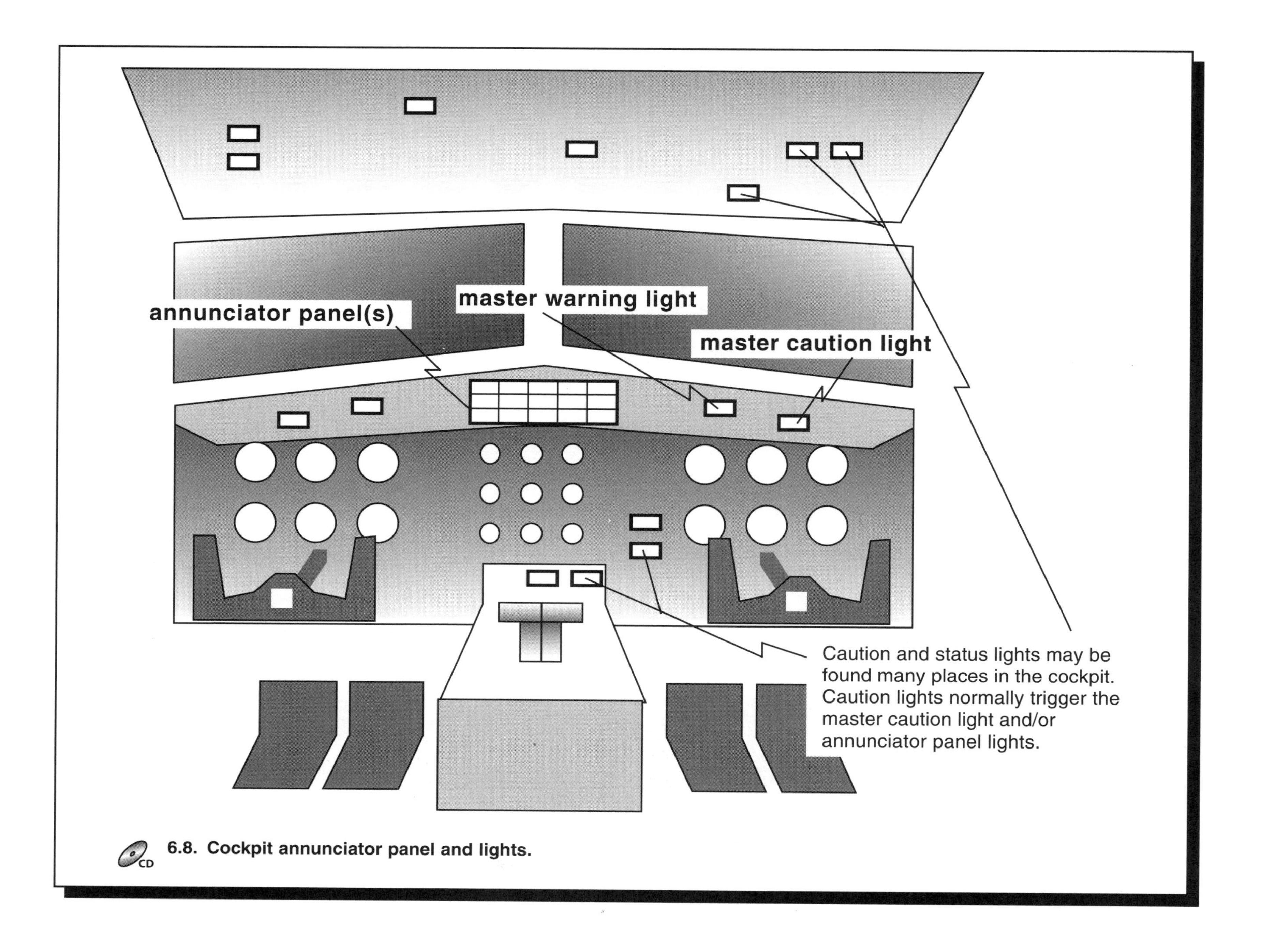

6.8. Cockpit annunciator panel and lights.

Status Lights

Status lights are normally green, blue, or white. These indicate operation of systems that should be monitored or recognized by the pilots but that in themselves do not represent abnormal situations. Ice vane deployment on turboprops and bleed-powered ice protection systems on jets fall into this category. Fuel crossfeed is another such condition.

Airplane systems vary so much by aircraft type that what justifies a status light on one airplane may dictate a caution light on another and may be unindicated on a third. Status lights are sometimes installed on the same panel with caution and warning annunciators or may otherwise be scattered around the cockpit, depending on aircraft type.

Audio Advisory and Warning Annunciation

Every turbine aircraft you fly will likely have some sort of an audio advisory and warning system used in conjunction with annunciator and warning systems. Like annunciator lights, these audio systems are designed to advise the flight crew of the status or operation of certain aircraft systems and to warn the flight crew in the event of an impending emergency situation. A variety of audio signals may be found on typical turbine aircraft: bells, horns, chimes, clackers, verbal speech, and C-chord signals, to name a few. Some turbine aircraft may have dozens of different aural signals, and you will be required to know them all.

One popular turboprop associates all warning and caution lights with specific audio chimes. If a red warning light illuminates, it is accompanied by a triple chime, while an amber caution light is accompanied by a single chime. The audio warning chime is repeated every five seconds until the flight crew cancels it by pressing the appropriate master warning or master caution light switch.

Takeoff Configuration Warning System

One specific type of warning system often found in turbine aircraft is the *takeoff configuration warning system (TOCWS)*. This system is designed to warn pilots when an aircraft is not properly configured for takeoff. The TOCWS monitors flap and spoiler positions, trim settings, parking brake status, and a number of other parameters, depending on the airplane (for example, gust locks or condition lever settings in turboprops). If any monitored controls or flying surfaces are in positions not conducive to safe takeoff, a horn sounds and a caution light illuminates upon application of takeoff power by the pilots.

The takeoff configuration warning system can be reset by retarding throttles to idle and reconfiguring the aircraft properly for takeoff.

Fire Protection Systems

A fire is absolutely one of the gravest emergencies that can occur in an airplane. Therefore, numerous systems are installed to combat them. Preflight and before-start checklists always call for fire system tests, as do APU start procedures. Any in-flight fire dictates immediate emergency action, followed by landing as soon as possible.

As you might expect, most larger aircraft have built-in fire detection and/or extinguishing systems. These systems vary tremendously in detail, based on aircraft type, but there are a number of commonalities.

Fire Detection and Extinguishing Systems

Turbine engines are virtually always monitored for fire. Obviously, the combined presence of heat, fuel, and existing combustion makes engine nacelles potentially dangerous.

Fire detection in engine compartments may be accomplished in several manners. The most common is via a system of *fire loops,* installed around the engines so as to pass through the most likely areas where an uncontained fire could develop. Fire loops operate on the principle that the electrical resistance of a material changes with temperature. Excessive heating of any area on a fire loop signals a nearby detector that electrical resistance has changed and, therefore, that a fire may exist.

Another type of fire detection system operates optically. Infrared light (heat) that exceeds a certain threshold triggers a fire warning. (Optical fire detection systems are subject to false alarms in some aircraft, where sunlight at certain angles can trigger the detectors.)

To fight fires, remotely operated fire extinguisher bottles may be mounted in or near the engine compartment, or they may be remotely located in the fuselage. In some aircraft, multiple extinguisher bottles are cross-plumbed, so pilots can elect to fire multiple bottles sequentially into one engine to snuff a persistent fire.

Aside from the engines, fire detectors may be located in any number of other vulnerable locations. Any plenum through which hot, high-pressure bleed air lines pass is a good candidate for monitoring, as are locations of major electrical components. Wheel wells are often monitored, due to the possibility of brake fires. (Tires on large aircraft are often filled with nitrogen in order to reduce the oxygen available to support brake fires.)

Remote extinguishers may or may not be installed in every one of these various locations. In the case of a fire alarm in a bleed line plenum, the checklist may simply call for shutting off the appropriate bleed source for that area and then proceeding to land.

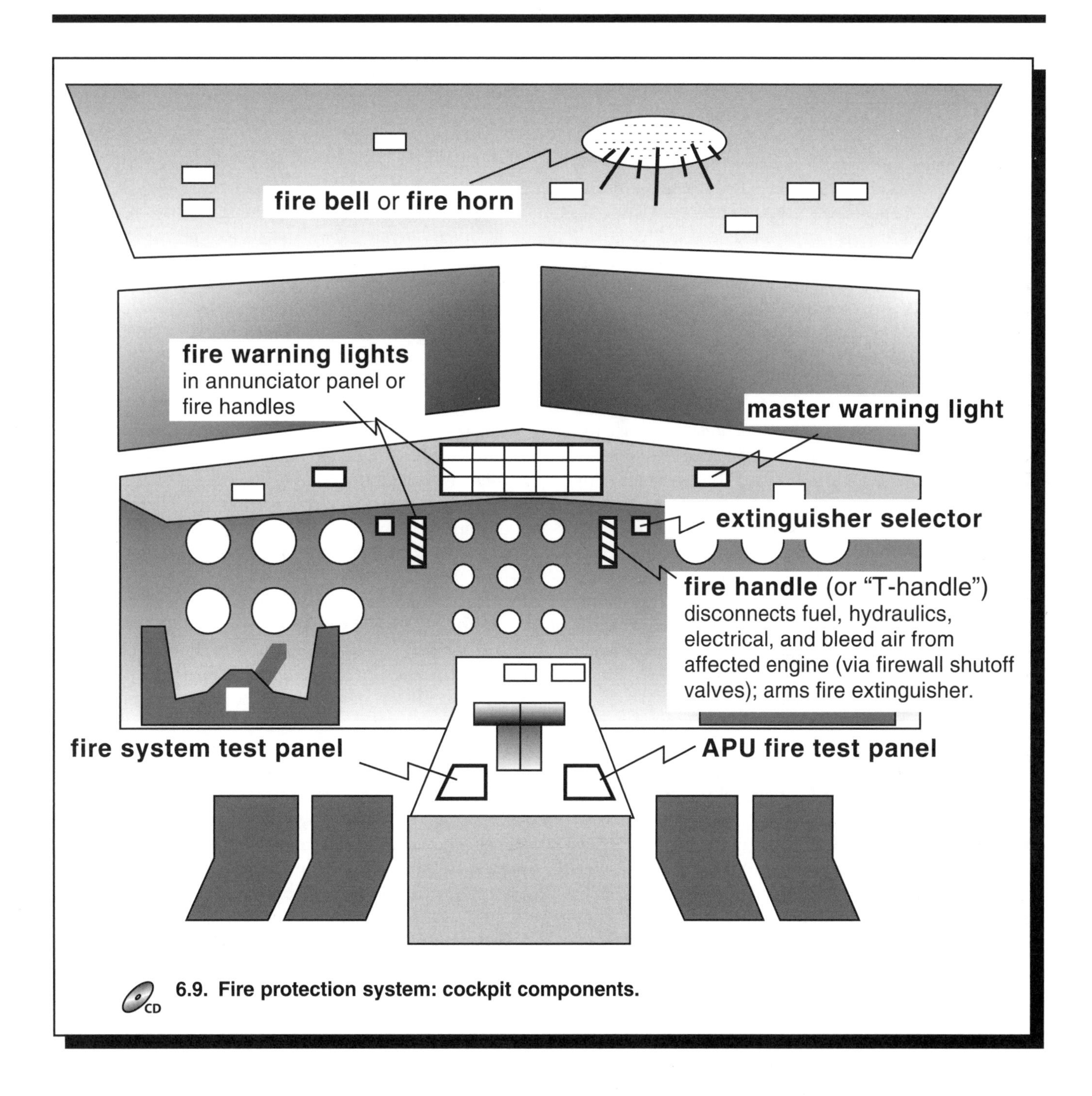

6.9. Fire protection system: cockpit components.

Pilot Actions and Cockpit Controls

Pilots must learn specific emergency procedures for fires, based upon aircraft type. These sequences usually include verifying the emergency and its location, shutting down the affected engine, pulling the "fire handle," and then selecting and firing built-in extinguishers, if installed.

Fire handles (or *T-handles*) are prominently located in the cockpit and often have fire warning lights installed directly in them to eliminate any confusion about which engine is burning. They are designed to completely isolate the engine with a single pilot action. Pulling a fire handle normally shuts off fuel plus hydraulic, electrical, bleed air, and sometimes oil connections to the affected engine. This action also arms the engine fire extinguisher, if installed, for pilot activation (Fig. 6.9).

Electrical Considerations

If any aspects of fire handle or extinguisher operation are electrically powered, they're normally supplied off the "hot battery bus" so as to be operational even under limited electrical power.

Any critical electrical fire protection items on other electrical buses are normally powered by the generator of a different engine. For example, on a twin-engined aircraft, emergency electrical items for the left engine are normally powered off the right engine's generator bus, and vice versa. Obviously, if a left engine fire occurred, it would likely disable the left generator in short order. Therefore, it's best that left engine emergency electrical needs be filled from the right side (unaffected) generator.

Cabin and Cockpit Protection

Cabins of larger aircraft are normally equipped with *smoke detectors,* as well as with *overheat detectors* in environmental air ducts. Cabin and cockpit fires are normally addressed using the hand-held fire extinguishers required by the FARs. Crew members are required to be trained in the use of these extinguishers, and equipment maintenance must be monitored and logged. In addition, pilots are trained to respond appropriately to various types of cabin and cockpit smoke. Procedures for dealing with electrical fires, for example, are different from those addressing environmental system smoke. Training includes proper in-flight donning and operation of *personal breathing equipment* (*PBE,* see below) for use in these situations.

On commercial aircraft, smoke detectors are also required in passenger lavatories due to a history of lavatory fires started by smoking passengers. Lavatory smoke detectors are often monitored at the flight attendants' panel, as well as in the cockpit. Frequently there's also an automatically triggered extinguishing system in the lavatory wastebasket. During training you will be required to demonstrate your knowledge of the location and use of each of these types of emergency equipment.

Portable Fire Extinguishers

Fire extinguishers are classified into four different types, with each designed for a different kind of fire. It is important (and an FAA requirement) that flight crew members know the locations and types of all fire extinguishers on board and which to use on a specific fire.

Fire Extinguishers:

- Type A: designed for fires containing ordinary combustibles such as wood or paper.
- Type B: designed for fires containing flammable liquids such as hydraulic fluid.
- Type C: designed for electrical fires.
- Type D: designed for combustible metals such as magnesium, which is found in some aircraft structures. (Magnesium is so combustible that it's used in making flares.)

This memorization task is no longer as daunting as it used to be, since most aircraft are now equipped with Halon fire extinguishers, which can be used to extinguish Type A, B, and C fires, thereby covering most fires one might confront on an aircraft. Many aircraft also carry water extinguishers, which are only used on Type A fires, so be careful when using this type—you wouldn't want to dump water on an electrical or magnesium fire and make conditions worse. A simple way to remember which type of extinguisher is used for which type of fire is:

- Type A = *A*sh: used on a fire that makes an ash (paper, wood)
- Type B = *B*ang: used on a fire that makes a bang (flammable liquid)
- Type C = *C*harge: used on a fire having a charge (electrical fire)
- Type D = *D*ent: used on a fire that makes a dent (fire involving metal)

Pilot Masks and Goggles

A challenge you'll face during training, but hopefully never in real life, is donning pilot oxygen masks and smoke goggles under simulated emergency conditions and shooting an instrument approach to minimums while wearing them. Seeing out of smoke goggles is restrictive even under the best of conditions, and as a result of recent smoke- and fire-related accidents, new designs are being tested. Oversized smoke hoods under development reach almost to the instrument panel, in an effort to improve visibilty and, therefore, pilot performance under dense smoke conditions.

Portable Breathing Equipment (PBE)

Portable breathing equipment, or PBEs, are basically "see-through" hooded bags designed to protect flight and cabin crew members from the effects of smoke or harmful gases. Small chemical oxygen generators in the rear of each hood may be activated by the user, providing about a 10 minute supply of oxygen and allowing the crew member to fight an in-flight fire or otherwise manage an emergency without being overcome by hazardous gases. One or more PBEs must be readily accessible in the cockpit, with additional units sometimes located elsewhere in the cabin.

Auxiliary Power Unit Fire Protection

Since auxiliary power units (APUs) are remotely located engines with their own fuel lines, they are always monitored for fire and normally have automatic self-extinguishing systems. Upon fire detection, these automatic systems perform basically all the functions of the fire handles associated with the main powerplants. They cut off fuel to the APU engine, close air intake doors in many cases, and trigger warning lights. Many aircraft are equipped with external APU fire control panels so that ground crew members can suppress APU fires without running to the cockpit.

CHAPTER 7

Limitations

AS YOU KNOW, every airplane is designed to operate under a set of limitations: airspeeds, engine parameters, temperatures, fuels, weight and balance, and various structural and system limitations. Many are depicted in the cockpit via placards and instrument markings, and others are found in the AFM (Aircraft Flight Manual). When "checking out" in turbine aircraft, many limitations for each aircraft type must ultimately be memorized by the pilot. In this section, we take a general look at some of the most common types of limitations, with emphasis on those uniquely found in turbine aircraft.

Airspeeds

Limiting airspeeds for turbine aircraft are, in many cases, similar to those of their piston-engined relatives. Best angle and rate of climb speeds are defined similarly (though using different terms), as are stall and minimum single-engine control speeds. However, there are some interesting variations at the higher end of the speed range. There is no redline on the airspeed indicator of a turbine aircraft. These planes are limited by V_{MO} and M_{MO} (both lumped under *maximum operating limit speed*), rather than the V_{NE} ("never exceed speed") found in piston aircraft.

V_{MO} is a structural limit designed to prevent airframe damage from excess dynamic pressure. M_{MO} protects the aircraft from shock wave damage as the aircraft approaches the speed of sound. (See "Aerodynamics of High-Speed/High-Altitude Aircraft" in Chapter 15.) The values of both V_{MO} and M_{MO} vary with altitude. V_{MO} governs as maximum operating speed at lower altitudes, while M_{MO} is most restrictive at higher altitudes. Since it would be impossible to depict all this information on the airspeed indicator using a fixed redline, maximum operating speed is shown on turbine aircraft by a red and white striped needle commonly referred to as the barber pole. On the airspeed indicator of a turbine aircraft, the barber pole moves automatically to display correct maximum operating limit speed for conditions at any given moment (Fig. 7.1).

At jet speeds, cruise is described by Mach number, rather than in knots or miles per hour. Mach number describes airspeed relative to the speed of sound, which is represented as Mach 1. An aircraft traveling at Mach 2 is traveling at twice the speed of sound, while one traveling at 0.85 Mach has attained 85 percent of the speed of sound. (See "Aerodynamics of High-Speed/High-Altitude Aircraft" in Chapter 15.) Aircraft Mach speed is displayed on a Machmeter, usually associated with the airspeed indicator.

Like its piston counterparts, each turbine aircraft is assigned a *maneuvering speed* (V_A), which limits the airspeed at which controls may be fully and rapidly deflected. However, a turbine aircraft additionally has a specific *turbulent air penetration speed* (V_B), which offers maximum-value gust protection.

In addition, many aircraft have specific airspeed limits for operation of various systems. For example, some Beech King Airs have limiting airspeeds pertaining to operation in icing conditions. There's a maximum airspeed for use of engine ice vanes and one for effective use of windshield heat. The same aircraft are also assigned a minimum airspeed for climbs in icing conditions, in order to prevent underwing icing.

Engine Limits

There are many limitations to learn for turbine engines, even though most are also depicted on the faces of engine instruments. Like other aircraft, normal operating ranges are

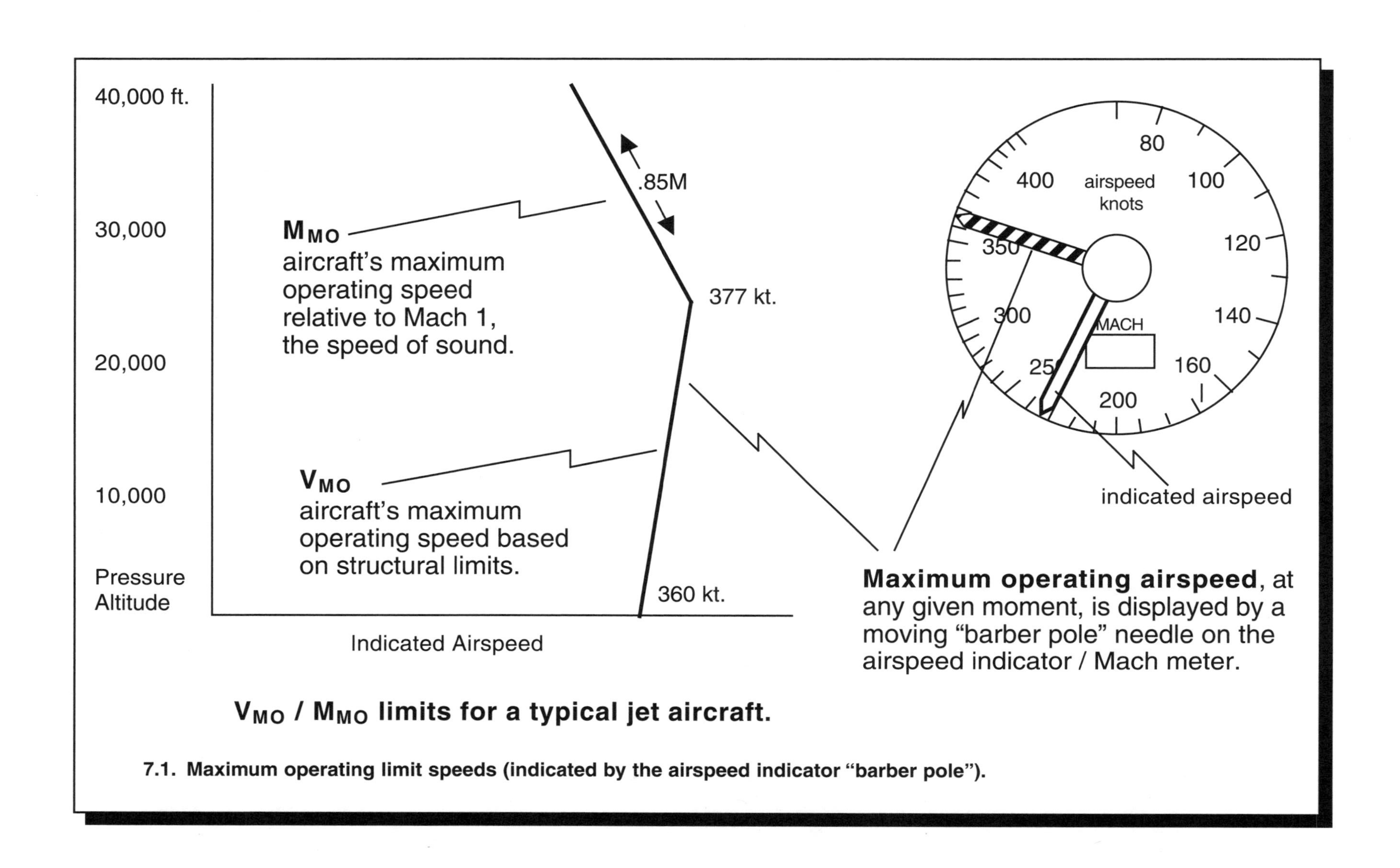

7.1. Maximum operating limit speeds (indicated by the airspeed indicator "barber pole").

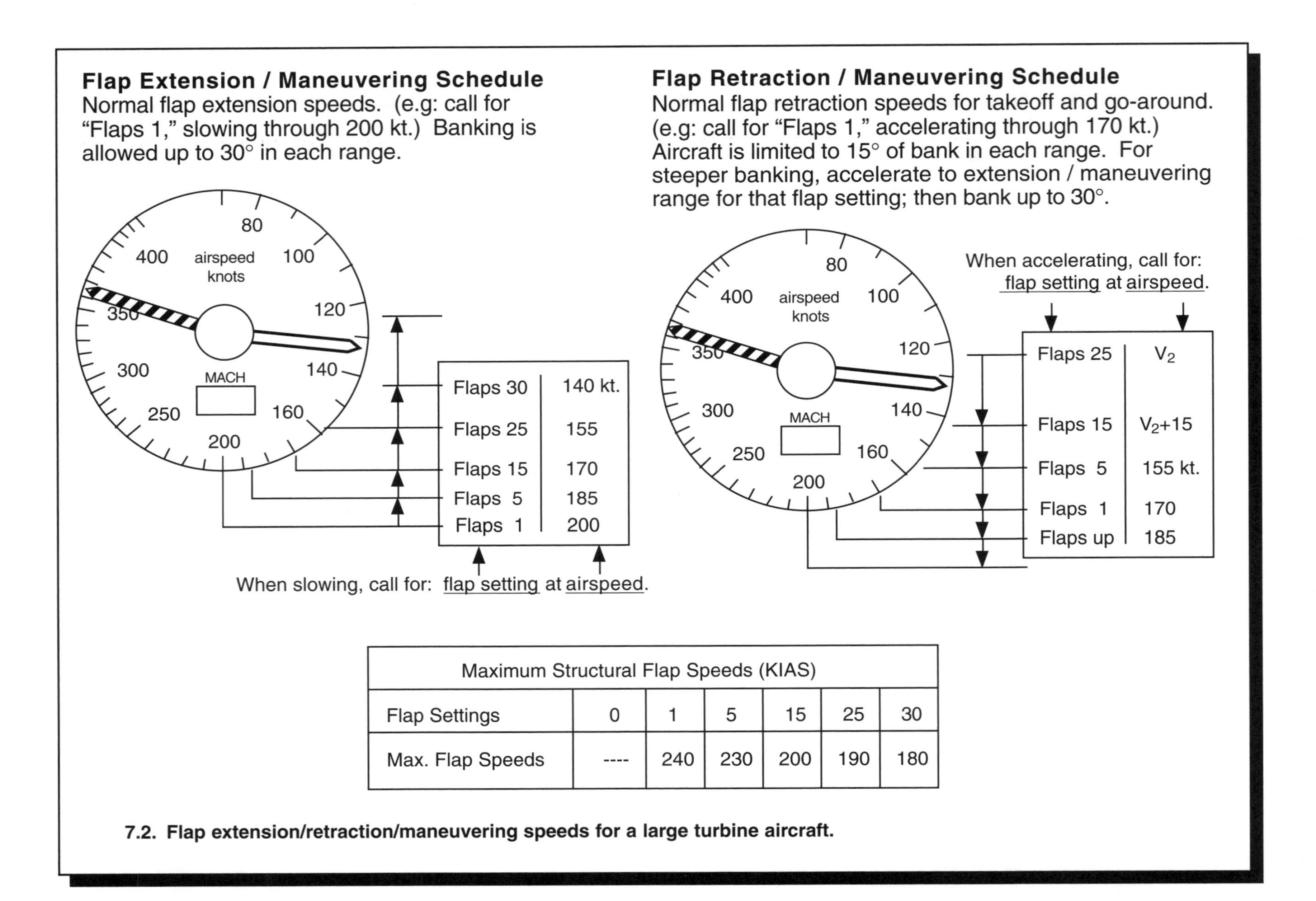

Maximum Structural Flap Speeds (KIAS)						
Flap Settings	0	1	5	15	25	30
Max. Flap Speeds	----	240	230	200	190	180

7.2. Flap extension/retraction/maneuvering speeds for a large turbine aircraft.

indicated by green arcs (or sometimes, no arcs). Yellow arcs depict caution or limited operating ranges, and red lines or arcs depict absolute limits and prohibited ranges. Along with oil pressures and oil temperatures, there are some new types of limitations to learn.

The most telling turbine engine temperature, depending on aircraft, is EGT (exhaust gas temperature) or ITT (interstage turbine temperature). In ground school, you'll have to memorize acceptable EGT or ITT values for your aircraft, including values covering all phases of flight from takeoff and cruise to reverse thrust. Engine temperatures in some types of aircraft bear watching during high summertime temperatures in order to avoid overheating.

Engine temperatures, along with engine rpm parameters, require exceptionally careful monitoring during engine start. A "hot start" is indicated when EGT or ITT shoots past normal start-up values and heads toward redline. The captain must be right on top of things if this happens. Failure to shut down an engine before exceeding temperature redline can in seconds convert a million dollars worth of engine to junk. For that matter, whenever any redline is exceeded in a turbine aircraft, even for a moment, maintenance must be called in before any further operations.

Other System Limitations

Every aircraft system has limitations, and when operating a turbine aircraft you'll get to know lots of them. To give you an idea of what you'll learn in ground school, here are just a few examples.

Most anti-icing systems are to be activated in visible moisture at temperatures below +10°C. A related parameter of particular interest to turbine operators in cold weather is fuel temperature. As you remember, absorbed water in turbine fuel can crystalize at low temperatures, clogging the fuel system and threatening flameout. Fuel system anti-ice should be activated long before that can happen, and you'll need to know exactly when. (See "Fuel Systems" in Chapter 5.) Generator amperage, hydraulic system pressures, oxygen and fire extinguisher dispatch pressures, and system fluid capacities are other items to memorize.

Some aircraft systems, with which you're familiar from light aircraft, have additional limitations that must be dealt with in larger turbine vehicles. A particularly interesting example is flap operations. On smaller turbine aircraft, flap speed limitations are based strictly on structural limits; as with piston aircraft, for each step of flap extension, there's a single structural limiting speed.

Large and high-speed aircraft, however, have additional flap speed limitations besides those based on their structural limits. As you remember, large aircraft flap systems are designed to radically improve lift at lower airspeeds. Therefore, stall speeds and drag characteristics can differ greatly between flap settings, so each flap setting has its own airspeed range for operation. Furthermore, consider that stall speeds increase in steep turns. Therefore, flap speed ranges on large aircraft often have additional limitations for bank angles greater than 15° (see Fig. 7.2).

CHAPTER 8

Normal Procedures

Crew Coordination

With only a few exceptions, turbine-powered aircraft used in business and commercial operations require two or more crew members. While it's true that many smaller turboprops and a few light jets are certified for single-pilot operations by the FAA, the fact is that two-member crews are still often necessary due to requirements of scheduled commercial operations, insurance, corporate policies, or just plain prudent operations under busy or adverse conditions. For our purposes, therefore, we will discuss the roles only of a two-pilot crew. (Interactions with other professional crew members, such as flight engineers and flight attendants, are extensions of the same principles.)

Captain and First Officer/Copilot

There are two ways in which a two-pilot crew is categorized for a flight. The first is by command. The *captain,* of course, is in command at all times and "owns" the responsibility for all aspects of a flight. The *copilot* (term used in private and corporate operations) or *first officer* (most often used in airline and commuter operations) is basically a professional assistant to the captain and under his or her command. Since the largest numbers of pilots hired these days go to regional or major airlines, from now on we'll use the term "first officer," or "FO."

Flying Pilot and Nonflying Pilot

In most operations, the captain and FO alternate flight legs in order to share the flying, keep sharp in various duties, and maintain variety. In this relationship, we refer to the *flying pilot* (*FP*) and the *nonflying pilot* (*NFP*). (Some operators use the terms "pilot flying," or "PF," and "pilot not flying," or "PNF.") While command rests at all times with the captain, the duties of the two pilots, in most cases, are defined by who's flying.

The primary job of the flying pilot, at all times, is to fly and maintain safe control of the aircraft. Any duties that may distract attention away from this responsibility are assigned to the nonflying pilot. The FP also calls for checklists at the appropriate times and double-checks their execution by the NFP when called for (Fig. 8.1).

The role of the nonflying pilot is to perform all other duties that must be accomplished. This includes reading and performing checklist items, operating radios, handling systems abnormalities, obtaining weather and flight information, calling out critical flight information such as altitudes and fixes, and assisting the FP in any other manner called for.

There is a certain rhythm or cadence to properly executed multipilot operations that takes some getting used to for fighter pilots and others accustomed to single-pilot operations. Statistics show that aircraft operated by a properly trained, multipilot crew can be significantly safer than single-pilot operations. At the same time, the presence of two pilots who are not working together in a coordinated manner is a contributing factor in many accidents. Hence, today's emphasis by business and commercial operators on CRM, or crew resource management, training.

Crew Resource Management

For more than a decade the airline industry has been active in the study of *crew resource management* (CRM). Crew resource management focuses on the relationships and coordination between crew members in flight. As a flight officer applicant, you will be expected to know the meaning of "CRM" and be somewhat familiar with its basic goals and objectives.

In a three-pilot cockpit the flight engineer reads the checklist, and the flying pilot and nonflying pilot respond accordingly.

8.1. Crew coordination in a multipilot cockpit.

The primary goal of crew resource management was initially to reduce the number of accidents and incidents attributable to "pilot error." As the airline industry matured, it became evident that while fewer airline accidents were being caused by mechanical problems the number caused by flight crews was not declining. The initial CRM goal of accident prevention has evolved into a number of relevant objectives. Crew resource management training is designed to teach pilots to make use of all resources available to the flight deck. This includes not only cockpit resources but also information and expertise offered by cabin crews, maintenance technicians, and company dispatchers.

All major airlines and many regional/commuter and corporate departments have some type of crew resource management training. It may be called "crew coordination training," "cockpit resource management," "command-leadership resource management," or something similar. Whatever the name, the goals are the same: improved communication between crew members, better overall flight management skills, development of a team performance concept, and improved mesh of crew members' technical knowledge and skills.

Optimizing Crew Communication

Everyone knows the old saying, "two heads are better than one," but it's only true about pilots when they're working properly as a team. Through decades of research, poor communication between crew members has been found to be a primary cause in many aircraft accidents and incidents. Therefore, a major focus of CRM training is to *optimize communication among crew members.*

An often discussed accident demonstrates the dangers of poor crew member intercommunication. United Airlines Flight 173, a DC-8-61 on approach to Portland International Airport on December 28, 1978, had experienced a landing gear malfunction. The crew held for one hour while dealing with the problem. During the last thirty minutes, the first officer and flight engineer made several comments regarding low fuel supply. Eventually, the aircraft ran out of fuel and went down 20 miles from the airport. The captain never acknowledged the low fuel situation and in fact sounded surprised when the engines began to flame out, one by one.

The probable cause of the accident was determined by the National Transportation Safety Board (NTSB) to be the captain's failure to monitor the aircraft's fuel state and to properly respond to the crew members' advisories about the fuel. The NTSB also cited failure of the first officer and flight engineer to assertively communicate their concerns to the captain.

There are a number of practices that, when used in conjunction with standard operating procedures, promote better communication between crew members. Crew coordination is enhanced when, at the beginning of a flight, there is a thorough briefing of all crew members, including cabin crew. The briefing should identify possible problems such as weather, ATC delays, or any mechanical abnormalities affecting the operation. Captains should call for questions or comments. Each crew member should share relevant information with the rest of the crew.

Each crew member should state intended course of action during various phases of flight, thereby keeping the entire flight crew in the loop. If there is disagreement, crew members must feel free to state their concerns regarding crew actions and decisions.

The key player in improving communication between crew members is the captain, who sets the tone in the cockpit. He or she must encourage open communication, while at the same time retaining decisive command. An intimidating environment seriously reduces input from other crew members. There are many documented cases of aircraft accidents that occurred because the FO or flight engineer hesitated to speak out about a problem or spoke out and the captain chose to ignore it. Accordingly, much CRM training is directed toward resolving cockpit ego and rank considerations for decision-making purposes. Ideally, both pilots should evaluate choices before making them final, if time permits. In any case, the relevant saying goes, "Focus on what's right, not who's right!"

Improving Overall Flight Management

Another CRM goal is *improving overall flight management.* This is done through preparation and planning and by maintaining situational awareness. Here again, the captain plays an important role by prioritizing cockpit tasks and dividing the labor under normal and abnormal situations. This includes delegating primary flight duties (flying the aircraft and dealing with emergency or abnormal events) and secondary flight duties (ATC, company, or passenger communications). In CRM programs, crew members learn to assess problems and avoid preoccupation with relatively minor issues. They also learn to recognize and report work overloads for themselves, as well as for other crew members.

The classic example of an aircrew's preoccupation with a relatively minor problem leading to an accident is Eastern Airlines Flight 401. Flight 401, a Lockheed L-1011, was on approach to Miami International Airport on December 29, 1972, when it experienced a landing gear malfunction. The nose gear indicating system did not show that the nose gear was down and locked. The crew circled for a few minutes trying to exchange landing gear indicator light bulbs, during which time the aircraft's autopilot became inadvertently disengaged. The flight crew, fixated on this relatively minor landing gear indicating problem, didn't notice that the aircraft had begun a gradual descent. Eventually the aircraft crashed into an Everglades swamp northwest of Miami.

The NTSB cited the flight crew's lack of situational awareness during the final minutes of the flight. The crew members' preoccupation with a malfunction of the nose gear's position indicating system distracted their attention from their primary duty of flying the aircraft.

Development of a Team Performance Concept

Another CRM objective is the *development of a team performance concept.* This training is designed to instill proper balance between respect for the captain's authority and the assertiveness other crew members must sometimes exhibit. The key is to maintain an open and comfortable relationship in the cockpit so that crew members are never intimidated into silence. While crew members don't necessarily have to like each other, they must be able to adapt to one another's characteristics and personalities. The result is a team that works together in a complementary manner.

A fine example of team performance was displayed by the crew of United Airlines Flight 232 on July 19, 1989. The DC-10 was at cruise altitude when it experienced a catastrophic failure of the number two (tail-mounted) engine. The explosion that accompanied the engine failure knocked out the aircraft's hydraulic systems, leaving the flight controls inoperative. The flight crew determined that the only way to control the aircraft was to adjust the thrust of wing-mounted engines one and three.

A flight attendant informed the captain that a company DC-10 training check airman was in the passenger compartment. The captain invited him to assist the flight crew. After inspecting the wings and finding no damage, the check airman was instructed by the captain to manipulate the thrust levers in order to maintain level flight. This allowed the captain and first officer to concentrate on regaining use of the flight controls.

The flight crew never regained use of the flight controls and was forced to make an emergency landing at Sioux Gateway Airport, Sioux City, Iowa. Using only thrust-lever movement to control the aircraft's attitude, the DC-10 was difficult to control. It landed wing-low, cartwheeling into an inverted position.

While a number of people were killed, many others survived. It is clear that the teamwork of cockpit crew, cabin crew, and the off duty check airman prevented what could have been a much larger catastrophe. The captain clearly delegated tasks and maintained situational awareness throughout the entire approach. He also listened to the inputs of his first officer, flight engineer, and flight attendants in order to develop a plan of action and get the aircraft safely on the ground. This represents crew resource management at its best.

Crew Resource Management Training

Crew resource management (CRM) techniques are normally taught in classroom and group workshop settings. Many airlines also include CRM training in their initial and recurrent simulator exercises. This *line orientated flight training* (*LOFT*) is designed to instill a team attitude on the flight deck. Crew members learn to mesh their joint technical skills and knowledge, thereby improving overall proficiency and efficiency.

Professional pilots are expected to possess high levels of basic stick and rudder flying skills. They also must demonstrate thorough knowledge of aircraft systems and procedures. While CRM cannot replace these skills, it encourages more effective individual and group responses to specific situations. Crew resource management improves the odds that, when a problem occurs, the proper course of action is taken and mistakes that might compromise safety are avoided.

Checklists and Callouts

Checklist Procedures

The primary function of checklists is to ensure that the flight crew safely and properly configures the aircraft before and during each flight. Three interrelated methods of conducting checklists are commonly used in turbine aircraft: flow checks, challenge-response, and do lists.

Flow Checks

Flow check is a checklist method where each pilot memorizes a sequential pattern for doing specific cockpit tasks. Each pilot starts at a specific panel location and works through every switch and indicator to confirm settings for various phases of flight. Typically, the captain will call for "after start flows," for example, and both pilots perform their sequences (Fig. 8.2). Some flow checks are then backed up with one of the following checklist methods to ensure the completion of critical items.

Challenge-Response

In civilian aviation, *challenge-response* checklists are often used to follow up flow checks (see Fig. 8.3). After the pilots have configured their aircraft for a particular phase of flight, the flying pilot calls for the checklist to verify that all items have been completed. The nonflying pilot "challenges" the FP by reading the items off the checklist to ensure completion; the FP then verifies that each item was accomplished properly by checking that item and then calling out the appropriate "response." (NASA recommends touching each item to ensure that it's checked.) As you might imagine, challenge-response checklists are used for critical items and phases of flight (such as "before landing") where double-checking and awareness by the FP are important.

You military transport pilots may be accustomed to a variation of challenge-response, where a checklist action is performed only after it is specifically called for. In that step-by-step methodology, jumping ahead of the checklist is unacceptable. Most commercial operators, however, use checklists to verify completion of required actions after they've been accomplished. (There are exceptions: both civilian and military checklists favor step-by-step challenge-response actions in abnormal or emergency situations.)

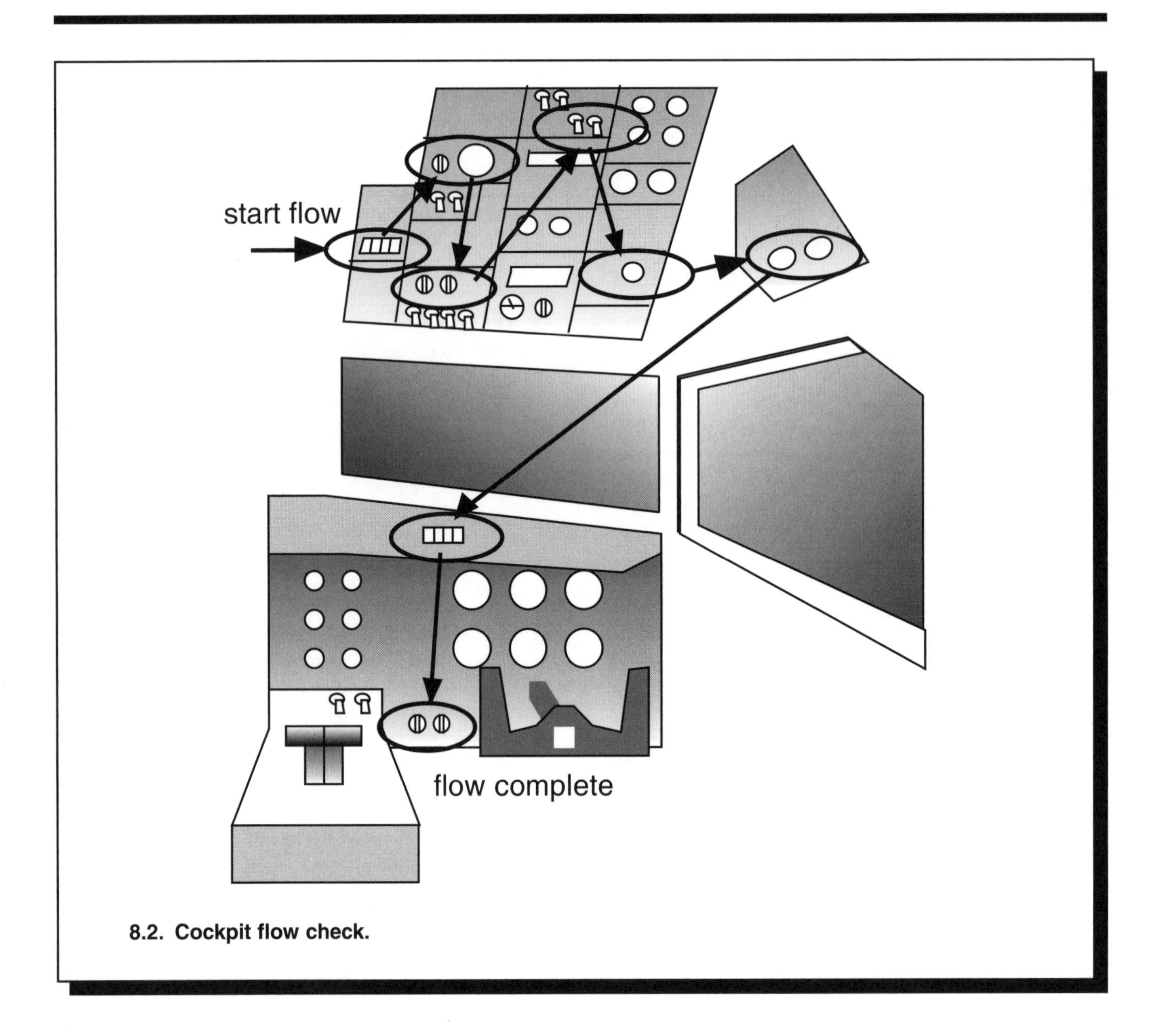

8.2. Cockpit flow check.

Do Lists

Do lists are just what they sound like. The NFP simply reads and accomplishes everything "on the list" for a given phase of flight. Do lists are commonly used for cleanup or housekeeping chores, such as "climb" and "cruise" checklists (Fig. 8.4).

Most airlines employ a combination of checklists. Flow check plus challenge-response is used for the bulk of the checklists. Flow check plus do lists are used for such rote tasks as "engine start," "after start," and "securing the aircraft."

Types of Checklists

While the preceding methods are basically the same for all types of flying, there are various ways of presenting checklists in aircraft. Paper checklists are probably still the type most commonly used by all sorts of aircraft operators. A number of air carriers have installed standardized mechanical or electromechanical checklists in their aircraft over the years. However, with the advent of so many computerized functions in the cockpit, smart electronic checklists (including computer-aided, voice synthesis, and even self-check systems) are certainly the way of the future.

BEFORE LANDING CHECKLIST

C = Captain
FO = First officer
FP = Flying pilot
NFP = Nonflying pilot

NFP... PRESSURIZATION........................ CHECK
C........ AUTO-FEATHER SWITCH............ARMED
FO......BRAKE ANTISKID...........................ON
NFP... FLAPS.. 27° POSITION
NFP... LANDING GEAR CONTROL.........DOWN (SPEED CHECK)
NFP... LANDING LIGHTS..........................ON
FP...... TAXI LT.(when cleared to land)....ON

crew member who performs the task.

task to be completed; **challenge** read from this checklist by the NFP.

task completed and confirmed; this **response** called out by the FP.

8.3. Challenge-response checklist for a turboprop aircraft.

CRUISE CHECKLIST (SILENT)

NFP... CRUISE POWER....................................... SET
NFP... ENGINE INSTRUMENTS..........................MONITOR / TREND CHECK
NFP... GALLEY POWER.......................................ON
NFP... FUEL QUANTITY..CHECKED
NFP... PASSENGER BRIEFING..........................COMPLETE

NFP (nonflying pilot) confirms that all checklist items have been completed, then notifies FP (flying pilot), "Cruise checks complete."

8.4. Do-list checklist.

Paper and Mechanical Checklists

Paper checklists simply list tasks on a paper card. This is the simplest and most common type of checklist (for now). A *scroll checklist* consists of a narrow strip of paper that scrolls vertically between two reels in a metal case. The box has a window on the front marked with a lubber line. The pilot scrolls through the checklist by rotating a knob on the side of the box. When a checklist item is completed, the pilot simply scrolls the next item up to the lubber line.

A *mechanical checklist* consists of a small panel containing several plastic slides moving over a list of checklist items. As the item is accomplished, the plastic slide is moved over that particular checklist item. The checklist is complete when all items are covered. Similar to the mechanical checklist, the *electromechanical checklist* consists of a small panel with an internally lighted list of items. Next to each particular item is a toggle switch. When that item is accomplished, its associated switch is toggled, turning on a colored light. The checklist is complete when all items are marked by illuminated lights.

Electronic and Computer-Aided Checklists

The introduction of the modern glass cockpit has made it possible to electronically display checklists on cockpit CRT (cathode-ray tube) displays such as EFIS (electronic flight instrumentation systems) and radar screens. Some computer-aided checklists are "passive," while others are "active." A passive computer-aided checklist simply displays checklist items on a screen to read from like a paper checklist. On active versions the computer verifies that specific checklist items have been completed.

Voice Synthesis Checklists use a computer-generated electronic voice to speak an audible checklist. The manufacturer or operator preprograms the unit with checklist items. There are two buttons, normally mounted on the yoke, marked "proceed" and "acknowledge." When a checklist item is accomplished, the acknowledge button is pressed, followed by the proceed for the next item. If "proceed" is pressed, without first depressing the acknowledge button for the previous item, the device will repeat the checklist item.

Normal Checklists

Normal Checklists are those used for routine portions of a flight. Depending on the operation, some Normal Checklists may be silent do lists, while others are called aloud with crew member response (challenge-response). In any case, each checklist is called for at the proper time by the flying pilot or the captain and executed as appropriate. For purposes of our discussion, we will assume that the FP calls for each checklist and the NFP completes it. (By the way, you'll always be the FP for an interview simulator check.)

The details of Normal Checklists vary tremendously, of course, by aircraft type and operator. At the same time, there are some basic similarities that can be summarized here. Most aircraft operated by multipilot crews have eight to twelve Normal Checklists, similar to the following.

Before Start Checklist

The *Before Start Checklist* is called for after aircraft and cockpit preflights are completed and once all flight deck crew members are in place and ready for engine start. (The start sequence itself is normally accomplished from memory by the captain.)

After Start Checklist

To confirm proper status of engines and systems and to prepare for taxi, an *After Start Checklist* is accomplished immediately after all engines are started.

Taxi and/or Before Takeoff Checklists

Taxi and/or *Before Takeoff Checklists* are normally accomplished enroute to or while holding at the runway and are used to complete all of the general pretakeoff tasks. One key item is the takeoff briefing by captain or FP. This briefing summarizes how the takeoff sequence will progress, how any emergencies or anomalies will be handled, the duties of the crew members in such situations, and a review of takeoff speeds.

Runway or Cleared-for-Takeoff Checklist

A *Runway* or *Cleared-for-Takeoff Checklist* is usually very brief, with a few key items handled as the aircraft taxis onto the runway. Lights on, arming of any emergency systems (such as auto-feather or antiskid), and final review of takeoff speeds are normally included.

After Takeoff or Climb Checklist

An *After Takeoff* or *Climb Checklist* is called for as soon as the aircraft is safely airborne and workload permits. This is often tied to a speed, altitude, or operation, so it won't be forgotten. For example, in some aircraft it might be tied to final flap retraction. "Flaps up, set climb power, After Takeoff Checklist."

Cruise Checklist

A *Cruise Checklist,* as you might suspect, is completed upon level-off at cruise altitude. It includes such items as the setting of cruise power, checking of engine gauges and systems, and engine trend monitoring, if appropriate.

Descent Checklist

A *Descent Checklist* is called for upon initiation of descent, or "through 18,000 feet," in order to complete items preparatory for landing, such as setting pressurization, reviewing landing speeds, and setting altimeter.

Initial or Approach Checklist

An *Initial* or *Approach Checklist* is accomplished before entering the final approach phase. Approach procedures are reviewed, safety systems are armed, landing briefings of crew and passengers are completed, and speed and configuration objectives are established.

Final or Before Landing Checklist

A *Final* or *Before Landing Checklist* is accomplished when the aircraft is established on the approach but is still outside the final approach fix (FAF). The idea here is to have everything accomplished and configured for a fully stabilized approach before crossing the FAF or outer marker. Note that "finals," as well as other checklists, may be divided for convenience, usually by a line on the checklist. For example, the aircraft may be fully configured "to the line" for landing. Upon glideslope intercept, the FP calls for "below the line," and the remaining tasks of that checklist are accomplished; "gear down, three green lights, and landing lights on."

After Landing Checklist

The *After Landing Checklist* is executed after the runway has been cleared, and it includes such cleanup items as flaps up, transponder off, various flight systems shut off, and APU started.

Shutdown Checklist

Upon taxiing into the gate for parking, the *Shutdown Checklist* is used to ensure that the proper sequence for shutdown is followed and that the aircraft cockpit is properly secured.

Note that long forms of the ground checklists, such as before start, after start, and shutdown, may be used for first and last flights of the day. Abbreviated versions are common when one crew is making multiple turnarounds in the same aircraft.

Standard Callouts

Standard callouts are the routine callouts made by multipilot crews. As with checklists, "standard calls" vary a bit by operator but are similar enough for summary here. Callouts serve (1) to ensure that important tasks are remembered during critical phases of flight and (2) to minimize the possibility of errors or problems going unnoticed.

Taxi Callouts

When taxiing, it is common for the NFP to look and call "clear" when turns are blind to the FP. The same is true when taxiing onto any runway or when making turns in flight.

Takeoff Callouts

On the takeoff roll, the NFP is to monitor speeds and engine instruments, to call out abnormalities, and to set power for the FP, if applicable. Typical callouts by the NFP on the takeoff roll include:

"Power (or thrust) is set."
"Auto-feather is armed" (turboprops).
"Engine instruments are normal."
"Airspeed is alive, both sides."
"V_1, V_R, rotate."

Once the aircraft is airborne:

NFP: "Positive rate."
FP: "Gear up."

(These may be called in either order, but in any case, the gear is not to be raised by the NFP until positive rate of climb is confirmed.)

Climb Callouts

Usually there is an additional callout early in the climb, at a predetermined altitude above ground level:

NFP: "One thousand feet."
FP: "Flaps up, bleeds on, climb power."

(The Climb Checklist may also be called for at this time.)

In-Flight Callouts

In flight, the FP usually repeats instructions assigned by ATC (to which the NFP has responded on the radio). Altitude calls are particularly important in this regard.

NFP: "One thousand feet to go" (or, "through 4000 feet for 5000 feet").

Incidentally, many corporate and all Parts 121 and scheduled 135 operators require *sterile cockpit* below 10,000 feet or cruise altitude (whichever is lower) and on the ground whenever the aircraft is in motion. No nonessential conversation or activities are permitted under sterile cockpit conditions.

Approach Callouts

On instrument approaches, the NFP normally calls out passing of key fixes and target altitudes: "localizer (course) alive," and "glideslope alive." In addition, calls are made at 1000 feet above decision height (DH) or minimum descent altitude (MDA) and then at every 100 feet from 500 feet down to the minimum. Under IMC (instrument meteorological conditions) most operators consider it standard operating procedure for the FP to stay on the gauges until the NFP calls, "runway environment in sight."

Final Approach and Landing Callouts

A "three green" wheels down call is common on final approach to landing, as are radar altitude calls down to 10 feet above ground level. (Radar altitude calls are particularly useful in large aircraft; they aid the FP in controlling descent rate and flare.)

Finally, landing reference speeds are often called out by the NFP on short final, usually relative to the precalculated landing reference speed: "V_{REF}+20," "V_{REF}+10," and "V_{REF}."

CHAPTER 9

Emergency and Abnormal Procedures

AS YOU'VE ALREADY learned in your flying career to date, a good deal of the training in any aircraft is in how to deal with emergencies. Several factors make emergency training "interesting." One is that emergencies occur relatively rarely, so pilots are faced with the challenge of staying sharp on procedures not often used. Another is that proper emergency training develops the confidence and clarity of thinking required to deal promptly and effectively with a problem, without making it worse. Finally, such training should provide the understanding required to deal with a "nonstandard" emergency, meaning one which is not specifically addressed by checklists and manuals and therefore requires independent action by the cockpit crew.

Emergency versus Abnormal Situations

Let's begin by defining a few terms, relative to aircraft operations. An *emergency* is a situation where immediate crew action is required in order to maintain the safety of the flight. In other words, failure of the crew to immediately exercise proper *emergency procedures* could jeopardize the vehicle and the lives of the people on board. Fire is such an emergency. Also, any situation that merits a red warning light on the annunciator panel (see "Annunciator and Warning Systems" in Chapter 6) should be treated as an emergency until confirmed otherwise by the crew.

Then there are those situations where something goes wrong that needs attention, but not with such immediacy. *Abnormal procedures* address these types of problems. An electrical bus failure, for example, is addressed by abnormal rather than emergency procedures on most aircraft. A bus failure is a serious problem and may affect operability of equipment critical to the flight. However, it is not generally considered an emergency because instant action is not required. The problem simply must be addressed in a methodical manner as soon as time permits. Note, however, that an abnormal situation can rapidly develop into an emergency if it's not dealt with promptly.

It's also important to understand that aircraft differ tremendously in aerodynamics and systems design. What is abnormal (or even routine) in one aircraft may very well be an emergency in another. That's another reason why thorough training is so important for pilots.

As with normal procedures, emergency and abnormal procedures are accomplished through the use of checklists. Again, the checklists serve to ensure that problems are dealt with thoroughly and in proper sequence.

Emergency Procedures

You learned the first step in dealing with emergencies way back during your first flight lesson: "Fly the airplane!" When an emergency occurs in a multipilot aircraft, the duty of the FP (flying pilot) is to concentrate on flying the airplane. Properly fulfilling this responsibility is important for several reasons. It ensures that the emergency doesn't become even more serious due to loss of control of the airplane. It also frees up the NFP (nonflying pilot) to handle the emergency.

You can see that, if either pilot gets overly involved in the responsibilities of the other, the emergency cannot be addressed efficiently. In some cases, the captain may feel the need to assume a specific role where he or she feels the most can be accomplished. In that case a reversal of roles may be assigned. "You fly the airplane, and I'll handle the emergency." In any case, each crew member must assume and fulfill one of the two roles of FP and NFP. The relationship between crew members is a critical safety factor. (See "Crew Resource Management" in Chapter 8.)

Many operators take the approach that an emergency may be so time critical that there's no time to look for the checklist until key memory items are completed. Others have concluded that it's better to let a few seconds pass—to get the checklist out first and to ensure that nothing is forgotten. We'll concentrate on the memory item type of emergency checklist here since it's most likely new to you.

Since an emergency requires immediate action, *emergency checklists* are often marked with striped or red borders for quick identification and often use large carefully laid out type for clarity (Fig. 9.1).

Prominent on many emergency checklists is the presence of *memory items.* These are laid out at the beginning of the checklist and are sometimes called "boxed items" due to their graphic treatment. In the event of an emergency, memory items are normally called out loud by the FP. Since lives may well depend on their proper execution, memory items must be learned to the degree of being second nature by every pilot.

The last memory item on every emergency checklist is "Call for the checklist." Memory items are the critical ones for getting the emergency under control. The checklist is used to confirm completion of the memory items and then to execute less-urgent follow-up procedures for properly configuring systems and continuing the flight to a safe landing site. Examples of some common emergency checklists include:

- Engine fire
- Engine failure during takeoff (at or below V_1)
- Engine failure during takeoff (at or above V_1)
- APU (auxiliary power unit) fire
- Electrical smoke or fire
- Bleed air overtemperature
- Environmental system smoke
- Environmental overtemperature
- Rapid depressurization
- Emergency descent
- Passenger evacuation

Let's follow the execution of the engine fire or engine failure in-flight emergency checklist from Figure 9.1b, in order to see how it's done. We'll assume that the aircraft is climbing to altitude when the FP detects a power loss.

FP: "We've lost the left engine! Max power, gear up!"
NFP: "Positive rate, gear's up."
FP: "Confirm left engine has failed."
NFP: "Left engine has failed. Confirmed."
FP: "Left condition lever."
NFP: (Puts hand on left condition lever.) "Left condition lever. Confirm."
FP: (Looks to make sure that the NFP's hand is on the correct control.) "Confirmed; left condition lever 'fuel cutoff.'"
NFP: "Left condition lever 'fuel cutoff.'"
FP: "Left propeller lever."
NFP: (Hand on left prop control.) "Left propeller lever. Confirm."
FP: (Checks NFP's selected control.) "Confirmed. Left propeller 'feather.'"
NFP: "Left prop feathered."
FP: "Left fire handle."
NFP: (Hand on left fire handle.) "Left fire handle. Confirm."
FP: (Checks NFP's selected control.) "Confirmed. Left fire handle 'pull.'"
NFP: "Left fire handle pulled."
FP: "Is the engine on fire?"
NFP: "Yes, the left engine is on fire."
FP: "Number one fire extinguisher."
NFP: (Hand on left fire extinguisher button.) "Number one fire extinguisher. Confirm."
FP: (Checks NFP's selected control.) "Confirmed. Activate number one fire extinguisher."
NFP: "Number one fire extinguisher activated."
FP: "Engine fire or engine failure in-flight checklist."
NFP: (Pulls out the named checklist, confirms that all memory items have been completed, and executes the remainder of the checklist.)

Note that, especially for engine failures and fires, every key action requires confirmation by the flying crew member. Doing this quickly and thoroughly takes practice and training. It's worth it, though. Consider the consequences of accidentally securing the good engine, rather than the dead one!

Abnormal Procedures

As we've discussed, abnormal procedures cover all of those things that can go wrong without creating an immediate emergency. A complete set of abnormal checklists is available in the cockpit to address most such problems. These checklists are once again set up on the premise that the flying pilot will concentrate on the flight mission, while

EMERGENCY DESCENT

EMERGENCY DESCENT........................	ANNOUNCE (Cabin and ATC)
IGNITERS...	ON
THRUST LEVERS.....................................	CLOSED
FLIGHT SPOILERS...................................	MAXIMUM
TARGET SPEED..	MAX. OP. SPEED (V_{MO} / M_{MO})
LEVEL-OFF ALTITUDE............................	Highest of: Lowest safe altitude or 10,000 ft.

FLIGHT SPOILERS...................................	RETRACT
IGNITERS...	AS REQUIRED
CREW OXYGEN	STBY.

9.1a

ENGINE FIRE OR ENGINE FAILURE IN FLIGHT

Affected Engine:

CONDITION LEVER..................................	FUEL CUT-OFF
PROPELLER LEVER.................................	FEATHER
FIRE HANDLE-AFFECTED ENGINE......	PULL
FIRE EXTINGUISHER...............................	ACTUATE (if necessary)

ENGINE AUTO IGNITION.........................	OFF
ELEC. HYD. PUMP-AFFECTED SIDE...	ON
GENERATOR...	OFF
PROPELLER SYNCHROPHASER..........	OFF
ELECTRICAL LOAD.................................	SHED LOAD as necessary

9.1b

In each case, boxed items are "memory items." Once they have been performed, the FP (flying pilot) calls for the appropriate checklist. After confirmation that memory items have been completed, the crew executes remaining (nonboxed) checklist items.

9.1. Typical emergency checklists.

ILLUMINATION OF DEICE PRESS. CAUTION LIGHT

1. AUTO DEICE TIMER................. OFF
2. DEICE PRESS. indicator......... Compare deice pressures, both sides. If either side is below 12 psi,
3. ISOLATION VALVE.................. CLOSED
4. MANUAL selector..................... Limit operation of remaining system, as much as possible.
5. Exit icing conditions as soon as possible.
6. On landing: Max. flap angle 20°; see AFM for V_{REF} correction and landing field length.

9.2. Typical abnormal checklist (nonemergency).

the NFP works his or her way through the abnormal checklist and keeps the FP informed (Fig. 9.2).

Among other things, illumination of any yellow caution light on the annunciator panel (see "Annunciator and Warning Systems" in Chapter 6) generally indicates an abnormal situation requiring crew attention. (In many aircraft an abnormal checklist is directly associated with each caution light.) Other abnormal situations, such as engine-out approach and landing procedures, are also covered. There may be many abnormal checklists for any given aircraft. Common categories include:

- System failures (such as generator and electrical bus failures and hydraulic system failures)
- Engine-inoperative landings
- In-flight engine start
- Engine low oil pressure
- Abnormal flight controls (various)

CHAPTER 10

Performance

IF YOU'VE BEEN BUZZING around the flatlands 'til now in piston airplanes, you've probably reduced the whole issue of performance to just a couple of items: cruise speed and fuel planning. While everyone studies takeoff and landing performance, a close look is rarely required for low-elevation piston operations. Those of us who routinely fly in mountainous areas, or who operate at gross weight in piston twins, must more frequently consider takeoff, climb, engine-out, and landing performances.

In turbine-powered aircraft, these performance issues must be addressed more diligently than ever. Despite much greater installed power, most turbine aircraft tend to operate, in many respects, much closer to their limits than typical light piston aircraft. Runway length requirements, especially for jets, are much greater than for piston aircraft. Therefore, turbine aircraft tend to require a larger percentage of the average runway for operations. This calls for more careful planning, especially for the possibilities of engine failures, go-arounds, and aborted takeoffs.

Also, turbine aircraft are designed to operate routinely at high weights since their best economic performance is generally tied to full passenger or freight loads. Even with all that extra power, many turbine aircraft don't actually perform all that much better engine-out, at gross weight and high-density altitude, than many light piston twins.

A related factor is the large range of weights at which a typical turbine aircraft operates. As you know, the operating airspeeds of an aircraft vary, based upon weight. Therefore, when flying at gross weight in a light aircraft, you may decide to rotate at a higher airspeed and on final to "cross the fence" a little faster than normal. Due to the range of operating weights for larger turbine aircraft, along with density altitude and other considerations, the spreads of proper rotation, emergency, and landing speeds are much larger. "A little faster" simply is not accurate enough for safe operation of turbine aircraft.

Sure, cruise performance must be addressed on every flight, due to fuel planning requirements, but among the major performance planning issues for turbine aircraft are takeoff, climb, landing, and engine-out situations.

Takeoff, Climb, Landing, and Engine-Out Performances

If you've been flying light piston aircraft until now, you're used to learning a specific set of airspeeds for each plane. There was a standard takeoff rotation speed, best angle (V_X) and best rate of climb (V_Y) speeds, and recommended airspeeds for approach and for crossing the threshold on landing. These same parameters apply to turbine pilots, but they're described differently. Since turbine aircraft operate under broad variations of weight, configuration, and environment, a wide range of operating speeds must be calculated for takeoff, landing, and emergency or abnormal operations (such as engine-out and no-flap landings). Since pilots can't possibly memorize every speed variation for every possible situation, an entire airspeed terminology has been developed that is unique to turbine aircraft.

There's more than convenience involved in the use of turbine "V-speeds." Sharp multiengine piston pilots know the safety benefits of engine-out planning prior to takeoff. In the event of an aborted takeoff, will there be adequate runway for stopping? Can the trees be cleared in the event of an engine failure after rotation?

The FAA has carefully defined calculation of turbine V-speeds in order to ensure that engine-out aircraft performance meets minimum safety standards under such circumstances. "V_1," "V_R," and "V_2" are among the airspeed terms familiar to every turbine pilot. Let's first examine these terms in the context of required safety-related performance planning. Then we'll consider practical usages of these terms in day-to-day flying.

Takeoff and Climb Performances

All takeoff and initial climb performances in turbine aircraft is planned with one situation in mind: safe continued operation after an engine failure. Certain standard decision points are established for takeoff, with specific performance capabilities required at each. These decision points are described in the form of airspeeds attained by the accelerating airplane and are known as "V_1," "V_R," and "V_2." As safety speeds, these takeoff V-speeds allow the aircraft to achieve optimum or required performance for each particular segment of takeoff in the event of engine failure. Each is computed before every takeoff and varies with the aircraft's weight at the beginning of takeoff roll. (By the way, you ex-military types will miss the takeoff acceleration check; few civilian airports have runway distance markers.)

V_1: Takeoff Decision Speed

Let's start with *V_1, takeoff decision speed.* Simply put, it is "go or no-go" speed. If an abnormality occurs before V_1 is reached, takeoff is to be immediately aborted. If an engine failure or other abnormality occurs after V_1 is attained, takeoff is continued, and the problem is treated in flight. Most airline and corporate flight departments have historically based V_1 speeds upon "balanced field length" for their particular aircraft. If an engine failure occurs exactly at V_1, the distance required to abort the takeoff and stop is the same as the distance required to continue the takeoff. In familiar multiengine terms, this means that accelerate-go distance equals accelerate-stop distance.

However, many airlines with large transport-type aircraft, and some corporate operators, now adjust V_1 to a lower speed than the balanced field length V_1. Lowering V_1 has the effect of lowering the speed where an abort might be initiated; pilots have been doing this in response to numerous accidents that have resulted from high-speed aborts. With most multiengine turbine aircraft, it turns out to be safer to continue takeoff after engine failure than to abort and to try to safely stop the aircraft on the remaining runway. This is especially true for more massive transport category aircraft.

After all, balanced field length V_1 is determined during flight testing with a brand new plane, new brakes, a dry runway surface, and a well-rested super test pilot type in the left seat who's prepared for any emergency. With this in mind, it seems more than reasonable to reduce the V_1 speed slightly to compensate for the average pilot flying a used airplane in bad weather after seven flying legs.

For the same reasons, many captains now carefully define "abnormality" on the before takeoff briefing reference the V_1 "go/no-go" decision. "We'll abort for minor malfunctions or yellow caution lights up to 80 knots and for red warning lights or the fire bell up to V_1. Takeoff will be continued for any problem occurring after V_1, and we'll handle it in flight." Operators may require further adjustments to V_1 under special conditions, such as takeoff on wet or slippery runways (Fig. 10.1).

You may have heard the expression, *V_1 cut.* This term refers to the act, during pilot training or testing, of simulating an engine failure precisely at V_1 on the takeoff roll. It's easy to see both the training benefits and safety hazards of experiencing engine failure at such a critical moment. V_1 cuts are among those exercises best experienced in the simulator.

V_R: Takeoff Rotation Speed

V_R is *takeoff rotation speed* in turbine-powered aircraft. In the event of an engine failure, rotation at V_R to a specified pitch attitude is designed so the aircraft will attain V_2 by the end of the runway—at least 35 feet above the surface. V_R may be set equal to or higher than V_1.

V_2: Minimum Takeoff Safety Speed

V_2, or *minimum takeoff safety speed,* allows the aircraft, in the event of an engine failure, to maintain an FAA-required climb gradient in the climbout flight path. From a practical standpoint, consider V_2 as engine-out best rate of climb speed in takeoff configuration. V_2, depending on terrain, is normally maintained up to a company-mandated altitude, usually 400 to 600 feet above ground level (AGL). Then airspeed is allowed to increase to engine-inoperative best rate of climb speed up to some specified altitude, usually 1500 feet above ground level. Once above that altitude, airspeed is increased to normal climb speeds, and appropriate After Takeoff or Climb Checklist(s) are accomplished.

Those who've flown twin-engined piston aircraft know engine-inoperative best rate of climb speed as V_{YSE} (single-engine best rate of climb). For three- or four-engined airplanes and for some turbine twins other terms are used, such as "V_3," "V_2 + 10 (knots)" or "best-rate-remaining engines," depending on aircraft type. Because the terminology for this speed is not well standardized, we'll stick with "engine-inoperative best rate of climb speed."

Engine-Out Climb

Engine-out climb gradient must also be considered for departure. In many cases, this is the most restrictive of all turbine-powered aircraft performance factors, especially at

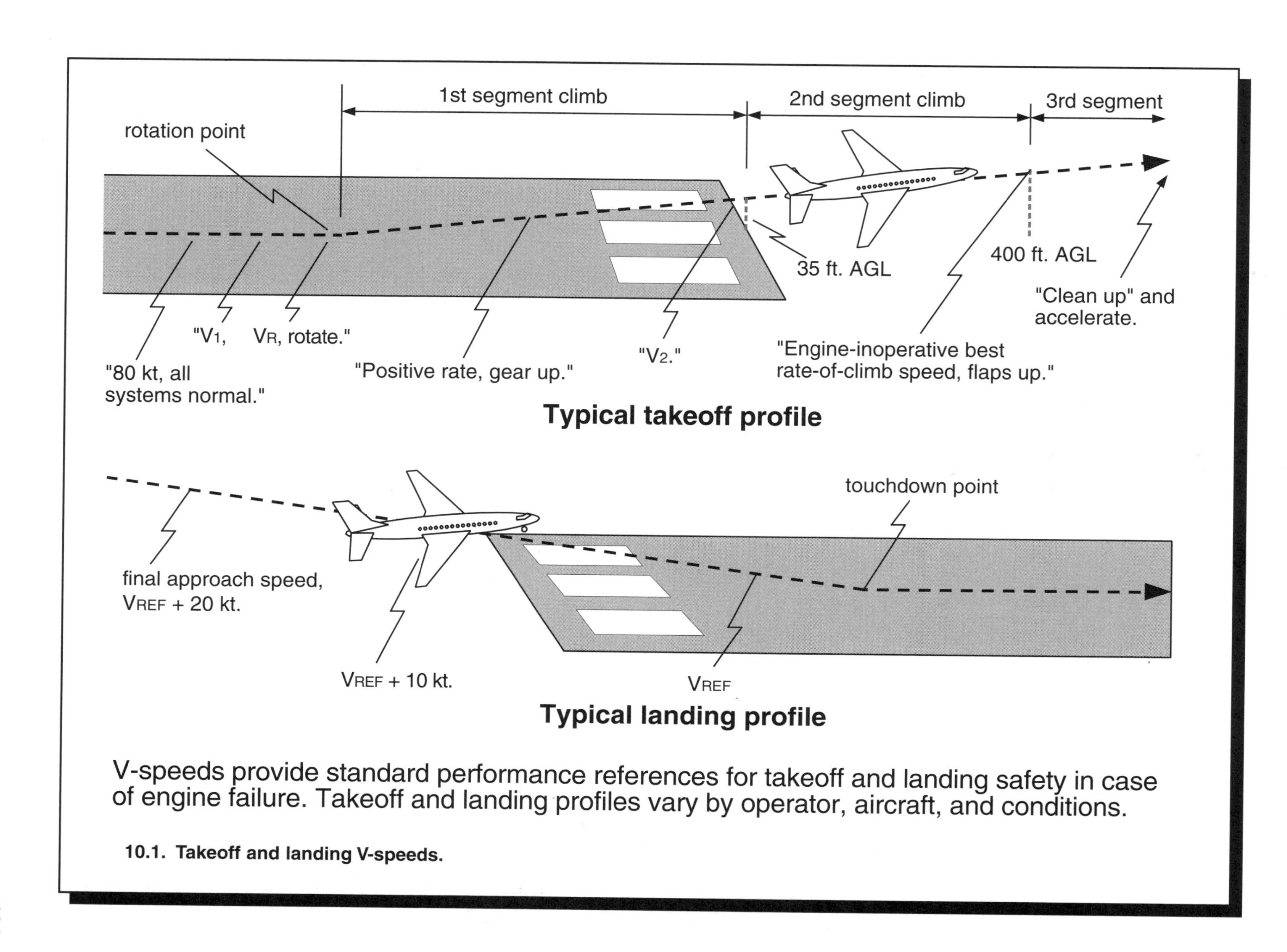

10.1. Takeoff and landing V-speeds.

high-density altitudes and in mountainous terrain. Reduced takeoff weight is the way to provide for adequate engine-out climb performance. Pilots are sometimes faced with tough choices on departure: minimize fuel or restrict loading of passengers and cargo.

For routine (all engines working) climb performance after the initial takeoff segments, most manufacturers publish a "Time, Fuel, and Distance to Climb" chart or something similar. These charts estimate fuel burn, distance, and elapsed time required to climb to cruising altitude. Turbine aircraft climb charts provide more information than those you may have used in the past. For example, flight crews may be offered several different climb speeds to choose from. Most flight departments instruct their pilots to use climb power settings slightly reduced from maximum continuous power.

Enroute Engine-Out Performance Planning

Enroute engine-out performance for turbine aircraft requires a little more planning than what you may be used to in piston aircraft. FAA certification requires multiengine aircraft to achieve certain engine-out performance during enroute flight. However, MEAs (minimum enroute altitudes) over mountainous areas are sometimes higher than the one-engine-inoperative service ceiling for a given aircraft. In order for operators to utilize their aircraft under these circumstances, the FAA allows use of a *drift-down procedure* for high-MEA route segments (Fig. 10.2). Cruising altitude must be high enough that, in the event of engine failure, the airplane can pass safely out of the critical airway segment before "drifting down" to the MEA. (Those of you flying in the Midwest needn't worry much about this one!)

Aircraft performance during cruise depends upon many of the same factors you dealt with flying piston aircraft. Engine power settings, aircraft weight, and weather conditions of wind and temperature all affect performance.

Landing Performance

Landing performance calculations for turbine aircraft are similar to those used for piston aircraft. The greatest difference is that turbine aircraft generally operate closer to their performance limits. With their higher final approach speeds and much heavier weights, turbine aircraft require more runway. Therefore, landing performance planning requires plenty of attention in turbine aircraft.

V_{REF}: Landing Reference Speed

As discussed earlier, the large range of turbine aircraft operating weights results in a corresponding spread of landing speeds. Therefore, there's no single final approach speed used in large turbine aircraft. Rather, the term V_{REF} is used as *landing reference speed.* Lest all these new terms sound too mystifying, let's consider where V_{REF} comes from. Remember your old rule of thumb for calculating final approach speed? V_{REF} for many turbine aircraft is similarly calculated at 1.3 V_{SO} (1.3 × stall speed in landing configuration). V_{REF} accordingly increases with aircraft weight. A Boeing 727, for example, may weigh anywhere from around 110,000 pounds to 180,000 pounds on final approach. Based upon landing weight, its final approach speed varies from around 112 knots to 147 knots. As with the takeoff V-speeds, pilots must compute V_{REF} for every landing.

V_2 and engine-inoperative best rate of climb are also recomputed for every landing to allow for engine failure in case of a go-around. Approach climb gradient (climb performance during single-engine go-around) must also be considered for such situations; it's usually the most restrictive landing performance issue. To determine approach climb gradient, it is necessary to consider the outside air temperature (OAT), airport density altitude, aircraft configuration, and aircraft weight.

When determining landing performance, the FAA again requires that minimum standards be met for operations onto any particular runway. Airport density, altitude, wind components, runway surface conditions, and slope must be considered, along with aircraft configuration and weight, when determining required field length and maximum landing weight. The FAA requires most categories of aircraft to be able to land safely within 60 percent of the landing distance available at the primary destination airport. (Seventy percent is allowed for filed alternate airports.)

Braking Performance

One additional factor to consider on both takeoff and landing performance calculations is the aircraft's braking system. This is very important because all deceleration and stopping performance is predicated on use of wheel braking only. (Thrust reverser performance does not enter into the calculations, even though the effects may be great. Consider reversers as adding to the margin of safety.) Braking charts are provided for crew reference on larger turbine aircraft.

Routine Performance Planning

We've now covered most of the performance issues unique to turbine aircraft operations. The performance charts and graphs themselves, which you'll see in ground school, are similar to those you've used in the past for piston or military aircraft. But all of those performance calculations can be a lot of work!

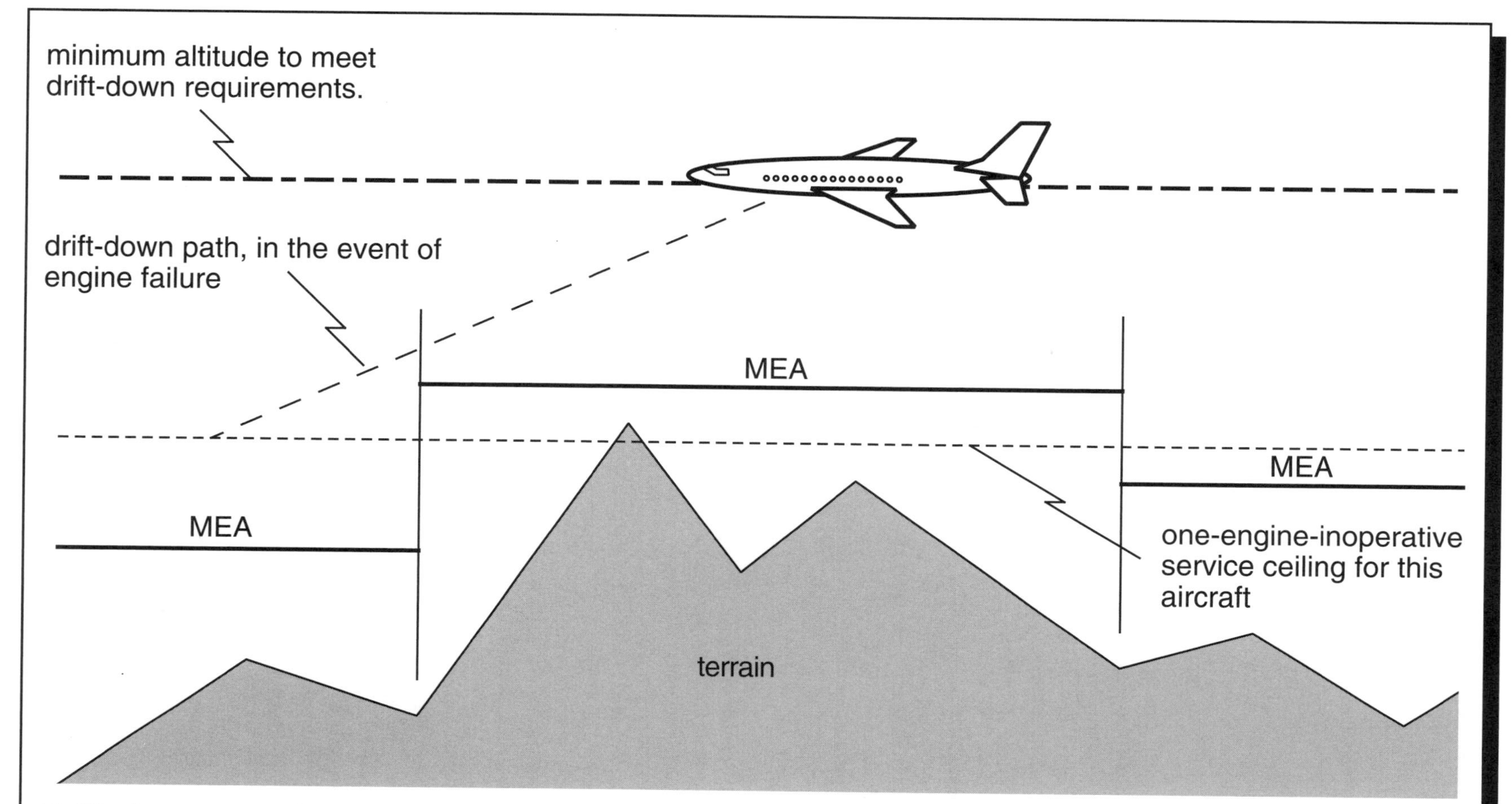

Drift-down requirements apply when an aircraft flies a route segment where the MEA (minimum enroute altitude) is higher than its one-engine-inoperative service ceiling. The aircraft must fly high enough so that, in the event of engine failure, it will reach a lower route segment before "drifting down" to its one-engine-inoperative ceiling.

10.2. Engine-out drift-down requirements.

When aircraft are delivered to corporate or airline flight departments, they come with manufacturers' performance charts. These charts tend to be very complex and too time-consuming for flight crews to use routinely. Therefore, most flight departments develop quick reference performance charts for their flight crews. Quick reference charts are particularly valuable for regularly scheduled operations, which often may include "ten-minute quick turns" and nine or more flying legs per day.

TOLD Cards

Two types of quick reference charts are particularly common among operators. Most flight departments construct simple quick reference *Takeoff and Landing Data Cards* (*TOLD Cards*), allowing flight crews to quickly determine aircraft V_1, V_R, V_2, and engine-inoperative best rate of climb for takeoff, as well as V_{REF}, V_2, and engine-inoperative best rate of climb for landing (Fig. 10.3). Using TOLD cards, these airspeeds may be quickly calculated for a given aircraft based upon gross weight, flap position, and sometimes other conditions. On new technology aircraft, computers are often used to calculate these airspeeds, along with power settings and other flight parameters.

Airport Analysis Tables

To deal with airport-specific performance computations in a safe and timely manner, all air carriers and many corporate flight departments enlist the aid of specialized companies or computer programs to prepare performance data for every likely situation. This information is provided to flight crews in the form of an FAA-approved *Airport Analysis,* a customized book of tables covering every authorized runway at every airport to which an operator flies. The *Airport Analysis* is a quick source of required aircraft performance under current conditions on any given day. It may include various options that the company deems necessary, such as intersection departures, rolling takeoff restrictions, company-preferred IFR departure procedures, or maximum operating weights for particular runways.

As you may imagine, it's often impractical for a flight department to include an analysis for every runway and every intersection at every airport where the company might operate. So be careful when considering intersection departures! Depending on specific FAA approval, most commercial aircraft may operate only from runways and intersections published in their own companies' *Airport Analysis* material, and then only under listed conditions.

Every company instructs its pilots on the use of its particular *Airport Analysis* charts during ground school. When used properly, this information keeps pilots safe and in compliance with FAA regulations without the labor of hand calculations. It is required to be onboard an operator's airplane every time it is flown.

It is possible, under unusual circumstances, to get approval from company dispatch for situations certified by the manufacturer but not covered in the company *Airport Analysis.* However, the company must then furnish the flight crew with all required performance material.

An interesting story related to this occurred in Phoenix, Arizona, during June 1990, when outside air temperature reached 122°F (50°C). Boeing aircraft were grounded because their published performance charts included OATs up to but *not including 50°C.* McDonnell Douglas aircraft, on the other hand, were flying that day. They carry certified performance data up to and *including 50°C*!

Cruise Performance: Fuel Planning

One of the great contributions of the turbine engine to aviation is its power-to-weight ratio. Tremendous power is available from relatively small turbine engines, resulting in excellent high-speed performance by most of today's turbine aircraft. The bad news is that turbine engines consume lots of fuel in delivering such terrific power. Older turbojets are among the biggest fuel burners. Turboprops "sip" fuel, by comparison, but are significantly slower. The latest generation turbofan engines are providing big advances in fuel economy, but trip lengths have been extended at the same time.

One way or another, lots of fuel must be carried by turbine aircraft, and careful fuel planning is always required. This is especially true among jets, with their higher hourly fuel consumption and extended missions. Let's examine some of the reasons why turbine fuel planning is so challenging.

First of all, turbine aircraft carry and burn a lot of fuel. To give you an idea of the numbers involved, an MD-11 carries a little over 256,800 pounds of fuel. Sounds like a lot until you consider the fuel burn of 16,000 pounds per hour (based on cruise fuel flow at FL 350 and aircraft weight of 460,000 pounds). Fuel planning for these aircraft must be very thorough in order to execute a flight within safe fuel limits.

Then there's the fact that fuel efficiency varies tremendously by altitude; turbine engines burn much more fuel down low than at high altitudes. That same MD-11 burns more like 26,000 pounds per hour in cruise at 11,000 feet and achieves a much slower true airspeed. You can see that unanticipated assignment of a low cruising altitude can completely change fuel consumption for a flight.

Just for comparison, let's also consider the performance of a popular business jet. At 33,000 feet, the "bizjet" cruises at a true airspeed of 415 knots. Fuel flow under those conditions is around 1200 pounds per hour. Down at 15,000 feet, the same aircraft cruises at 370 knots, burning

Aircraft Weight:	25,000 pounds		
TAKEOFF	Flaps 10°	5°	0°
$V_1=V_R$	82	89	-
V_2	88	98	105
V_{YSE}	-	-	107
LANDING	Flaps 30°	10°	0°
V_{REF}	86	92	111

Takeoff and landing data cards provide quick pilot reference for takeoff and landing V-speeds. There are two general types of TOLD Cards. The type shown is a part of a series of cards, each offering specific speeds for a given aircraft weight. Correction factors may be provided to compensate for such factors as field elevation and outside air temperature. The other type of TOLD Card (not shown) simply provides blanks to be filled out using "longhand" methods. The TOLD Card, in that case, just serves as a convenient place to display calculated speeds.

10.3. Takeoff and landing data card (TOLD Card).

1800-2000 pounds per hour of fuel. Consider the combined effects of a 60 percent increase in fuel flow, with a 10 percent decrease in true airspeed. Now you know why jet pilots don't like to hear "early descent" requests from ATC!

Jets accomplish fuel efficiency partially through ground speed. They burn lots of fuel per hour but travel a great distance in that hour. Therefore, they are relatively efficient in range (distance per fuel load) but less so in endurance (flying time per fuel load). Anything that adds unexpectedly to the duration of a flight, such as holding for weather, can be costly or even potentially hazardous.

Since turbine aircraft operate at middle to high altitudes, winds can be very strong, dramatically impacting ground speed. An extra hour in the air costs a lot of money in that MD-11. (Don't forget to add all of those other hourly operating costs to the extra 16,000 pounds of fuel burned.)

Finally, long-range flights often require huge fuel reserves and careful "what if" planning throughout the mission. Where must the LA to Hong Kong flight land if Hong Kong socks in? The same question might be asked of a Chicago to Boston turboprop driver. Where to, if the entire Northeast fogs in? (Not an uncommon situation!)

The answer may seem as simple as carrying lots of fuel. Unfortunately, it's not that easy. Since fuel consumption is so high in jets, lots of fuel capacity is required for longer missions. Therefore, there's a huge weight difference between full and partial tanks. It is very expensive and costs payload to "tanker around" more fuel than is necessary for a given flight. First there are the lost revenues when cargo and passengers are left behind in lieu of fuel. Then there are the performance penalties of carrying excess weight (ironically including reduced fuel economy).

Balancing all of these issues is a tremendous concern for pilots and aircraft operators. Safety must be first, but pilots who routinely carry too much fuel cost the company money. True turbine pros develop the planning skills to operate with no more than prudent fuel reserves.

CHAPTER 11

Weight and Balance

WHETHER A CESSNA 152 or a Boeing 747, every aircraft must be flown within certain certified limits of weight and balance. This helps to ensure that the aircraft performs adequately, remains controllable, and is not overstressed in flight or on the ground. Weight and balance principles for airline and corporate aircraft are identical to those of smaller general aviation aircraft. Some of the terms, however, are different, and of course the numbers are a lot larger.

The Weight in "Weight and Balance"

As with any aircraft, it's important that turbine-powered airplanes not exceed certain maximum weight values. In light piston aircraft there is often only a single maximum gross weight. Larger airplanes, however, are certified with several weight limits to allow for the large amounts of fuel carried and consumed during the course of a flight.

Maximum ramp weight is the most a given aircraft may weigh while parked, taxiing, or running up before takeoff. This value is slightly greater than takeoff weight, allowing for the weight of fuel burned during taxi and run-up.

Maximum zero-fuel weight (*MZFW*) is maximum allowable aircraft weight excluding fuel. This is a structural limitation; fuel in the tanks favorably redistributes aircraft structural loads (Fig. 11.1).

Maximum takeoff weight (*MTOW*) is the most an aircraft is certified to weigh for takeoff. MTOW, itself, is a structural limitation. However, there are usually performance-limited takeoff weights for a given aircraft. These further restrict takeoff weight under certain conditions of density altitude (airport elevation and temperature), climb requirements, runway length and conditions, or aircraft systems in use (such as bleed-powered engine anti-ice).

Maximum landing weight (*MLW*) is an aircraft's greatest allowable weight for landing. There is much more stress on an aircraft's structure during landing than on takeoff. Since most turbine aircraft fly for long distances, they burn off lots of fuel enroute. In order to get maximum utility from such aircraft, they are often certified for takeoff at greater weight than is allowed for landing. This greatly extends aircraft range by allowing takeoff with more fuel, but there's also a downside. The problem occurs when an aircraft lifts off at MTOW and then must return unexpectedly for landing. To address this problem many large aircraft have *fuel dump valves* to unload fuel overboard and thereby reduce weight below MLW for landing. Those aircraft without dump valves are faced with flying off the extra fuel before landing, or (like the DC-9) have special low descent rate landing procedures for heavy landings.

Aircraft Weight Categories

The FAA divides aircraft by weight for purposes of wake turbulence separation. *Heavy aircraft* are those certificated at 255,000 pounds or more maximum takeoff weight (MTOW). (Note that aircraft are always classified based on maximum certificated weight, whether or not they actually operate at that weight.) *Large aircraft* have maximum certificated takeoff weights from 41,000 pounds up to 255,000 pounds. *Small aircraft* have certified MTOWs of less than 41,000 pounds and basically include turboprop aircraft carrying thirty or fewer passengers. As might be expected, ATC provides increased separation behind a heavy aircraft and those that follow it, based on these classifications.

A different FAA weight definition concerns pilot certification: type ratings are required to pilot all "large aircraft" having maximum certificated takeoff weights greater than 12,500 pounds. (This rule effectively applies only to large

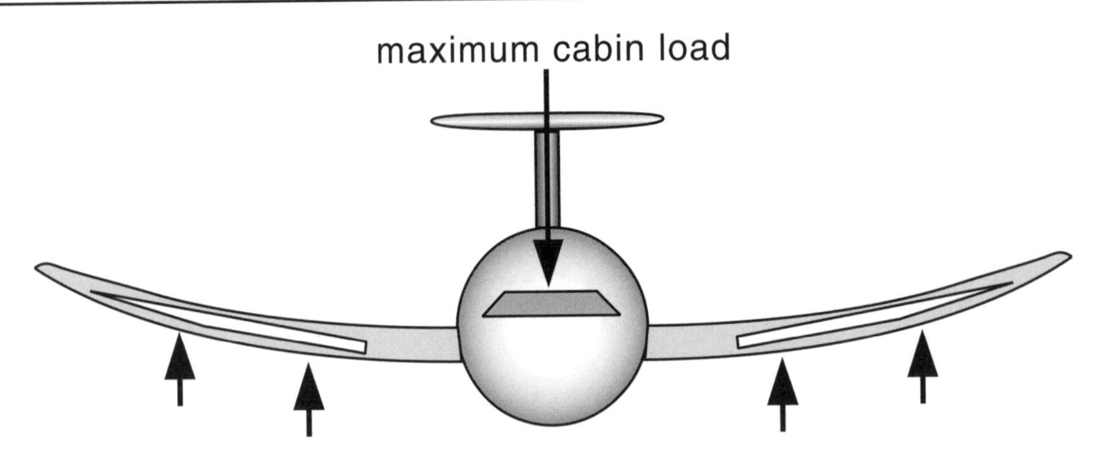

With a big cabin load and not much fuel the wings can be overstressed.

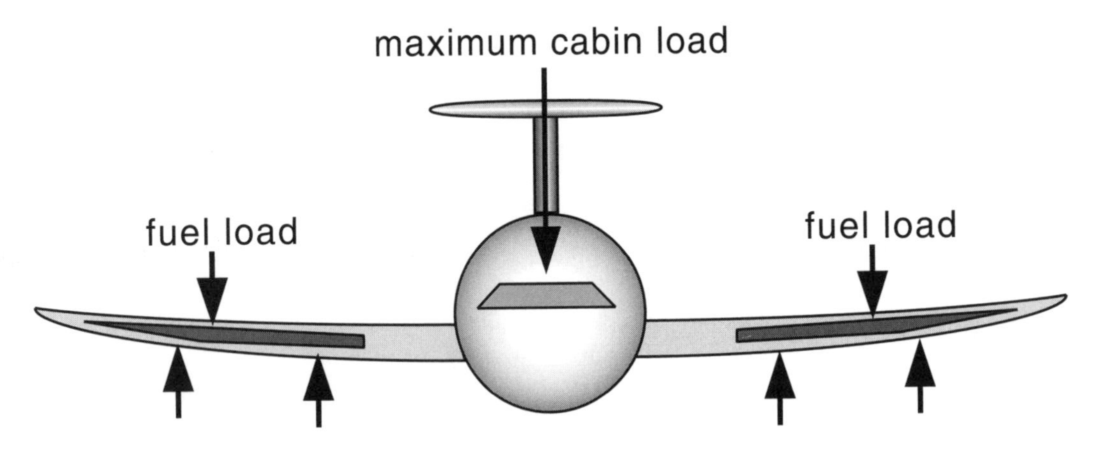

MZFW limits ensure that cabin loads do not overstress the wings in bending. Fuel may be added beyond MZFW, however, because it loads the wings outboard and thereby actually reduces wing bending moments.

11.1. Maximum zero fuel weight (MZFW).

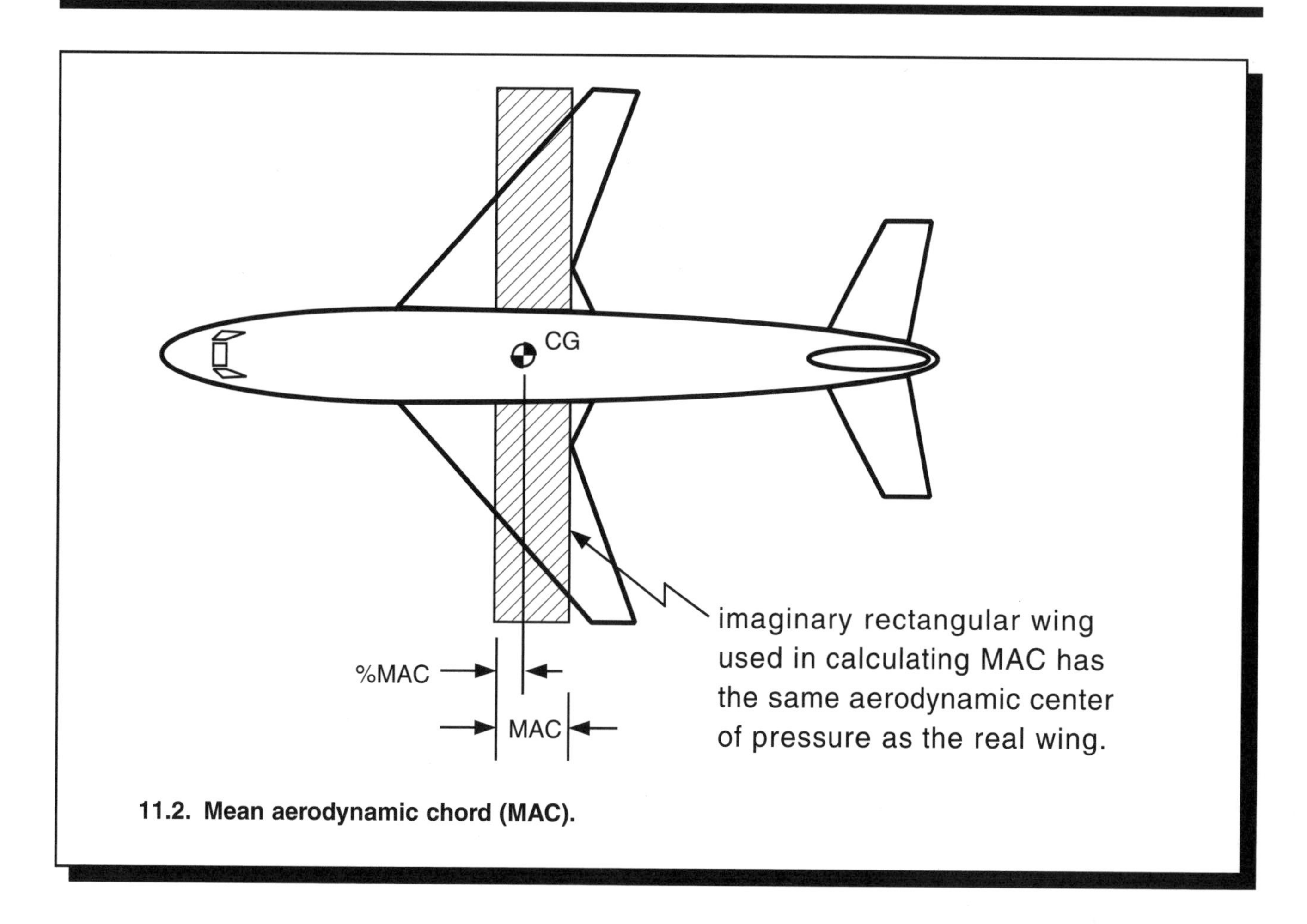

11.2. Mean aerodynamic chord (MAC).

turboprop and piston aircraft, as type ratings are required to pilot all turbojet-powered aircraft, including "pure jets" of any size or weight.)

Balance Considerations

In order for an aircraft to fly safely and be controllable, its *center of gravity* (*CG*) must lie within a certified range. While this is important in light aircraft, it becomes even more critical in large aircraft due to the huge weights and loading ranges involved.

CG as Percentage of MAC

CG computations for light turbine aircraft are usually accomplished in the same general manner you've used before. Many large aircraft operators, however, identify CG location not in terms of inches from datum but as a percentage of *mean aerodynamic chord* (*MAC*). The only difference is in how the CG location is described. (Whether in inches or percentage of MAC, the actual CG position is the same.)

MAC is simply the chord of an imaginary rectangular wing that has the same aerodynamic characteristics as the actual airfoil of the airplane. This simple imaginary wing represents in performance what otherwise might be a very complex swept wing planform. MAC is used primarily as a reference for longitudinal stability characteristics (Fig. 11.2).

Calculating CG position as a *percentage of MAC* (%MAC) is relatively straightforward. CG location is first calculated in the conventional manner. It may be described as inches from datum or in terms of "loading stations." Loading stations are simply measuring units made up by the manufacturer. The calculations, in these cases, are the same as what you're used to; just use the loading stations as arm values, instead of inches.

The length of MAC is calculated by subtracting LE MAC (leading edge MAC) from TE MAC (trailing edge MAC). These values are supplied with the aircraft weight and balance data. LE MAC is the distance from datum to the leading edge of mean aerodynamic chord. TE MAC is the distance from datum to the trailing edge of mean aerodynamic chord. Once the length of MAC is known, it's simple to determine where the CG falls on MAC. The CG's distance aft of LE MAC, divided by MAC, and multiplied by 100 yields percentage of MAC (Fig. 11.3).

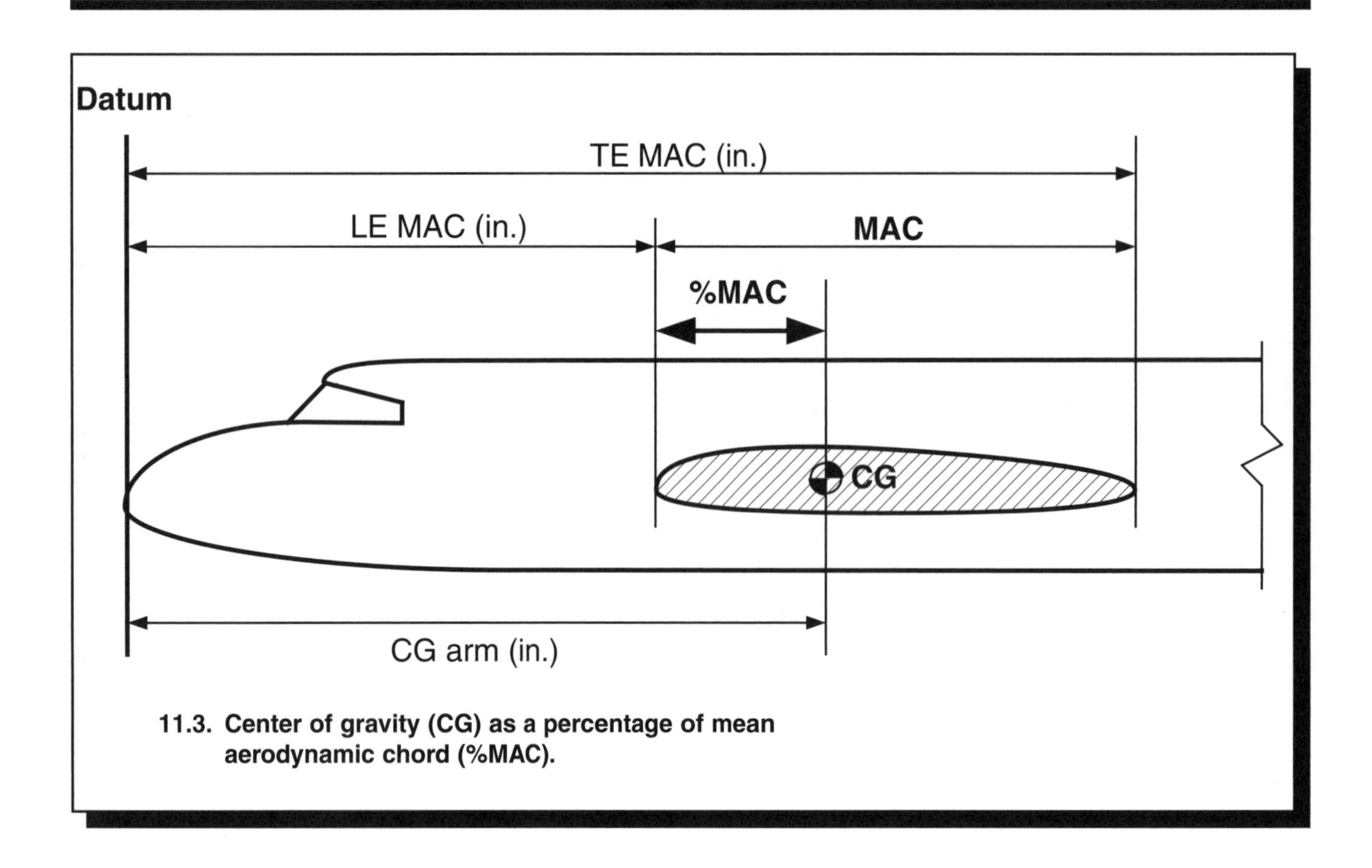

11.3. Center of gravity (CG) as a percentage of mean aerodynamic chord (%MAC).

An aircraft's CG range may be described using the same terminology. Most aircraft typically have CG ranges of around 10 percent to 30 percent MAC, though this varies significantly. (Beechcraft's 1900C commuter, for example, has an unusually wide CG range of 4 percent to 40 percent MAC.)

Performance Benefit of an Aft CG

You may remember from commercial ground school that there's a benefit to loading a plane near aft CG limits—faster cruise. This is because tail-down force, while only an aerodynamic load, must be carried by the airplane just as if it were another container of baggage. When the aircraft is loaded aft, less tail-down force is required to achieve aircraft balance. Therefore, an aft-loaded airplane requires less lift than one loaded forward. This translates to faster cruise and better climb rate. While the speed advantage may be minor in small slow airplanes, it can be significant in big ones. One airline captain won a large cost-saving award several years ago for suggesting that company aircraft be routinely loaded aft for increased performance. The resulting higher cruise speeds generated significant savings in time and therefore costs.

In-Flight CG Movement

In-flight CG movement can also enter balance computations for large turbine-powered aircraft. Most noticeably, the large fuel burns and fuel movements in turbine aircraft can significantly impact CG location between takeoff and landing. Such factors require computation of both takeoff and landing CGs in order to ensure that an airplane will stay within limits throughout the flight. A few aircraft use fuel movement to their performance advantage. The MD-11, for example, is equipped with an automatic "tail fuel management" system. Fuel is shifted aft into horizontal stabilizer tanks during climb and cruise, then back forward during descent for a "normal CG" landing. Tricks like these have helped make modern jet transports incredibly fuel efficient when compared with older models.

Operators and aircraft designers must even allow for movements of passengers, flight attendants, and food and beverage carts around their aircraft. Consider, for a moment, that each airline beverage cart weighs about 240 pounds and is operated by two flight attendants. Four carts are often found on a large airplane. This means that, during beverage service, some 2500 pounds of carts and people may be rolling along the length of the airplane, not including

LOAD MANIFEST / WEIGHT & BALANCE

Flight		Date	
N		Capt.	
From		F/O	
To			

Alt	#1		#2		T/O	

Performance						
T/O	Temp		Rwy		MATW	
Ldg	Temp		Rwy		MALW	

Item			**Computations**	**Adjustments**
BOW				
Pax		x165/170		
Children		x 80		
Lap Children			N/A	N/A
Bags		x 23.5		
Cargo		Items		
ZFW. 14000				
T/O Fuel				
T/O Weight				
Fuel Burn				
Ldg. Weight				

Balance Form Attached (if required)

CG at T/O	MOM		CG Ldg	MOM	

I certify that this aircraft is loaded in accordance with the loading schedule and CG is within limits

Captain's Signature:

Random loading rules

Less than five pax:
Passengers may sit only in Rows 3 through 6.

Five to eleven pax:
Passengers may sit anywhere except the last three rows.

Eleven to nineteen pax:
Passengers may be seated anywhere.

Maximum Weights - BE 1900

ZFW	14000
MTOW	16600
MLW	16100
MRW	16710

Key

BOW = basic operating weight
CG = center of gravity
MALW = maximum allowable landing weight
MATW = maximum allowable takeoff weight
MLW = maximum landing weight
MOM = moment
MRW = maximum ramp weight
MTOW = maximum takeoff weight
ZFW = zero fuel weight

11.4. Typical commuter aircraft load manifest and random loading rules.

passengers. Does it affect trim? You bet it does; no wonder airline pilots rarely hand-fly during cruise!

How It's Done in the Real World

Obviously, weight and balance computations can be time-consuming. While many corporate operators perform them manually, others, along with virtually all commuter and airline operators, use simplified methods to save time. Some aircraft manufacturers, for example, provide computers or quick reference charts for graphical weight and balance determinations.

Average Passenger Weights

The FAA also makes life easier by allowing operators to use standard *average passenger weights.* Instead of weighing every passenger boarding an airplane, an average adult weight of 160 pounds may be used during "summer" months (May 1 through October 31) while 165 pounds may be used during "winter." (Some operators opt to use higher weights.) For domestic operations, the FAA allows a (minimum) average of 23.5 pounds for each passenger carry-on bag.

Using average weights for passengers and baggage on big airplanes makes weight and balance computations easier and therefore saves time. (Besides, the passengers get nervous if asked for their weights!) Of course, this approach must be accompanied with some common sense. If your aircraft is carrying the entire Japanese national sumo wrestling team, it is wiser (and the FAA expects you) to do a "long-form weight and balance" using true weights for your computations.

Random Loading Programs

Many aircraft operators use FAA-approved *random loading programs.* To design these, weight and balance are computed for every possible loading configuration. Then, so that weights can be calculated quickly, a series of *load manifests* is designed for use by pilots and/or ground crew. If necessary, random loading rules are established to keep the aircraft in balance (see Fig. 11.4). While such rules take a little getting used to, they allow the pilots to compute the weights quickly and simply (knowing that the aircraft is in balance) and to get on with other business. Every random loading program for commercial operators must be approved by the FAA.

The major airlines do all of their load planning and calculations by computer. Sometimes this is done at the station where the aircraft is located; at other times weight and balance is calculated for all aircraft from one central company office. In any case, load planning and calculations these days are normally accomplished somewhere off the aircraft. This information is then provided to the captain for inspection and approval, prior to departure.

Remember that no matter who actually does the calculations, the pilot in command still remains responsible for their accuracy. As pilot, you'll need to know all the procedures for your aircraft, whether you're lucky enough to escape the physical calculations or not.

CHAPTER 12

Airplane Handling, Service, and Maintenance

AS YOU ENTER the world of turbine aircraft, especially in scheduled operations, you'll find your interaction with maintenance professionals and ground service personnel changing. Charter and corporate pilots often still do much of the ground service of their airplanes, including aircraft cleaning and checking of oil and other fluids. They usually do their own flight planning, often arrange catering, and sometimes book hotels and ground transportation for their passengers.

Scheduled operators, on the other hand, generally keep their pilots so busy that much more ground support must be provided by their companies. Detailed maintenance records must be kept, and frequent mechanic sign-offs are required. The result is that pilots of scheduled aircraft usually spend far less time on routine ground tasks than their corporate and air-taxi counterparts.

Flight Dispatch

Under FAR Part 121 operations, the captain shares responsibility with a flight dispatcher to approve and release each flight. The *flight dispatcher* is an FAA-licensed professional, trained in virtually every aspect of aircraft operations except the flying itself. These people do all of the flight planning for an airline, including planning for weather conditions, fuel, cargo, and passenger loads. The *dispatch flight release,* which includes all information relevant to a given flight, is the official document releasing a Part 121 aircraft for its trip. Many Part 135 and corporate operators also use dispatchers in order to increase efficiency of their operations.

Fueling Procedures

At the "majors," and at large regional airlines, the flight dispatcher determines the proper amount of fuel to be onboard the aircraft based upon passenger load, cargo, and weather. The captain reviews the aircraft's dispatch release for the required amount of fuel and then ensures that the aircraft has been properly fueled. If the captain doesn't feel comfortable with the amount of fuel onboard, he or she may negotiate with the dispatcher on the fuel load. Some companies allow for "captain's fuel." That is, a certain amount of fuel may be added at the captain's discretion without negotiating with the flight dispatcher.

At smaller regional airlines, charter outfits, and corporate operators, the captain usually determines the amount of fuel required to be onboard for a flight. In these cases, the captain may also have to determine when and where to buy the fuel, and even at what price. While this may seem like added workload, there is a sense of more operational control by the captain, and some prefer it. Either way, the captain is ultimately responsible for ensuring that the required amount of fuel is on the plane for departure.

Most corporate and all commercial operators have developed their own specific aircraft fueling procedures. These procedures must be FAA-approved for commercial operators and are designed to protect against fuel contamination, fire, and explosion. Most importantly, they're designed to protect the flying public during fueling operations.

Among other rules, most companies require a flight crew member to be present during fueling procedures. (If this is required by your company, you can bet it'll probably be your duty as first officer to be there.) If passengers are on the aircraft during fueling, a qualified person trained in emergency evacuation procedures must be onboard during servicing. An aircraft door must remain open, and a clear evacuation route must be present.

Standard Preflight

Preflight of a turbine aircraft includes many more items than you may be used to, and the duties of each crew member vary by equipment and company. At the same time, turbine preflight procedures follow a commonsense pattern that will seem familiar to you almost at once. In this section we cover some general preflight procedures (with emphasis on commercial operations).

A standard preflight may be broken down into four basic categories: review of aircraft documents, cockpit check, emergency equipment check, and external check of aircraft.

Aircraft Documents Review

A standard preflight normally begins with a review of all aircraft documents required to be onboard. The following is a generic list of required documents found on turbine and/or commercial aircraft:

- Airworthiness Certificate
- Aircraft Registration Certificate
- Radio Station License
- Aircraft Flight Manual
- Aircraft Maintenance and Flight Records
- Normal, Abnormal, and Emergency Checklists
- Airport Analysis and/or Aircraft Performance Data
- Minimum Equipment List (MEL)/Configuration Deviation List (CDL)
- Takeoff and Landing Data Cards (TOLD Cards)
- Load Manifest
- Compass Deviation Cards

Often, it is the first officer's duty to ensure that all required aircraft documents are onboard the aircraft. The captain must review the Aircraft Maintenance and Flight Records in order to determine the aircraft's airworthiness prior to flight. (The captain is not to accept an aircraft having any "open write-ups," or unresolved maintenance issues.) Once satisfied that any maintenance issues have been properly addressed, he or she signs the records, signifying acceptance of the aircraft for flight. While the captain does this, the first officer (FO) completes the rest of the aircraft preflight.

Cockpit and Emergency Equipment Checks

Once the paperwork has been checked, the crew conducts a cockpit preflight flow check (Fig. 12.1). This ensures that switches, controls, and circuit breakers are in their proper positions. Depending upon aircraft, season, and whether or not it's the aircraft's first flight of the day, the FO may opt to turn on the APU (auxiliary power unit) or GPU (ground power unit). Environmental systems can then be activated to adjust cabin temperatures in preparation for passenger boarding. An emergency equipment check covers the presence and operability of all fire extinguishing equipment, supplemental oxygen equipment, and emergency exits.

Exterior Preflight Check

Exterior preflight check of a turbine aircraft includes many of the same items you're used to from piston models (Fig. 12.2). The condition of fuselage skin, windows, and maintenance and cargo access doors must be checked, along with antennas, aircraft lights, and deice and anti-ice equipment. You'll also be checking various system pressures, such as hydraulics, emergency oxygen, and fire-extinguishing bottles. (Gauges are often located in wheel wells or compartments where they're hard to read. Pilots quickly learn to carry a good-quality flashlight on preflight.)

Turbine aircraft tend to have more outside covers and locks than are normally found on light airplanes. External gust locks are sometimes installed on control surfaces for longer-term parking, as are engine inlet and pitot tube covers. Forgetting any of these things on preflight can kill you . . . and most are not visible from the cockpit! Propeller tiedowns (prop ties) are often installed on free-turbine turboprops for parking, in order to prevent propellers from "blowin' in the wind."

One other important preflight item is to check for removal of *gear pins.* These devices, which are installed for parking on many larger aircraft, are lock-down pins inserted into the strut or retraction mechanism of each landing gear. Taking off with gear pins installed is both embarrassing and dangerous since the gear won't retract when you select "up." At least one airline puts gear pins in the pocket behind the captain's seat for flight so that their presence (proof of removal) can be confirmed on the Before Start Checklist.

On some aircraft, the nose gear scissors must also be checked to confirm that they're mated. On these models, the scissors are disconnected for towing of the aircraft by a ground tug. If the scissors are not reconnected, the nose gear is left freewheeling and unsteerable.

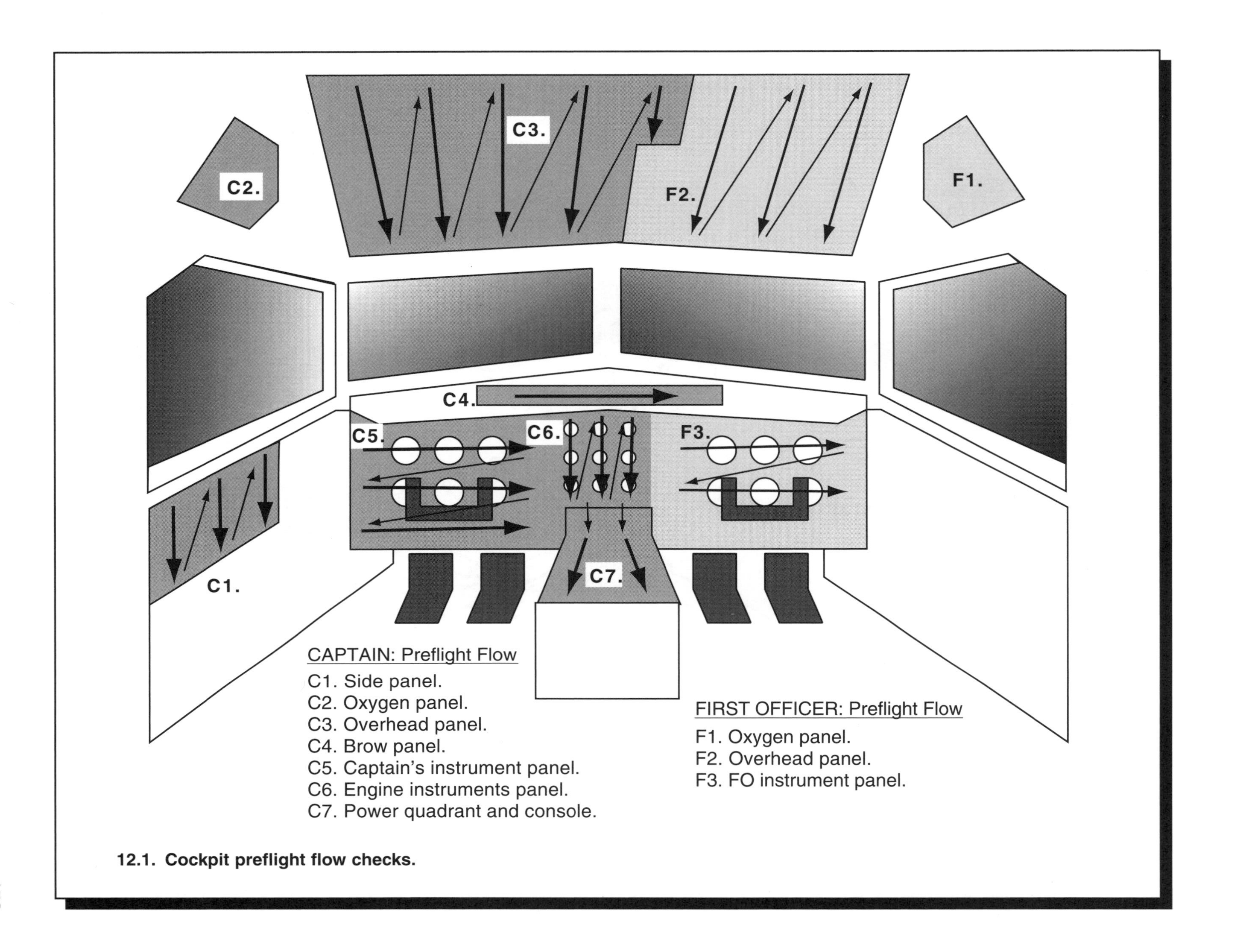

12.1. Cockpit preflight flow checks.

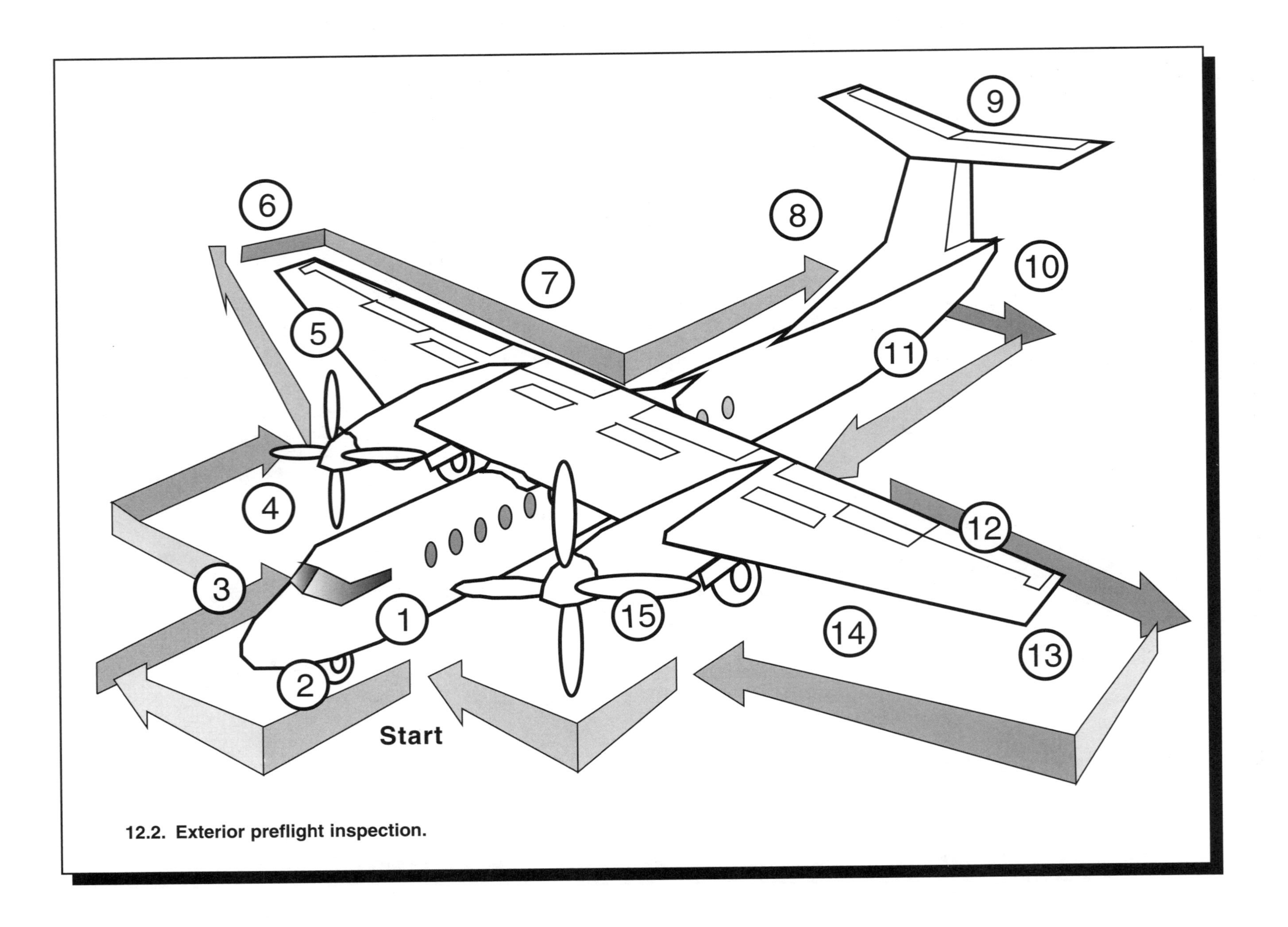

12.2. Exterior preflight inspection.

Final Preflight Preparations

While the FO is preflighting the aircraft, the captain may often be found in the *flight operations office* (*flight ops*) receiving and reviewing weather or, for companies with a professional dispatcher, the dispatch flight release. The dispatch release includes weather for the upcoming leg, the flight plan, and passenger bookings, along with the required fuel load.

Upon returning to the aircraft, the captain briefs the crew on the upcoming flight. Sometime during the preflight process, the first officer tunes in and copies ATIS (Automatic Terminal Information Service), then gets the ATC clearance. Although not always done, it's desirable for both flight crew members to listen to the clearance so both can confirm that it was properly received and recorded.

With this, the standard preflight is over. If it's a large aircraft, passengers have probably been boarded, and the plane is ready to roll. On smaller vehicles, the FO can now look forward to boarding the passengers in compliance with company and FAA policies, including random loading weight and balance requirements. (See Chapter 11, Weight and Balance.)

Minimum Equipment List (MEL)

One major difference between most general aviation piston aircraft, and turbine-powered corporate or airline aircraft, is the extensive redundancy of aircraft systems. In order to ensure safety and reliability, most major system components and communications and avionics equipment are duplicated. In many cases this allows corporate and airline aircraft to safely operate even after failure of individual aircraft components.

A *MEL* (FAA-approved *Minimum Equipment List*) contains items that are allowed to be inoperative on a given aircraft, while it is still considered airworthy. MELs are important because they allow flights to be conducted even though certain equipment is inoperative, while still being considered legal and "safe" by the FAA.

The MEL, which must always be onboard the aircraft, is normally in notebook form and is fairly simple to use. When an aircraft has an inoperative component, the pilots look it up in the MEL. The MEL states what flight conditions, performance limitations, crew operating procedures, maintenance procedures, placards, and duration limits are necessary in order to be legal for flight with that component inoperative. The idea is to ensure that acceptable safety levels are maintained, while allowing the aircraft operator maximum utility from its fleet.

When using a MEL, a few important things should be kept in mind. All equipment not listed in the MEL and related to the aircraft's airworthiness must be operative (that is, aircraft wings must be intact and in good condition). The airplane may not be operated if any OTS (out-of-service) components are not specifically listed in the MEL or if any MEL items are out of date. (MEL items are coded to show how long the aircraft may be operated before they are repaired.)

When multiple components are inoperative, pilots must consider the combined effects on aircraft safety and crew workload in order to establish the airplane's airworthiness. As with everything in flying, ultimate responsibility for maintaining acceptable levels of safety rests with the flight crew.

Configuration Deviation List (CDL)

Similar to the MEL, a *CDL* (FAA-approved *Configuration Deviation List*) contains additional items and limitations for operation without secondary airframe or engine parts, while still allowing the aircraft to be considered airworthy. Items such as missing inspection panels, access doors, or aerodynamic fairings are found in the CDL. The CDL is kept with the MEL, uses the same notebook format, and must always be onboard the aircraft. CDLs, like MELs, are important because they allow flights to be conducted that are still considered "safe" by the FAA even though certain equipment is inoperative, damaged, or missing.

The CDL must be referenced prior to flight for any performance penalties, limitations, or procedures to be applied for operation without some particular item. For example, a missing aerodynamic fairing may impose such a drag penalty that the aircraft must be dispatched with a greater than normal fuel load, thereby potentially displacing passengers or baggage.

CHAPTER 13

Navigation, Communication, and Electronic Flight Control Systems

PILOTS OF THE PAST, if they could return and visit modern aircraft, would probably be most astonished at the latest generation of avionics and flight control systems now found in turbine aircraft. Recent computer and electronic display advances have virtually revolutionized cockpit navigation and information systems.

Each corporate and airline flight department equips its aircraft with different navigation and communications options. Therefore, we will focus in this section on the operational principles of each system, rather than explaining details of use. (Be prepared for lots of acronyms—alphabet soup aficionados will love this chapter.)

Horizontal Situation Indicator

Depending on the sort of flying you've done, you may have already used a *horizontal situation indicator* (*HSI*) similar to those found on most turbine aircraft. HSIs range from analog "round-dial" models to the latest digital electronic displays. While the navigational information displayed on an HSI may be presented in different ways, the basic appearance and principles of pilot operation are the same.

An HSI simply combines a heading indicator and a *course deviation indicator* (*CDI*) into one instrument (Fig. 13.1). (In case you've forgotten, a CDI is just the course needle off your old VOR head.) On an HSI the CDI rotates with the heading indicator to reflect, in a simple graphic manner, the relationship of course to airplane. Glideslope or other vertical navigation (VNAV) information may also be displayed. Presenting all of this information on one indicator lessens pilot workload by reducing instrument scan. On electronic HSI indicators, the navigation presentation may also include an RMI (radio magnetic bearing pointer) and/or a DME (distance measuring equipment) range readout.

Traditionally, the CDI receives its navigation information from either a VOR or ILS (instrument landing system) transmitter. However, on many newer systems HSI course information is also received from other sources such as an inertial navigation system (INS), the global positioning system (GPS), or some other type of area navigation (RNAV) computer. (These terms are discussed later in this chapter.)

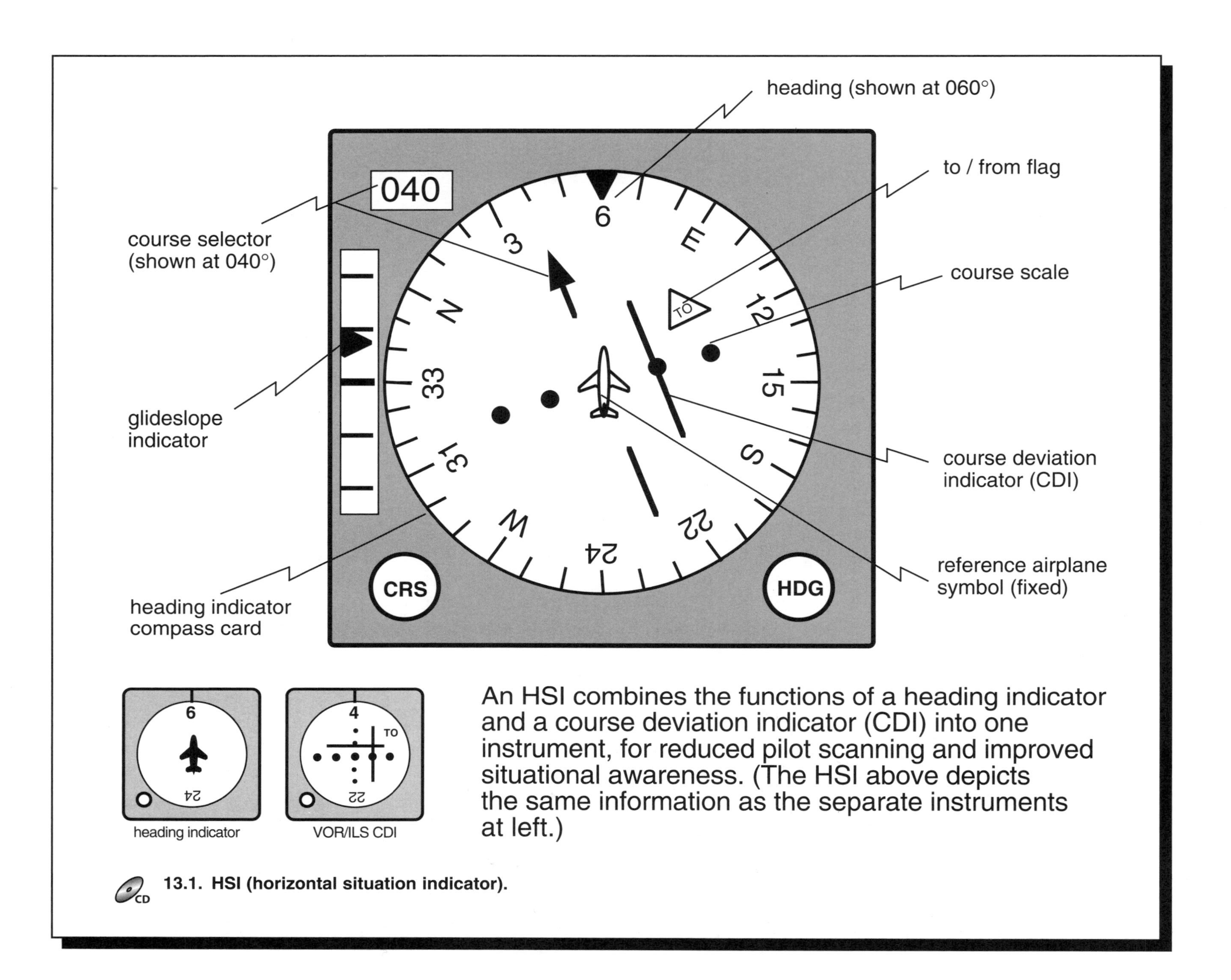

An HSI combines the functions of a heading indicator and a course deviation indicator (CDI) into one instrument, for reduced pilot scanning and improved situational awareness. (The HSI above depicts the same information as the separate instruments at left.)

13.1. HSI (horizontal situation indicator).

Autopilots

Autopilots, of course, are devices that automatically operate flight controls to fly the airplane. There are tremendous variations in autopilot systems, especially as a function of each unit's age. Today's turbine aircraft autopilots normally control aircraft attitude in all three axes: pitch, roll, and yaw. Even the most elementary units can normally perform at least the following functions: maintain altitude, hold a selected heading, track navigational courses from VOR or other nav equipment, and capture and shoot an ILS approach.

The autopilots commonly installed these days do a lot more: maintain preselected climb and descent rates and airspeeds, capture and level off at assigned altitudes, and, if a capable nav computer is installed, track multipoint navigational routes. Even descent parameters may be programmed into those systems equipped for VNAV (vertical navigation).

The sophistication of latest generation autopilots is in flying the airplane through a complex series of commands in all dimensions. This is due to the capability of today's flight management computers (FMC) to program and to store not only aircraft flight parameters but also complete RNAV routes and transition and instrument approach procedures. Integrated auto-throttle (airspeed control) and auto-land functions enhance these capabilities. (There is more information on FMCs later in this chapter.)

Aircraft certified for Category III "zero-zero" ILS approaches (see "IFR Operations in Turbine Aircraft" in Chapter 15) must have several independent autopilots installed, with "tie-breaking" capabilities. "Tie breaking" refers to the ability of the system, in the event of an autopilot malfunction, to identify which autopilot is working properly in a given situation and to assign it priority.

While autopilots are extremely capable, they can be complex to operate and require continual monitoring. The big liability of an autopilot is that it allows the crew to relax, thereby taking the pilots out of the loop of flying the airplane. Many accidents have occurred over the years due to unnoticed autopilot malfunctions or to poor programming and pilot inattention. In a surprising number of cases, pilots have been oblivious as the autopilot flew the airplane into terrain. Thorough training and continual monitoring are essential for safe autopilot use.

By the way, those of you flying autopilot-equipped piston singles or light twins may be in for a surprise. Given today's high pilot workloads, you might expect that autopilots would be installed in every one of those multimillion dollar turbine aircraft. However, this is not the case. Most nineteen-seat commuter aircraft are not equipped with autopilots. If you'll be flying those aircraft in your next job, plan on getting sharp at your IFR (instrument flight rules) "stick flying!"

Flight Director

A *flight director* is similar in many respects to an autopilot, except it does not have the ability to control the aircraft in any manner. Rather, it provides that information directly to the pilots for use in precise hand-flying. A flight director computer-processes heading, attitude, and navigation information and determines what pitch or steering commands are needed to fly the aircraft correctly. This information is presented to the pilot through use of one or more "command bars" superimposed over the attitude indicator. Command bars do exactly what the name implies, directing the pilot through control movements to meet preselected parameters. The pilot follows pitch and steering commands presented by the command bar in order to maintain course, heading, or altitude selected.

Two flight director arrangements are common. "Single-cue" systems use a triangular representation that fits into a V-shaped command bar (shown in Figure 13.2). "Double-cue" models incorporate a horizontal "pitch" bar and a vertical "bank director" bar; when on course or heading and altitude, the pilot sees centered needles having the appearance of ILS needles superimposed over an attitude indicator.

Flight directors can be tuned to nav radios for use in shooting instrument approaches; they are also normally tied to a go-around function, allowing pilots to conform to standard performance profiles in the event of a missed approach or go-around. (Go-around buttons for triggering this function are normally found on the thrust levers or power levers.)

While older aircraft generally have separate computers driving the autopilot and flight director systems, newer models use the same computer system to operate both. In either case, pilots can let the autopilot operate the controls, or they can choose to hand-fly the aircraft following flight director commands. Much of the time the autopilot does the flying, and the flight crew uses the flight director command bars to monitor autopilot operation (see Fig. 13.3).

Electronic Flight Instrumentation Systems (EFIS)

You've likely heard the term "glass cockpit" used to describe today's modern cockpits, which are packed with the latest computer wizardry. With the advance of computer and display technologies, electronic cockpit displays have become commonplace in today's airline and corporate aircraft. These systems are known as *electronic flight instrumentation systems,* or *EFIS.* EFIS systems replace conventional electromechanical instruments with CRTs (cathode-ray tubes, like TV screens) or with flat panel displays. Depending on instal-

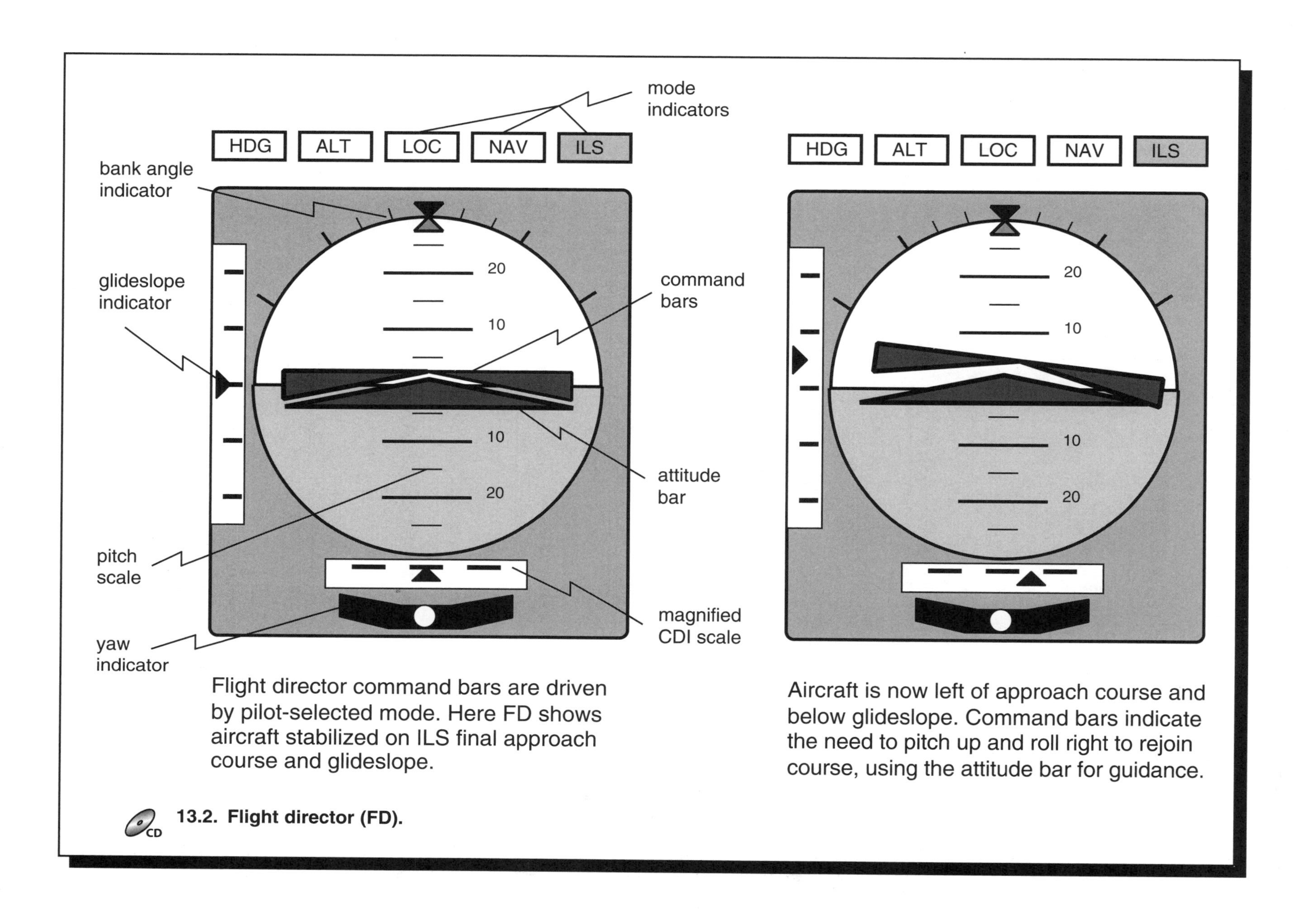

Flight director command bars are driven by pilot-selected mode. Here FD shows aircraft stabilized on ILS final approach course and glideslope.

Aircraft is now left of approach course and below glideslope. Command bars indicate the need to pitch up and roll right to rejoin course, using the attitude bar for guidance.

13.2. Flight director (FD).

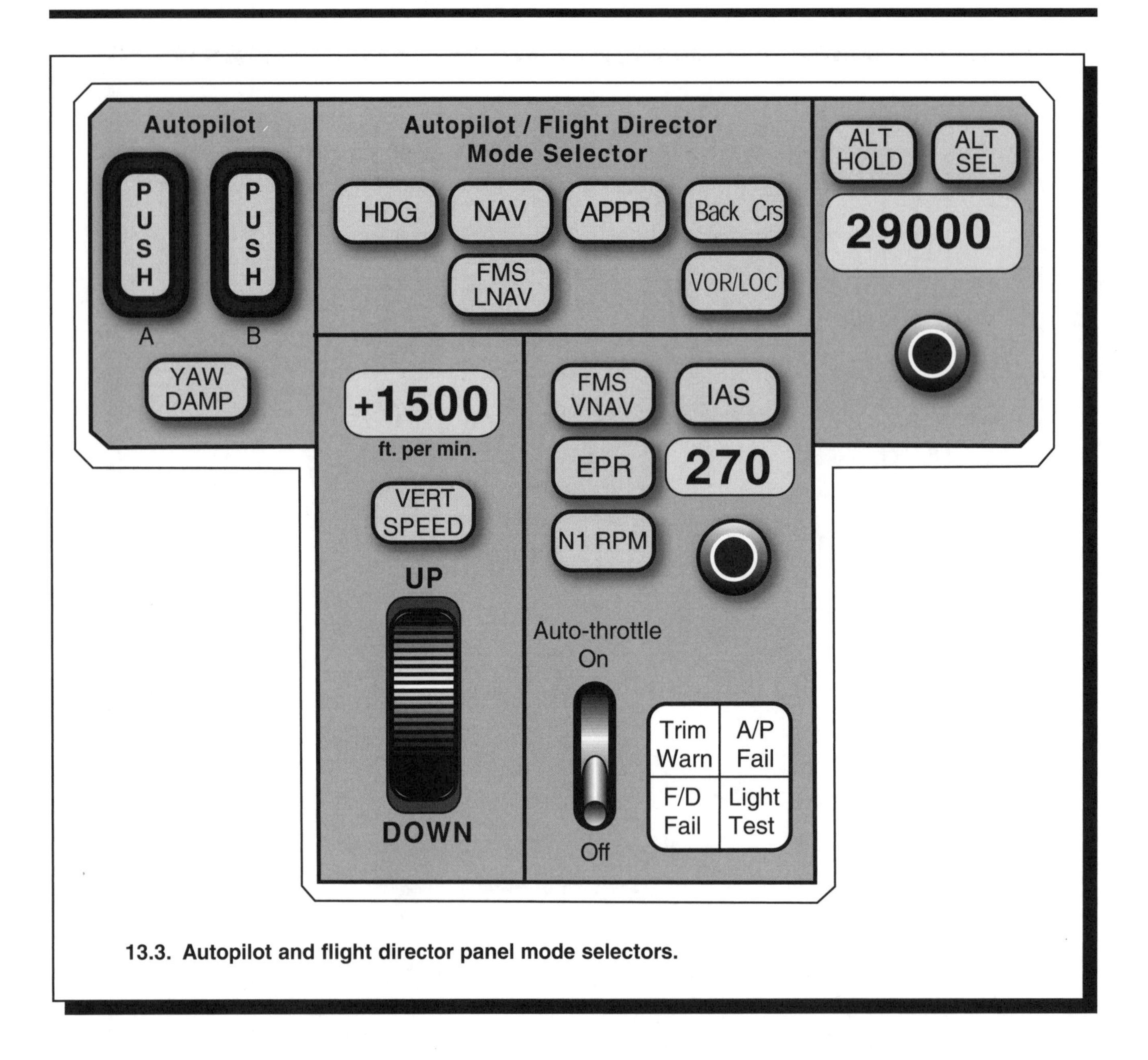

13.3. Autopilot and flight director panel mode selectors.

lation, most or all instrument indications are computer generated and displayed on these screens.

Levels of EFIS utilization in the cockpit vary tremendously by aircraft and operator. Some aircraft use only an electronic attitude indicator or HSI, while others, like the Airbus A-320, Boeing 777, McDonnell Douglas MD-11, and for that matter most new turbine aircraft, are "all glass." In many cases the airline or corporate cockpit you fly will contain a mixture of older-style analog instruments and newer EFIS displays. EFIS components have been added to many older aircraft midway through their service lives and are standard on many newer models. On smaller turboprops, in particular, EFIS is often installed only on the captain's side, while the first officer makes do with analog instruments. (Yet another privilege of command!)

Because EFIS systems come in many permutations, it's difficult to generalize about them. Full-cockpit EFIS systems commonly include anywhere from four to six displays, presenting virtually all cockpit information. (You may hear expressions such as "five-tube EFIS," to indicate how many displays are included.) Most all-EFIS cockpits incorporate variations on three types of displays (Fig. 13.4).

The *primary flight display* (*PFD*) is the most standard of EFIS instruments and incorporates the traditional functions of the "standard T" of flight instruments. (Think of it as the attitude display.)

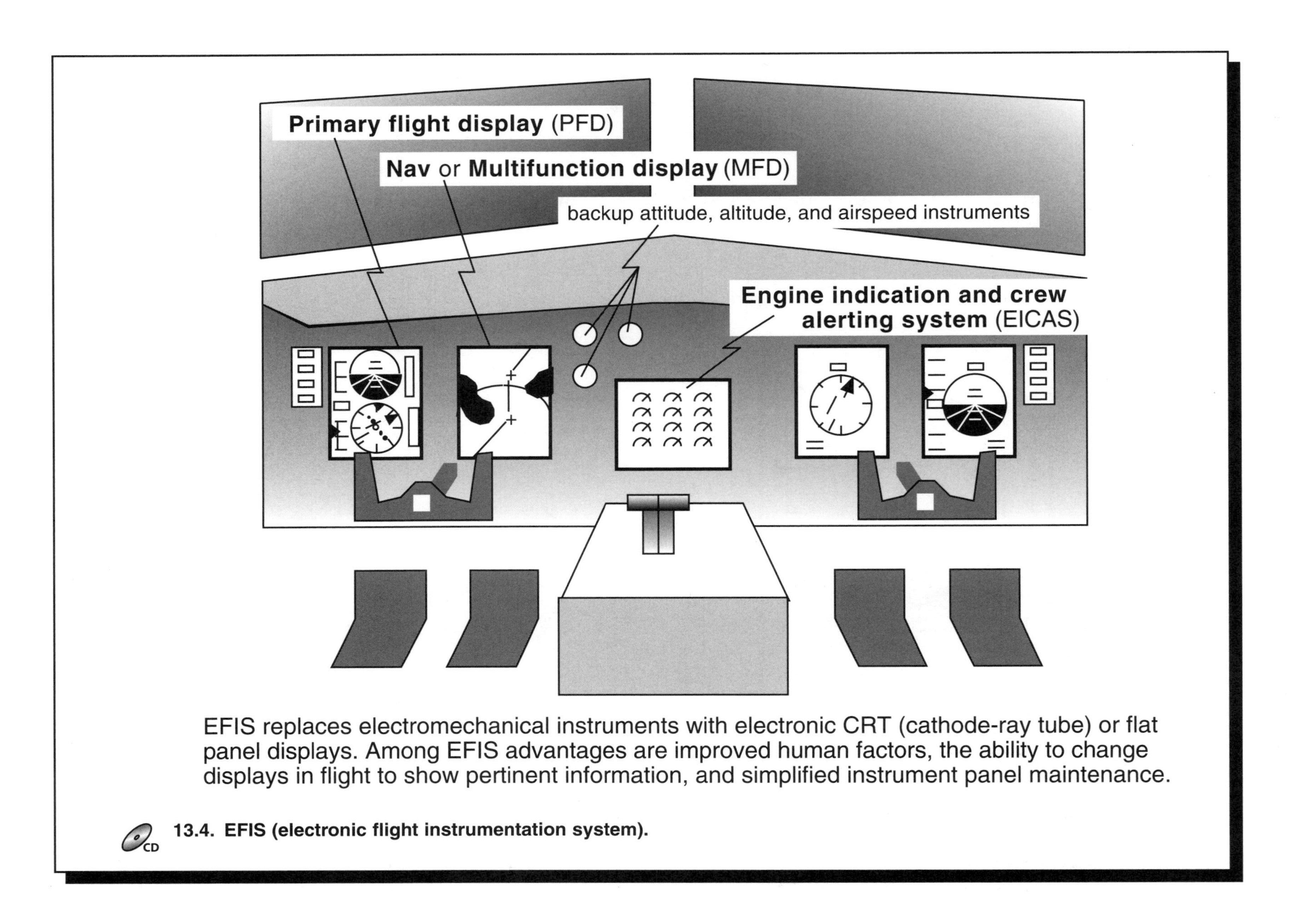

EFIS replaces electromechanical instruments with electronic CRT (cathode-ray tube) or flat panel displays. Among EFIS advantages are improved human factors, the ability to change displays in flight to show pertinent information, and simplified instrument panel maintenance.

13.4. EFIS (electronic flight instrumentation system).

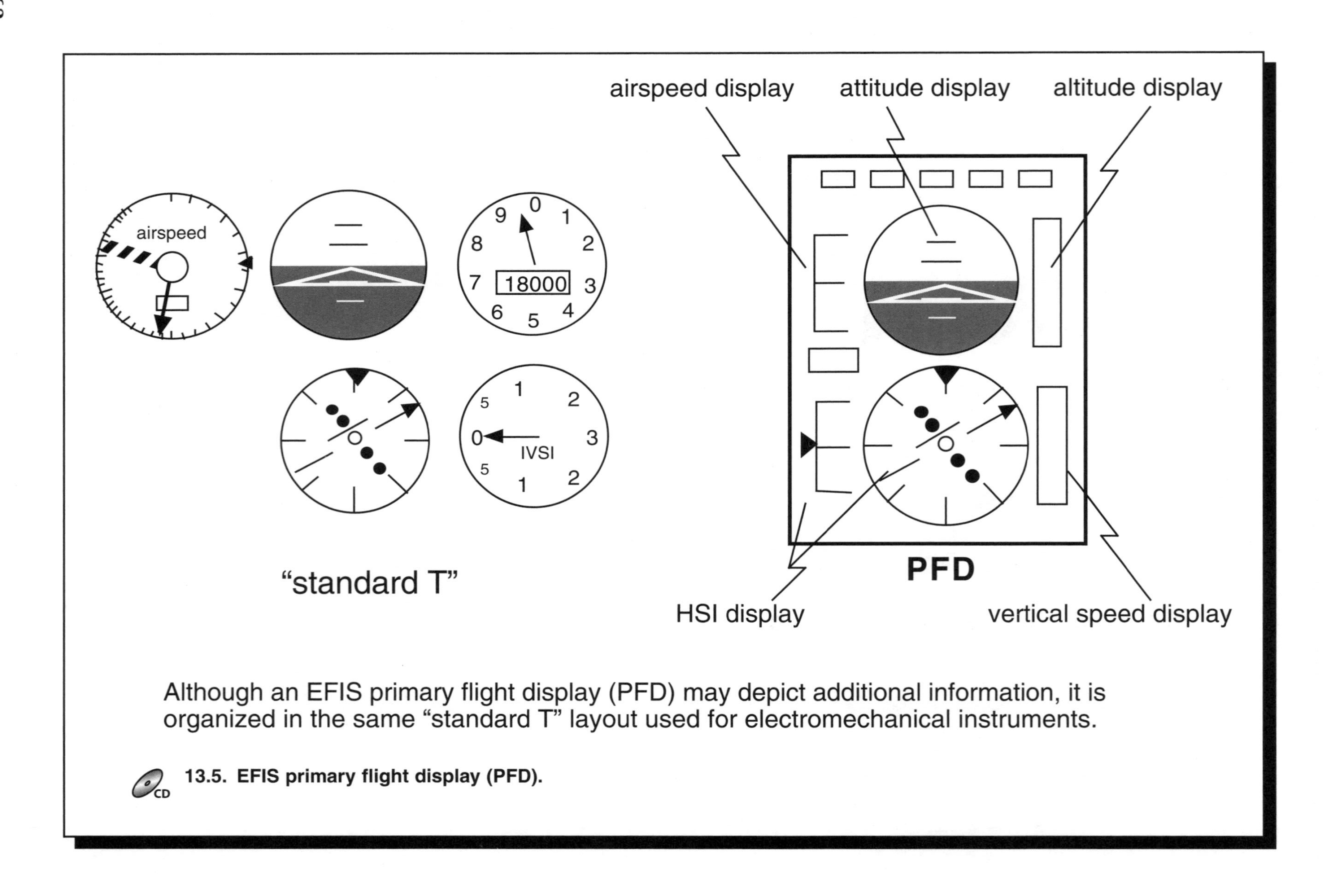

Although an EFIS primary flight display (PFD) may depict additional information, it is organized in the same "standard T" layout used for electromechanical instruments.

13.5. EFIS primary flight display (PFD).

The *nav display* or *multifunction display* (*MFD*) presents some combination of navigation, radar, TCAS (traffic alert and collision avoidance system), and other flight information. (Think of the MFD as the map display, incorporating HSI.) The most variable type of display is sometimes known as EICAS (engine instrumentation and crew alerting system) and presents engine and systems information.

EFIS offers a number of advantages over older-style analog electromechanical instruments. Perhaps the most important to pilots is flexible information display. An EFIS HSI, for example, can selectively present all sorts of useful information. A pilot can enlarge the display for easier reading. Depending on database and interfaced nav equipment, a "moving map" may then be superimposed behind the normal HSI display. Most such systems depict the locations of waypoints, navaids, airports, and even weather radar and lightning detector data. Some EICAS systems can display aircraft systems diagrams, with inoperative components identified, for pilot troubleshooting.

While a good bit of training is required for glass cockpit operations, most EFIS graphics are modeled after analog instruments and therefore are not difficult to learn. The biggest challenge is in learning to deal with (and, when appropriate, ignore) the tremendous amounts of information that can be displayed. After using EFIS a few times, most pilots wonder how they ever flew without it.

ACARS

ACARS (*aircraft communications addressing and reporting system*) is an onboard computerized communications system that provides a digital, voiceless radio "data link" between an aircraft and its company operations center. The operations center has a large computer that monitors and communicates with all aircraft in the operator's fleet.

ACARS acts as the electronic medium for all company messages, which are displayed as text on a cockpit screen, rather than being transmitted by voice in the traditional manner. A built-in cockpit printer generates hard copies of information the crew wants to save. Flight crews obtain weather, flight routings, loads, and even takeoff power settings through the system (Fig. 13.6).

The ACARS is typically linked to the aircraft's flight management computer (FMC), allowing direct uploads of flight plans. Departure and arrival times are automatically transmitted back to the company by the ACARS, as well as other messages the flight crew manually enters by keypad. ATC clearances can also be transmitted via ACARS to flight crews, reducing frequency congestion and the likelihood of errors.

Finally, ACARS can be programmed to transmit other types of flight or environmental information automatically. It is now being used, for example, to transmit winds-aloft information collected by aircraft navigation systems to ground stations for weather forecasting purposes. Several airlines have also linked their ACARS to SATCOM (satellite communications systems) for continuous contact on long overwater flights.

Keep in mind that appropriate ground equipment must be installed for pilots to use ACARS at any given airport—therefore, it is not yet available at many less-traveled destinations.

Head-Up Displays

A growing number of aircraft are equipped with a *head-up display* (*HUD*), also known as a head-up guidance system (HGS), which incorporates a special transparent plate, known as a "combining glass," mounted directly in the pilot's field of vision just inside the windshield. On the combining glass is displayed flight information and guidance cues for instrument approach and landing. Effectively, this allows the pilot to shoot an instrument approach while at the same time looking out the window for the runway (Fig. 13.7).

Originally, HUDs were developed by the military for weapons delivery. More recently, they have proven extremely useful for reduced visibility instrument approaches, especially during transition from instrument to visual flight. Many carriers have obtained FAA approval for lower than standard approach minima through use of HUD devices. Some operators are now authorized to conduct hand-flown CAT IIIa approaches using HUDs and for takeoffs with as little as 300 feet visibility.

Future versions of head-up guidance displays will integrate windshear, TCAS traffic alert warnings, and enhanced ground proximity warning systems (EGPWS). There's even talk of including an infrared sensor with the HGS units, as a visual aid. Infrared sensors detect heat sources and will allow pilots operating under poor visibility to see otherwise invisible aircraft tailpipes, runways, and even people walking around the ramp. Pretty neat stuff!

Area Navigation (RNAV)

RNAV, or *area navigation,* equipment offers navigational flexibility by allowing aircraft to proceed directly from any location to any other desired point, without having first to fly directly to or from any radio navaid. As a result, pilots with RNAV may fly directly from departure airport to destination airport or from waypoint to waypoint. A *waypoint* is a geographical location that can be described by

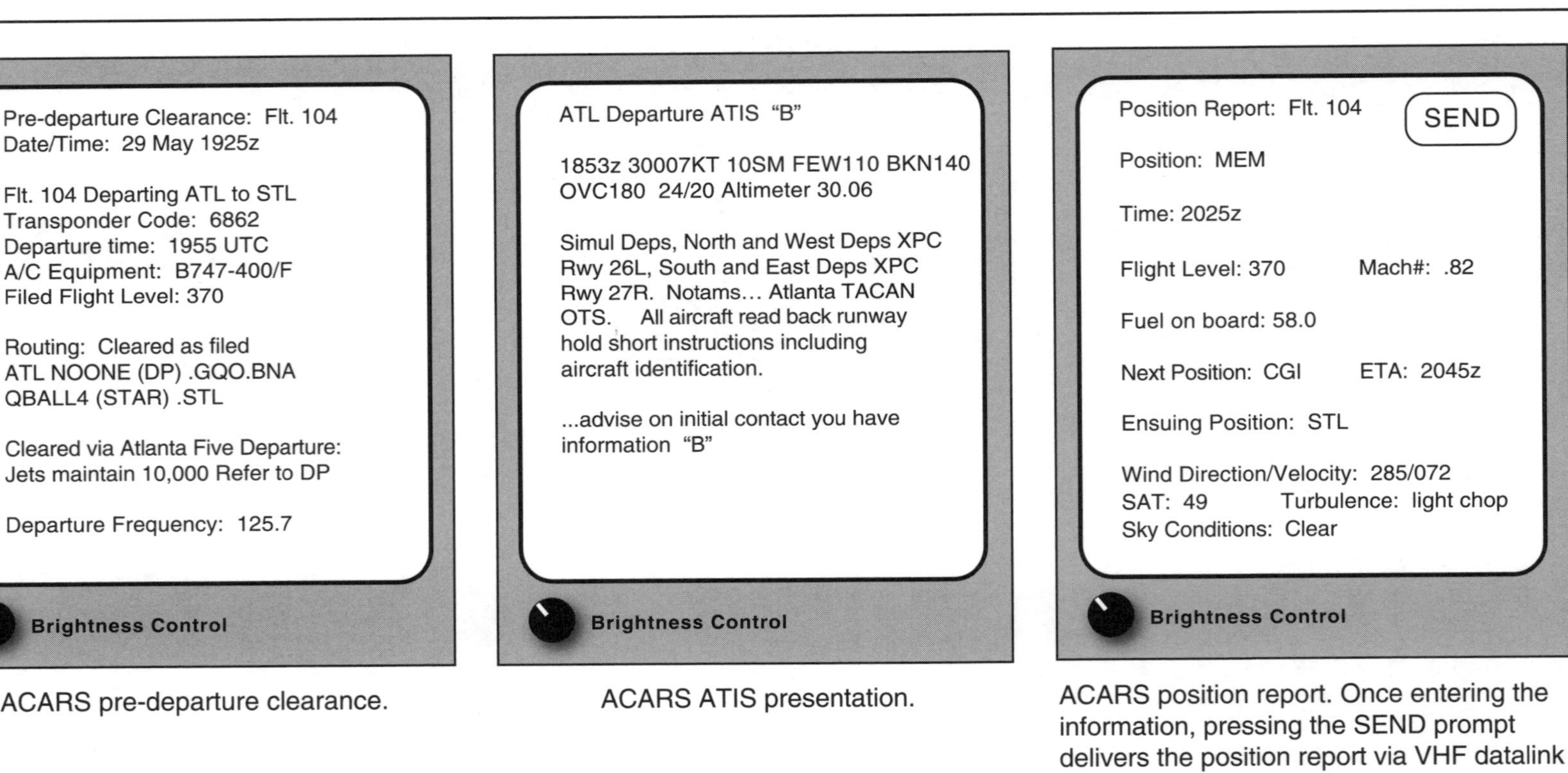

ACARS pre-departure clearance.

ACARS ATIS presentation.

ACARS position report. Once entering the information, pressing the SEND prompt delivers the position report via VHF datalink to the company operations center.

13.6. Typical ACARS communications pages.

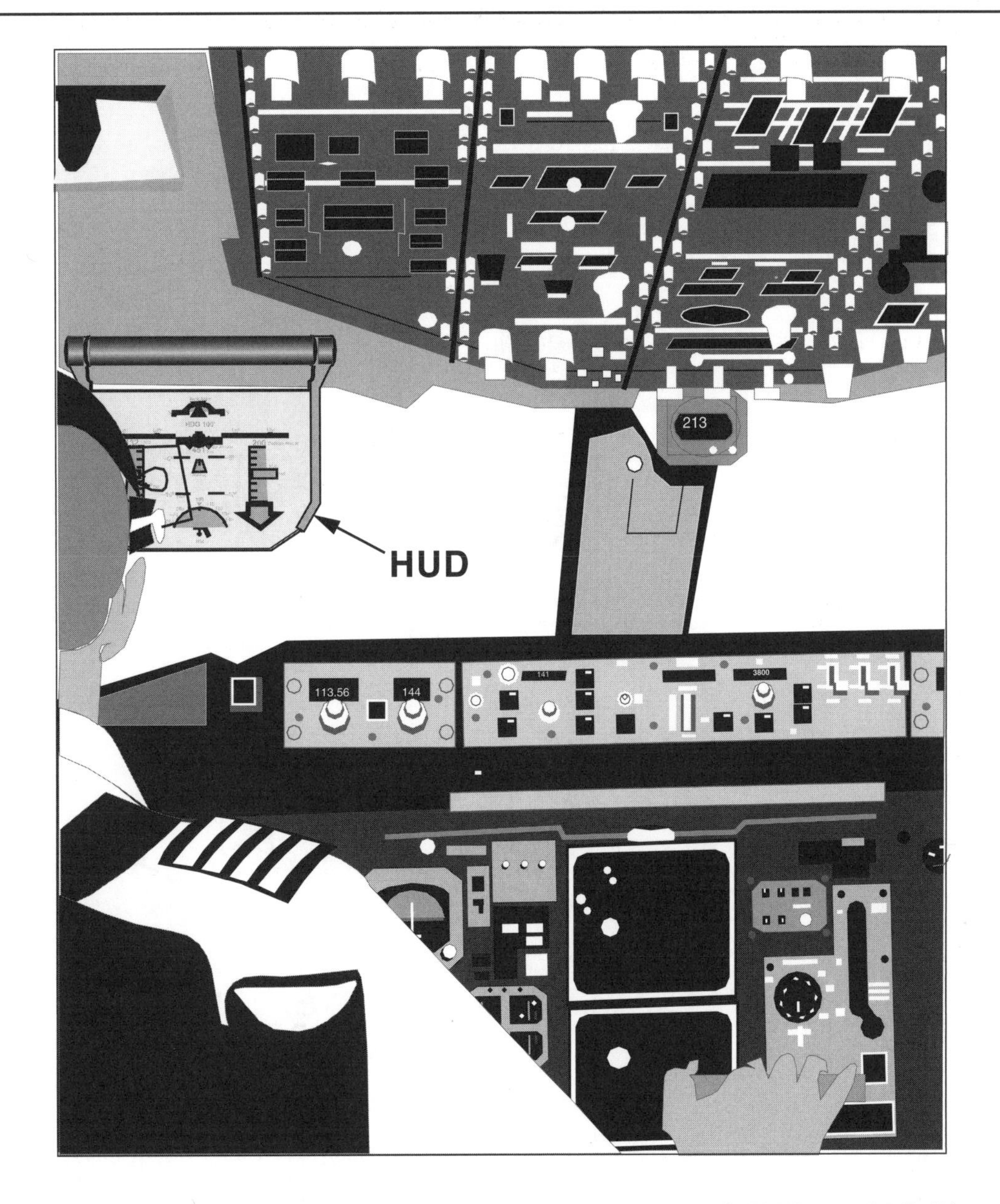

A **head-up guidance display system (HUD** or **HGS)** projects instrument data onto a transparent combining glass inside the windshield. Graphic cues show runway location before it's in sight, as well as instrument approach parameters. A HUD allows the pilot to shoot the approach while looking out the window.

13.7. HUD or HGS.

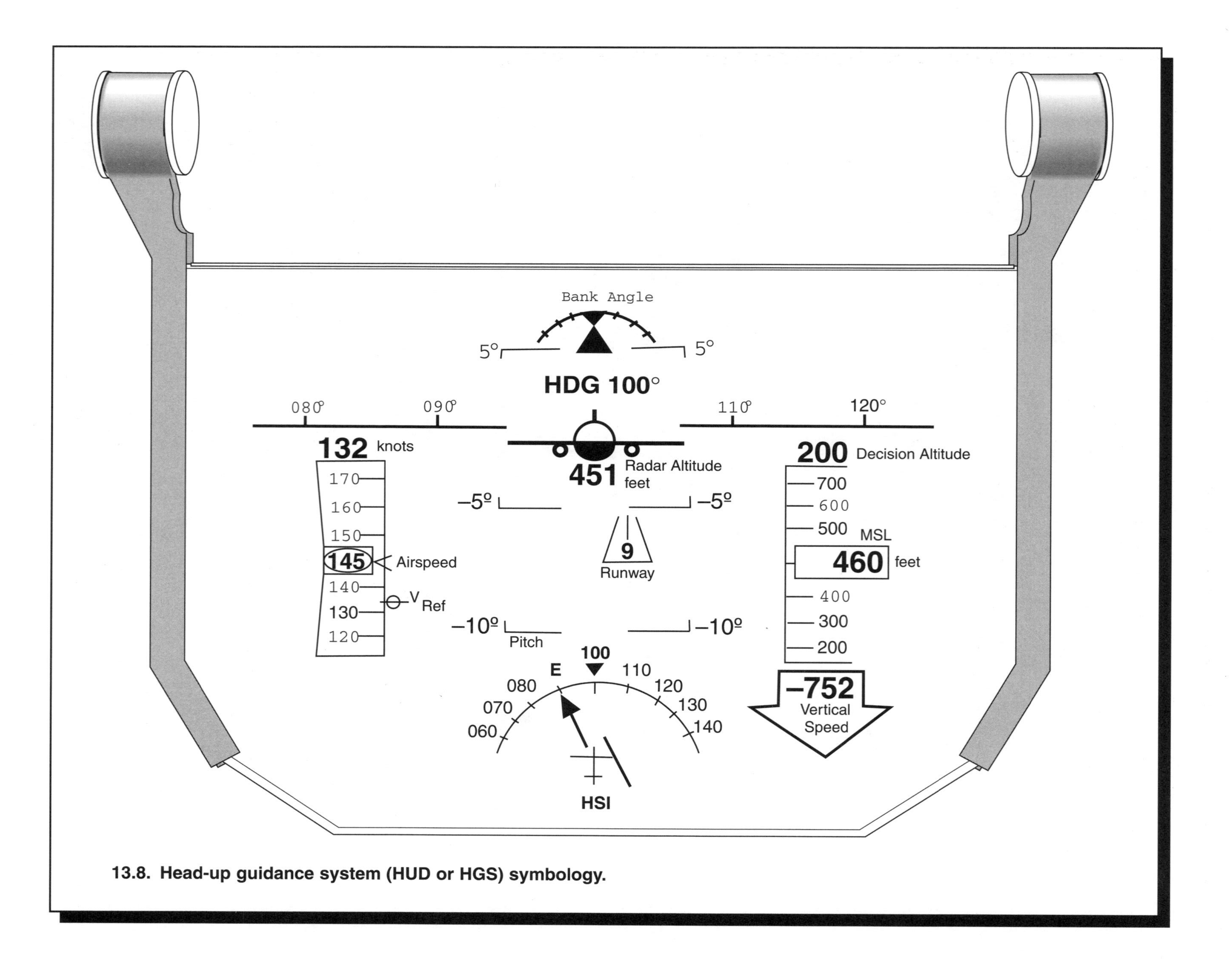

13.8. Head-up guidance system (HUD or HGS) symbology.

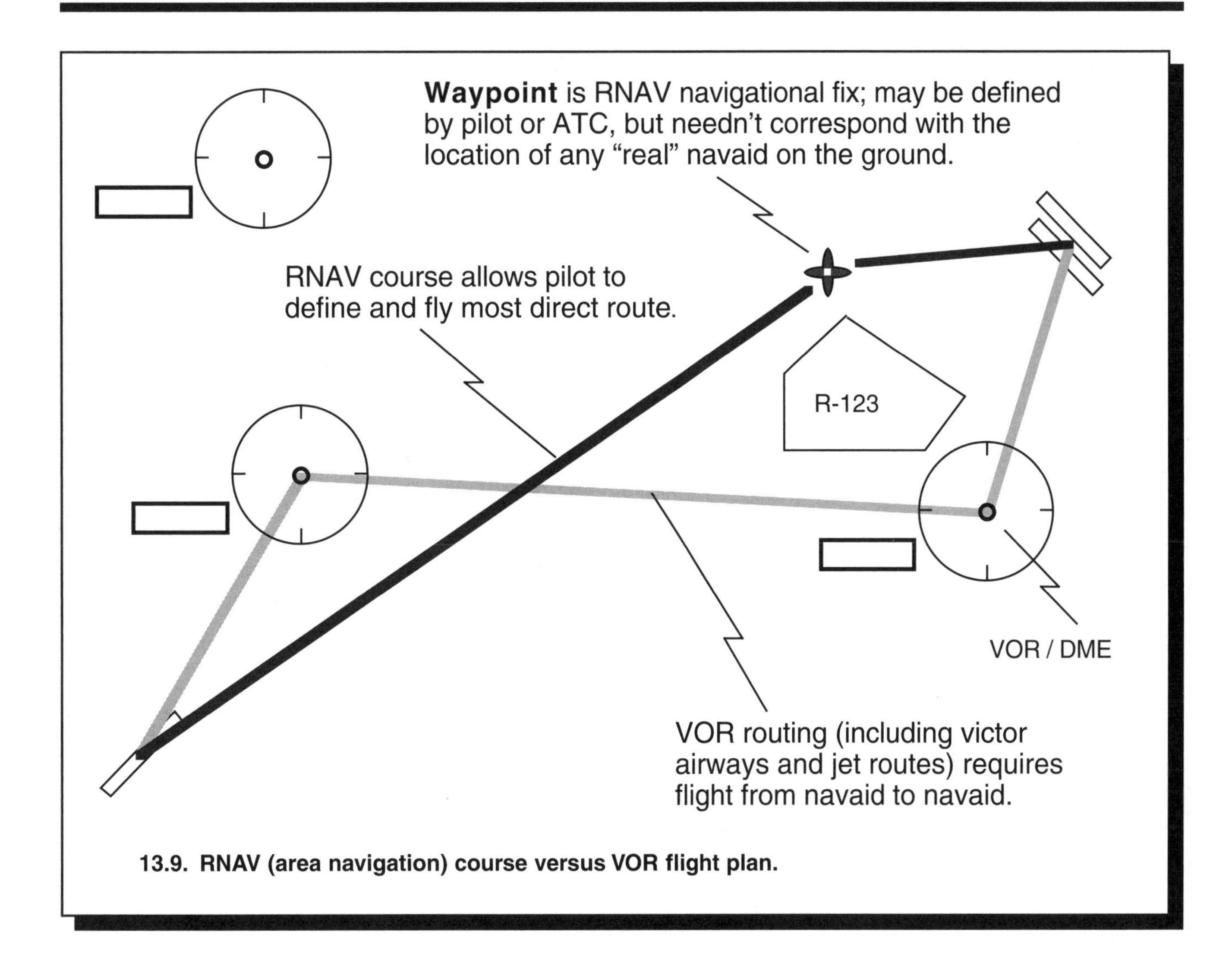

13.9. RNAV (area navigation) course versus VOR flight plan.

latitude and longitude or, when associated with a VOR station, by radial and DME distance (Fig. 13.9).

Other RNAV benefits include the construction of departure and arrival procedures from any point air traffic control managers may choose. This allows for building more efficient procedures that are tailor-made for local requirements, with the objective of relieving traffic congestion. RNAV also makes it possible to develop and certify instrument approach procedures at airports without local radio navaids.

RNAV offers the added benefit of relieving enroute traffic congestion in many areas, because it effectively makes more airspace available for navigation.

Depending on the installed system, an onboard RNAV computer may receive external or internal navigational inputs from a variety of sources. Following are brief descriptions of the different major types of RNAV systems.

VOR/DME-Based RNAV

With VOR/DME RNAV systems, waypoints are selected by radial and DME (distance measuring equipment) from a "parent" VOR. Waypoints may be placed anywhere, provided they are within the operational service volume of the parent VOR. It's as if you put a VOR station on a flatbed truck and moved it to a specific radial and distance from its old location that would be more convenient for your particular flight plan. Some RNAV computers use multiple VOR and DME inputs to achieve seamless coverage through areas where some signals may be interrupted.

LORAN

LORAN, or *long-range area navigation,* receives navigational signals from a number of ground-based,

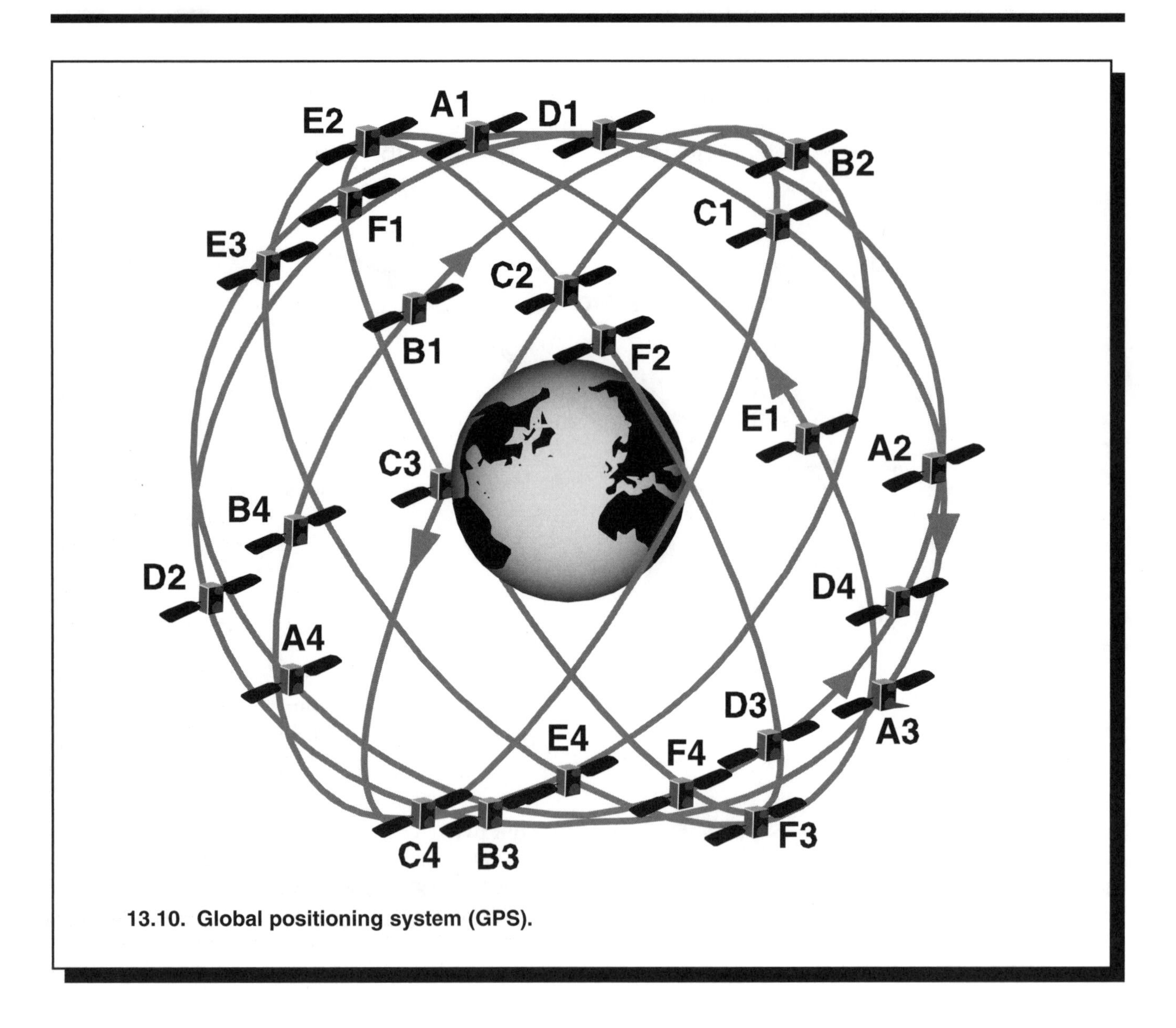

13.10. Global positioning system (GPS).

low-frequency, long-range transmitting stations separated by thousands of miles. A computer in the aircraft's system locates position by comparing time differences between synchronized pulses transmitted from the LORAN ground stations. This system was originally developed for maritime applications and was expanded for aviation use. Although still installed in many aircraft, LORAN is of decreasing use for primary navigation; chances are good that the system will be decommissioned or relegated to backup status within the next several years.

Global Positioning System (GPS)

GPS (*global positioning system*) RNAV systems receive navigational inputs from a space-based constellation of twenty-four geostationary satellites. GPS is very accurate and particularly valuable since ground-based nav stations are not required for its operation, except as projected for precision approaches (Fig. 13.10).

The airplane's GPS computer determines aircraft location by timing radio signals received from four or more GPS satellites, calculating the aircraft's position and velocity three-dimensionally by comparing satellite positions and times of transmission, and synchronizing that information with an internal clock. Since precise timing of transmissions is key to accurate GPS navigation, GPS satellites incorporate onboard atomic clocks and broadcast time information, along with other data used by aircraft GPS receivers, to calculate position.

Although currently GPS navigation is used only for enroute navigation and nonprecision approaches, long-term prospects are good for GPS to eventually replace most other aircraft navigational systems. For more precise GPS navigation, a combined ground-based/space-based GPS system is under development. These "differential GPS" systems will be used to improve GPS accuracy at specific geographic locations, allowing precision approaches to be conducted using GPS navigation in place of traditional ILS approach systems.

Inertial Navigation System (INS)

Another type of RNAV system is completely internal to each aircraft. The *inertial navigation system* (*INS,* or "inertial nav") receives no navigational signals from outside the aircraft, whatsoever. Instead, an internal computer determines location, ground speed, heading, and altitude through use of a system of acceleration sensors (accelerometers). In other words, INS determines position by sensing changes in the aircraft's movement. As technology has improved, INS systems have become extremely accurate. Because INS is a completely internal system, it has some inherent advantages over other navigational systems. INS provides accuracy without being subject to interference from weather, electronic jamming, or failure of any ground-based navigational stations.

Using RNAV

Regardless of which RNAV system is used, from a pilot's standpoint they all operate similarly. The pilot inputs the planned route of flight into the RNAV computer, which then collects signals from VOR/DME stations, LORAN or satellite transmitters, or internal accelerometers. In some cases, inputs from several of these systems may be processed and cross-checked through a single *integrated nav system.* After processing, the nav computer drives a CDI in the cockpit, providing pilots with an indication of position relative to desired course and displaying distance to the next waypoint.

Some advanced RNAV systems also have VNAV (*vertical navigation*) capability. Pilots use VNAV to adjust flight path to meet crossing restrictions, to provide standardized glide paths for nonprecision approaches to landing, and for preprogrammed "profile descents." (Profile descents are instrument procedures used for transition from cruise flight to approach. See "IFR Operations in Turbine Aircraft" in Chapter 15.) Many RNAV installations also calculate and display wind direction and wind speed in flight.

Everyone knows that the shortest distance between two points is a straight line—right? Actually, that's not entirely true. The shortest distance between two points on the earth's surface is "great circle distance." Long-range RNAV systems compute their courses as great circle routes, optimizing aircraft efficiency. An interesting caution, in this regard, applies to use of RNAV when flying published airways. Both high- and low-altitude airways are straight-line routes. On long flight legs, pilots using RNAV gear to fly airways must carefully monitor position, as a great circle RNAV course may place the airplane outside of the airway limits.

RNAV systems give pilots the ability to take the most direct course to their destinations, saving fuel and time. While navigation technology may change, direct routing is the way of the future, and that's RNAV.

Latitude and Longitude

The key to understanding GPS and INS area navigation systems is learning to think in terms of Earth's latitude and longitude. Familiarity with latitude/longitude is necessary for two reasons. First, it will help you to determine position along your planned course by reference to a map. Second, it will help prevent mistakes when you are programming waypoint coordinates into the navigation computer's control display unit.

The globe is divided into a series of imaginary latitude and longitude lines, with lines of latitude running east and west and lines of longitude running north and south (see Fig. 13.11).

Latitude lines are used to determine position north or south of the equator, with the equator defined as the 0-degree reference. Parallel latitude lines north of the equator number from 0 degrees to 90 degrees north, with 90 degrees north designating the north pole. Lines of latitude south of the equator are accordingly numbered from 0 degrees to 90 degrees south, with 90 degrees south being the south pole.

Lines of longitude define position east and west and extend from true north pole to true south pole. Just as the equator is the 0-degree reference for latitude, there is also a 0-degree longitudinal reference line known as the prime meridian. This line passes from the north pole through Greenwich, England and then down to the south pole. A total of 360 degrees of longitude surround the globe, from 0 degrees to 180 degrees west and 0 degrees to 180 degrees east, with the 180-degree line crossing the opposite side of the globe from the prime meridian through the Pacific Ocean. (Note that the coordinated universal time used by aviators reflects time of day at the prime meridian, while the international dateline runs 180 degrees opposite—with geopolitical adjustments—through the Pacific Ocean.)

You've probably noticed the latitude and longitude lines printed on your aeronautical charts. Each degree of latitude and longitude is divided into sixty minutes, and each minute is divided into sixty seconds. By knowing the latitude/longitude coordinates you can precisely describe the position of any spot on Earth. For example Los Angeles International Airport's coordinates are north 33° 56′6″ latitude and west

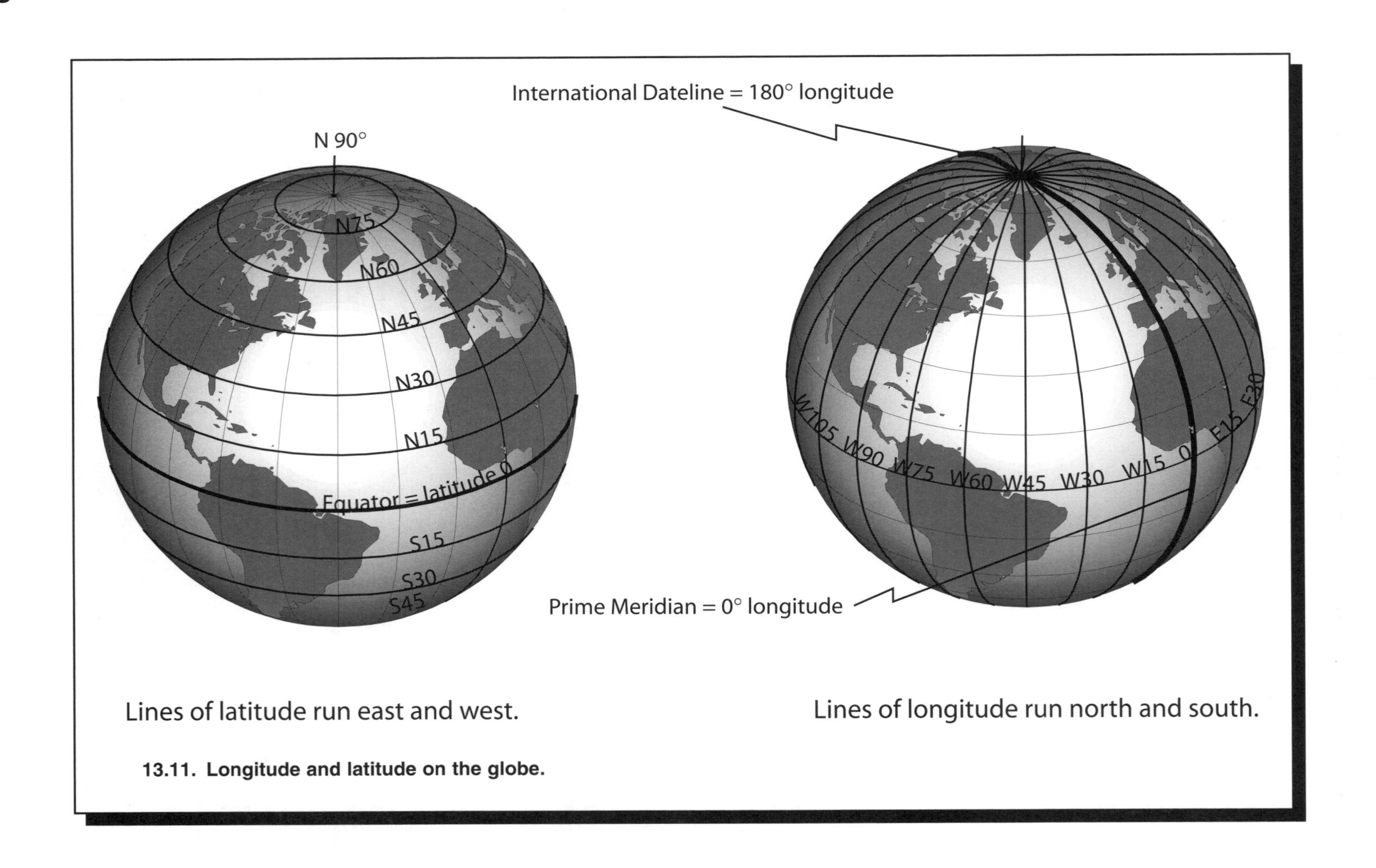

13.11. Longitude and latitude on the globe.

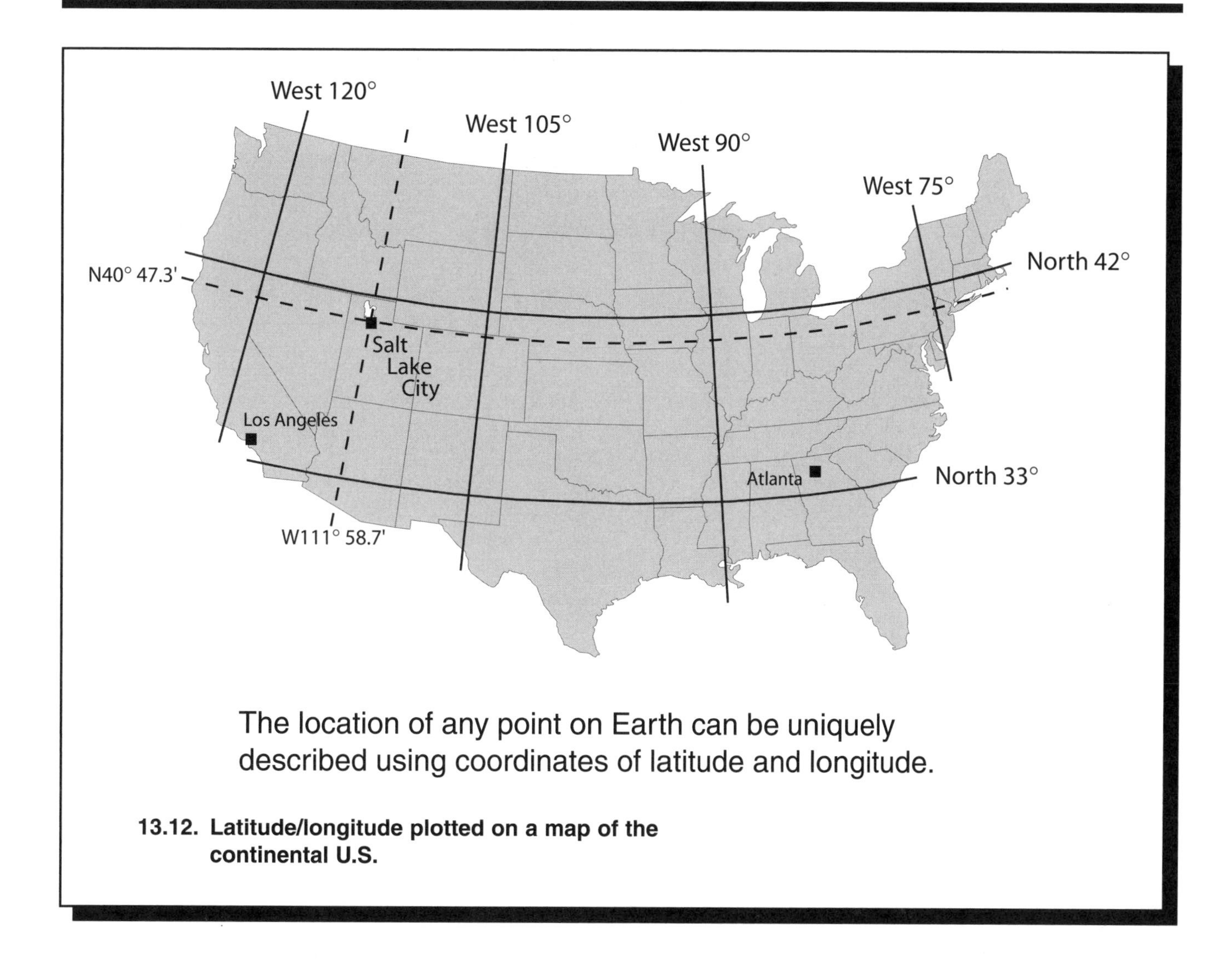

13.12. Latitude/longitude plotted on a map of the continental U.S.

118° 24′4″ longitude, and Atlanta Hartsfield Airport is located at N33° 38′4″ latitude and W84° 25′6″ (Fig. 13.12).

Flight Management System (FMS)

Virtually all new generation jets and larger turboprop aircraft produced today use a sophisticated system of computers to plan and automatically control aircraft path from the application of takeoff thrust to the landing rollout. This system of computers and associated aircraft control devices is called a *flight management system* or *FMS.*

Until recently, most piston pilots initially transitioned to light turboprop aircraft not equipped with flight management systems; instead, they had basic instrumentation and means of navigating similar to what they'd used in the past. However, with the widespread introduction of advanced turboprops and regional jets (RJs), chances are good that today's transitioning pilot will move directly into a turbine aircraft equipped with FMS. In this section we'll examine what FMS is designed to do, its basic components, and operation of a generic system.

Historically, FMS systems have not been particularly simple or intuitive to learn, which has often caused consternation among pilots new to them. It doesn't help that practically every airplane has an FMS unique to just that manufacturer or even the particular model of aircraft, and that all of these varying systems incorporate different keypads and display screens, different computers, different databases, and different sources of navigational information. Even within one airline, or corporate fleet, pilots may find very different software versions onboard otherwise identical FMS units.

The good news is that each new generation of FMS gets easier to use, and given today's much improved training programs and the average individual's better computer skills, most pilots now find learning FMS straightforward.

Basic FMS Components and Operating Principles

A flight management system uses computer technology and linked flight control devices, such as autopilot, flight director, and auto-throttles, to aid the flight crew in flying the aircraft as efficiently as possible. FMS does this by supplying data and automatically managing aircraft navigation and performance to achieve the most optimized flight path possible (Fig. 13.13).

The brains of an aircraft's FMS is a *flight management computer* (*FMC*), which processes inputs including navigational information, aircraft performance parameters, and pilot programming (see Fig. 13.14). Based on that information, the FMC directs the FMS's autopilot/flight director system and engine auto-throttles to automatically manage the aircraft's lateral route, or lateral flight path (LNAV), and vertical flight path (VNAV). In short, the FMC allows pilots to program an entire route and flight profile, including any altitude or airspeed restrictions, all before the airplane ever takes off.

Inside the flight management computer is stored a navigation database containing latitude and longitude coordinates for virtually every important navigational landmark around the world. The FMS uses this FMC navigation database to continuously update the aircraft's position. The FMC also provides for automatic navaid selection and tuning, relieving the flight crew of manually entering that information for use by the autopilot or flight director to navigate the aircraft.

The FMC navigational database includes coordinates for charted waypoints, approach fixes, airports, VOR and NDB navaids (frequencies too), elevations for airports, runways and radio nav stations, and even important airport landmarks such as airport gates and hangars!

Increasingly, detailed terrain information is also being incorporated into such systems for purposes of maintaining ground separation (see "Ground Proximity Warning Systems and Enhanced Ground Proximity Warning Systems" in Chapter 14). FMC terrain databases are becoming so sophisticated that pilots will soon be able to watch their real-time progress over ground terrain on a computer-generated cockpit display, much like what's done today on recreational flight simulator programs.

Along with making navigation even easier, such technology should increase pilot situational awareness, hopefully reducing the threat of controlled flight into terrain (CFIT) accidents.

The flight crew interacts with the FMS computer through a *control display unit* (*CDU*) (see Fig. 13.15). The CDU is basically a minicomputer terminal that allows pilots to input navigation and performance data, select navigation and performance options, and display navigation and performance information. The CDU simultaneously displays system status, version of operating system, and information about any malfunctions.

FMS guidance information is displayed for the flight crew in various ways. On glass cockpit aircraft having EFIS screens, electronically generated "moving maps" display aircraft position. Flight director bars on the primary flight display (PFD) present LNAV and VNAV flight guidance (see Fig. 13.16).

On most aircraft the FMS also displays "command markers"—FMC-controlled airspeed command bugs (pointers) and EPR or N_1 markers that assist pilots in efficiently hand-flying the airplane by indicating desired speed and thrust targets directly on the instruments. Pilots simply adjust controls to match the markers (see Fig. 13.17).

On nonglass aircraft the FMS presents flight guidance information to pilots using flight director command bars on the attitude indicator and analog HSI and DME displays for aircraft position.

Basic Operation of a Generic FMS

Let's consider basic operation of a generic FMS. Even with all the different types of FMS in use, they do share related operating principles. Here we will attempt to cover basic characteristics common to most.

The first step in preparing an FMS for flight is to perform an "FMS preflight." Like an aircraft preflight, where pilots check to make sure the aircraft is airworthy, an FMS preflight involves the flight crew ensuring that the flight management system is fully operational.

To do this, pilots select a "preflight contents page" on the CDU. This screen presents an index of all pages requiring data entry by the flight crew, prior to every flight. FMC pages are presented in a logical sequence, starting from preflight setup through takeoff performance, route, cruise, descent planning, and finally approach and landing data at the destination. On most CDUs you don't need to memorize or search for the next page requiring data input—at the bottom of the display a simple prompt directs you to the next page requiring data (see Fig. 13.18).

FMS Identification Page

The first page to review is what we'll call the aircraft "identification page," where the crew checks that the proper aircraft database is installed. If you're flying an Airbus A-320 with CFM56 engines rated at 25,000 pounds thrust, that's what you should see displayed—if the wrong aircraft or engine model database is installed, all FMS performance computations will be wrong. Also included on this page are valid dates for the installed navigation database. Just like your aeronautical charts, if the FMS database is out of date navigation information could be wrong and should not be

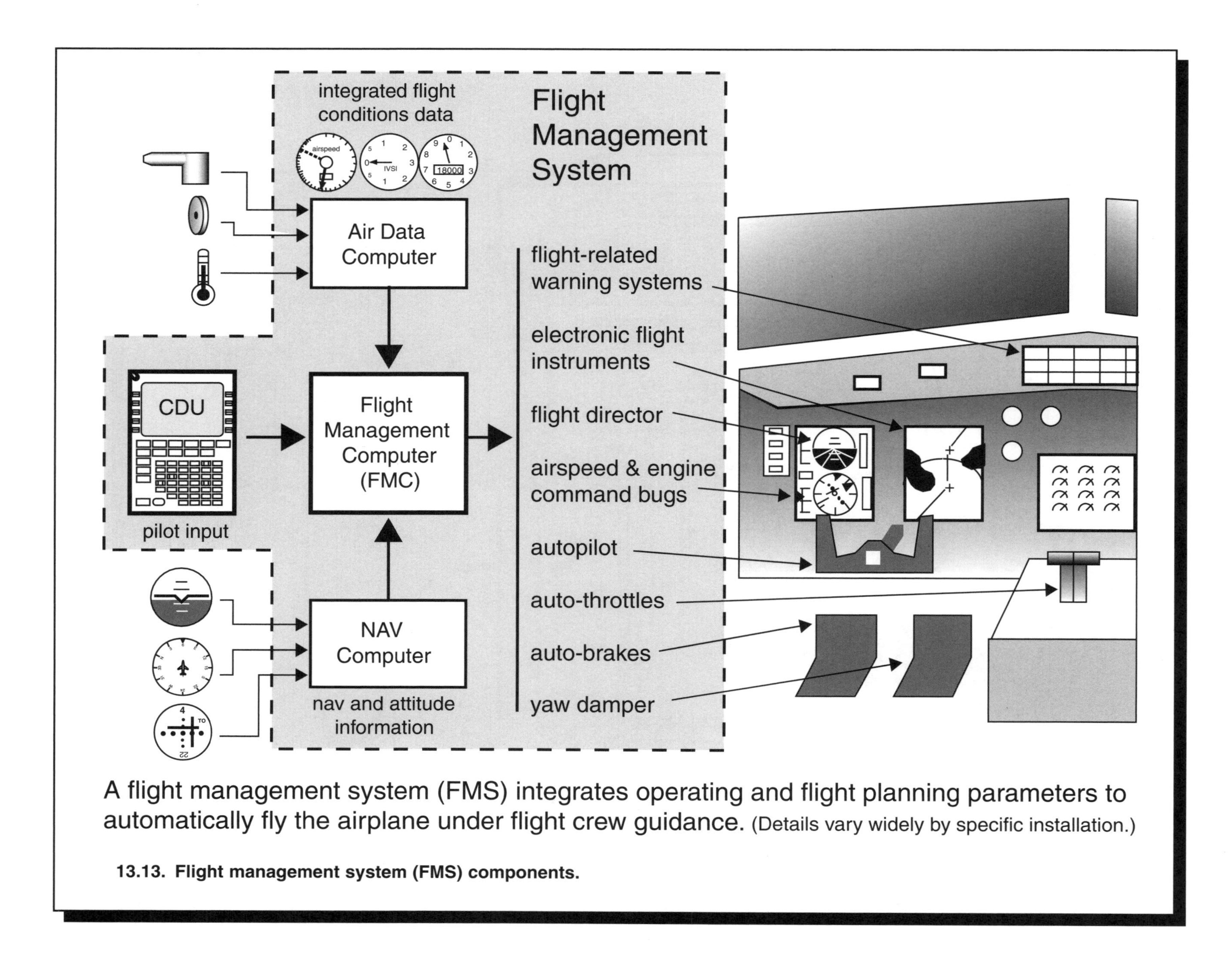

A flight management system (FMS) integrates operating and flight planning parameters to automatically fly the airplane under flight crew guidance. (Details vary widely by specific installation.)

13.13. Flight management system (FMS) components.

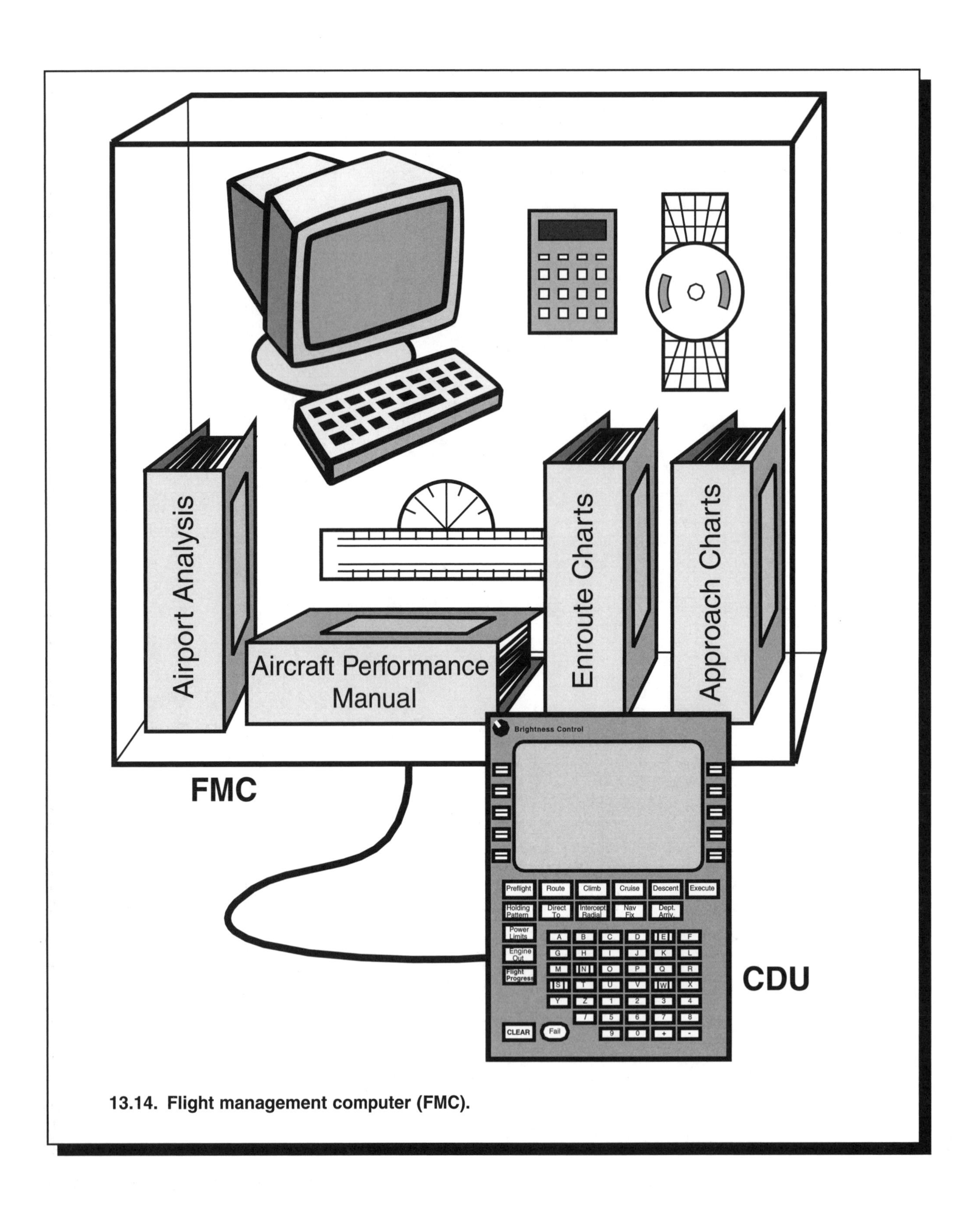

13.14. Flight management computer (FMC).

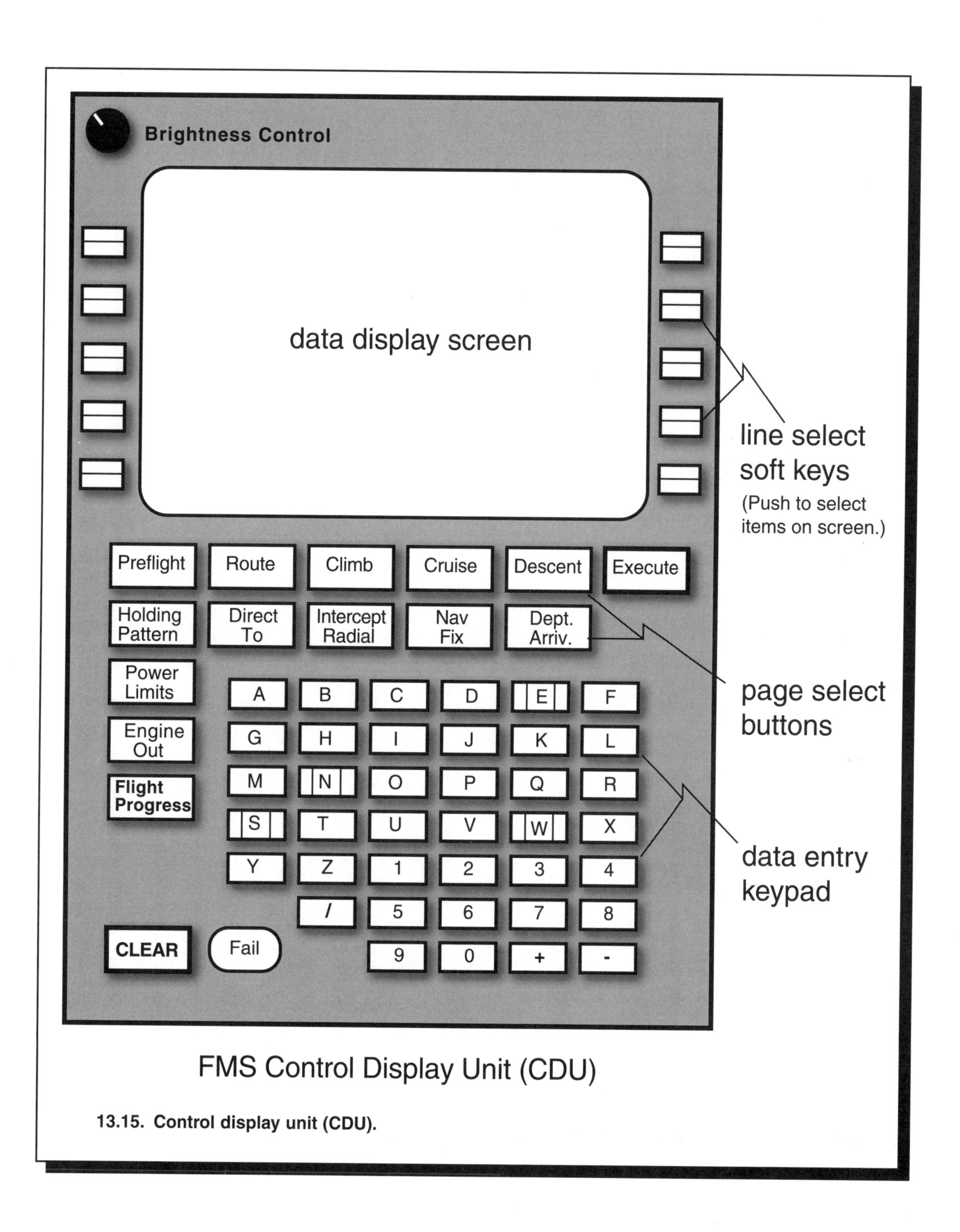

13.15. Control display unit (CDU).

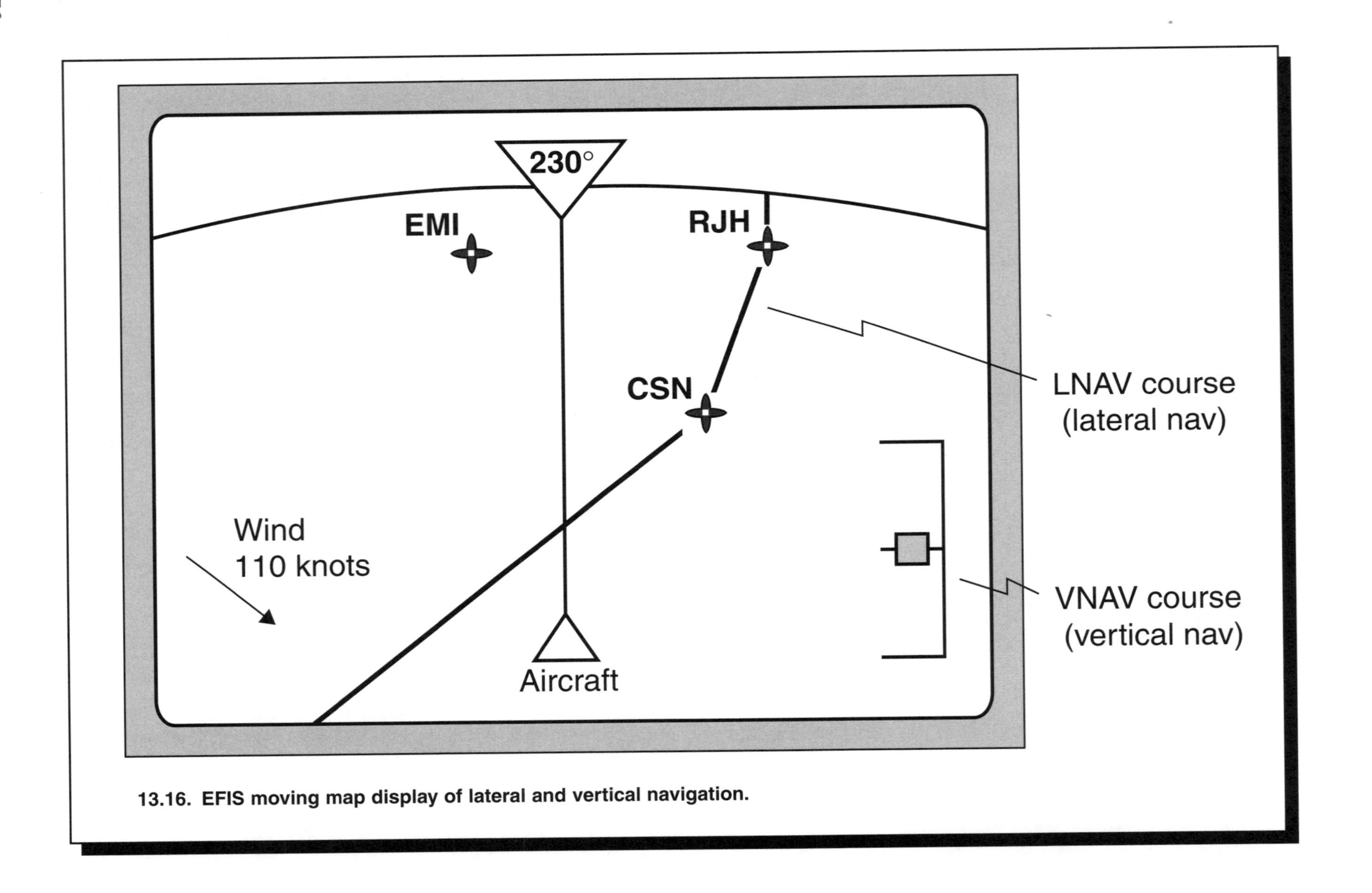

13.16. EFIS moving map display of lateral and vertical navigation.

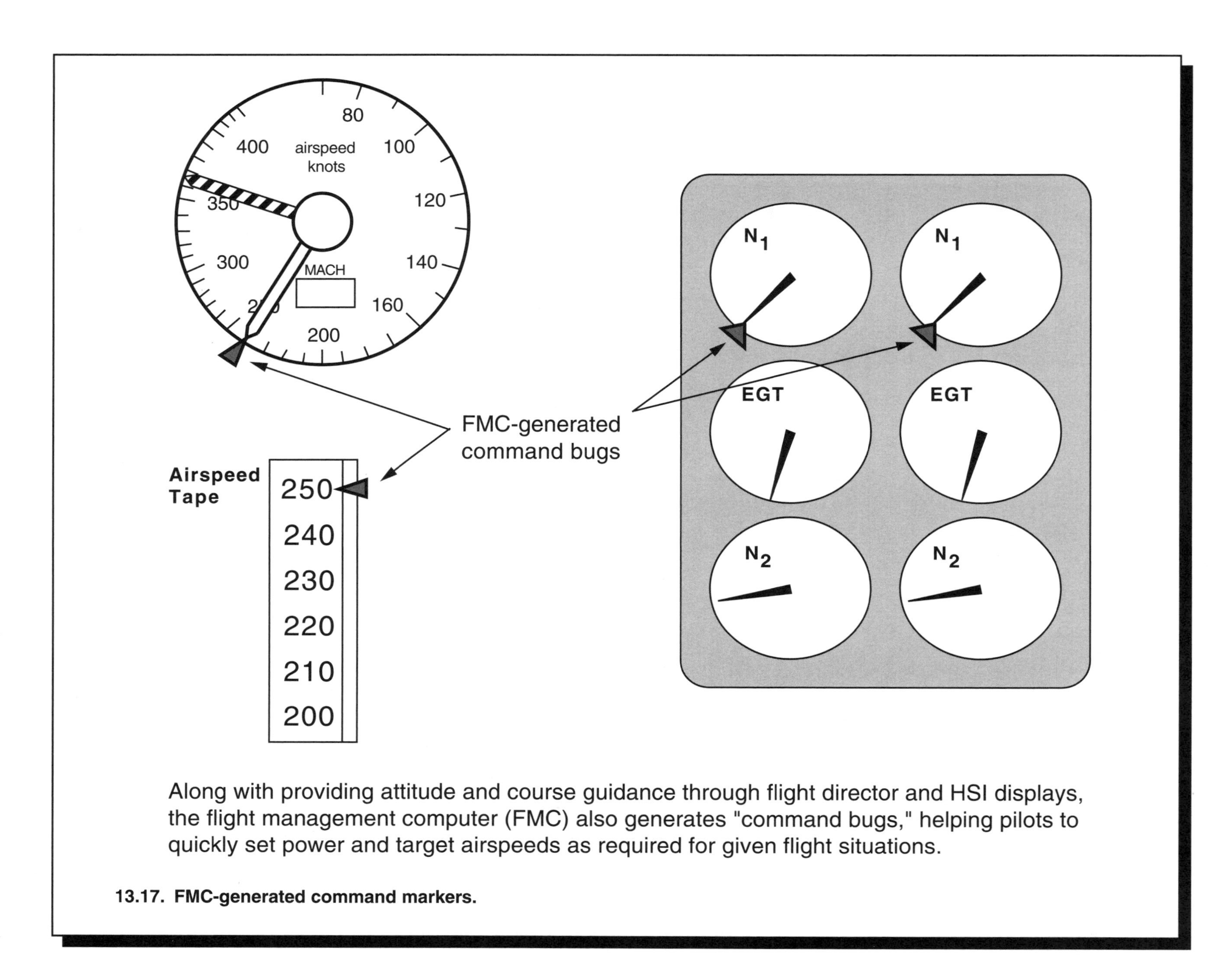

Along with providing attitude and course guidance through flight director and HSI displays, the flight management computer (FMC) also generates "command bugs," helping pilots to quickly set power and target airspeeds as required for given flight situations.

13.17. FMC-generated command markers.

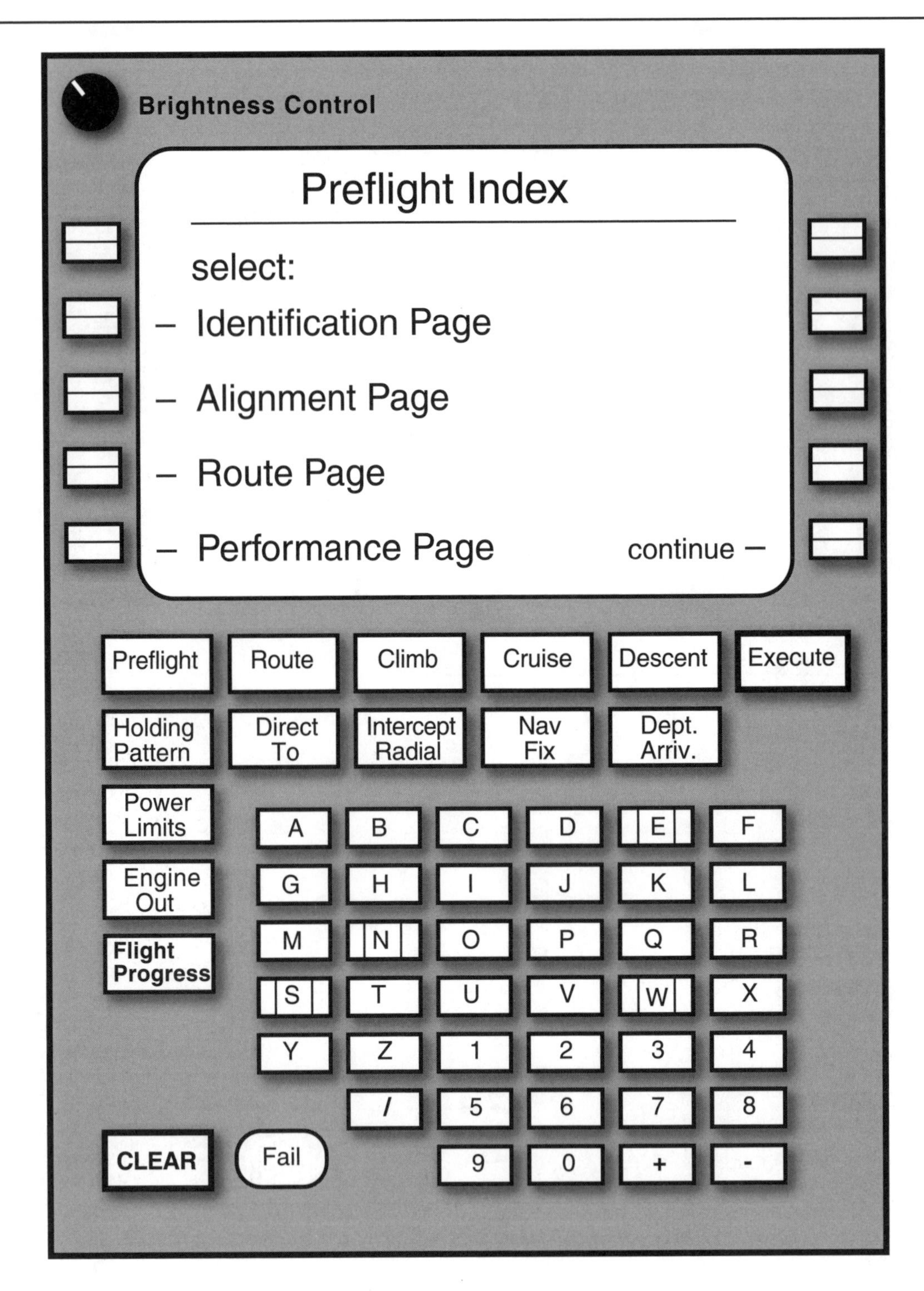

13.18. FMS preflight contents page.

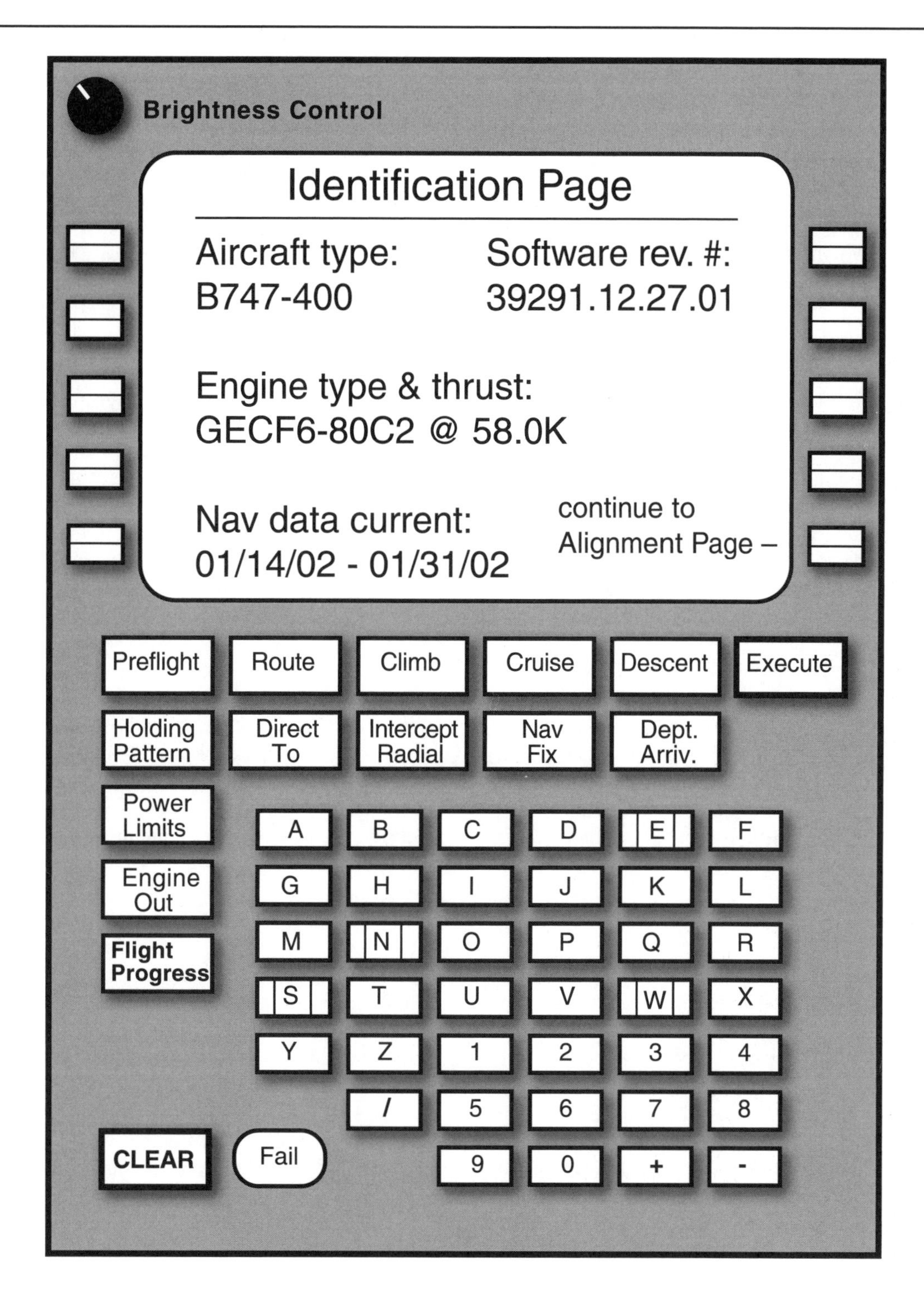

13.19. FMS identification page.

used for primary navigation. After reviewing the identification page to make sure all information is correct, you will be prompted to proceed to the alignment or initialization page (see Fig. 13.19).

Alignment Page

Before the FMS can begin navigating, you must first tell it where the aircraft is located. This process is called navigation alignment or initialization. Here the flight crew simply enters latitude/longitude coordinates for their current location (i.e., airport, gate, ramp, hangar, etc.) into the CDU. These coordinates are published in airway manuals and aeronautical charts and may sometimes even be found painted on the ramp, a hangar wall, or on a sign near the gate area (Fig. 13.20).

Once present-position coordinates are entered, some systems are ready to navigate; however, systems using inertial navigation system (INS) guidance may take more than ten minutes to get the internal gyros up to speed. With INS-based FMS navigation systems, any aircraft movement prior to completion of the alignment process will disable the navigation capability for the entire system.

Newer FMS installations rely solely on GPS information for determining aircraft present position. These systems are ready to go within minutes of pilots entering coordinates of where the aircraft is parked. Note that GPS navigational systems, when integrated with an FMS, are also commonly called global nav systems (GNS).

Once initial aircraft position has been defined, there will be a prompt to proceed to the flight planning route page.

FMS Route Page

The entire route for a planned flight may be entered in detail on the CDU's route page (see Fig. 13.21). Victor airways, jet routes, direct waypoint-to-waypoint routes, instrument departure procedures, arrival routes, and even specific approach and missed approach procedures may all be entered here. Once the route is entered, the FMC itself will check for completeness, assessing your flight plan to see that it travels in a logical sequence from departure airport to destination airport. The FMC will also ensure continuity along the entire route; that is, it will make sure that there are no gaps. Finally, the FMC will question any route information it cannot recognize from its database. Once route information has been entered, the flight crew will be prompted to go to the performance page.

FMS Performance Page

Once the FMS route page is complete, a performance page allows input of flight details, including specific airplane information, departure airport temperature, and top-of-climb wind data that is necessary for FMC performance calculations needed to enable the VNAV capability. The aircraft's gross weight, fuel load, planned cruise altitude, preferred climb and cruise airspeeds are also entered on the performance page. The flight crew can even input such details as climb power settings required to meet noise abatement or fuel conservation requirements. Finally, the FMC will calculate takeoff "V-speeds" and on most flight management systems even display command bugs (command markers) on the engine and airspeed gauges for the flight crew (see Fig. 13.22). Talk about a work-saver!

Once all data is entered, the FMC will display a message confirming that all required preflight data has been entered. The FMS is now ready for flight.

After Takeoff: FMS Flight Legs Page

When the aircraft begins moving, the FMS navigation system will automatically and continuously update the aircraft's position and continually direct the controls (or the pilots via flight director) to follow the planned course and profiles.

To navigate, flight management systems use some combination of INS or GPS inputs and VOR radial and DME information. During flight, the FMC automatically tunes in nearby VORs to update the aircraft's position. (This use of multiple sources to fix position is sometimes known as "integrated nav.") Newer FMS navigation systems rely solely on GPS to determine aircraft position and have no need to continuously monitor and "auto-tune" VOR stations during flight.

Once in flight, whatever LNAV and VNAV route is stored in the FMS during preflight is the path the aircraft will fly. To accommodate changes enroute, the FMC offers a duplicate route page we'll call the "flight legs page," where pilots can alter previously entered settings to meet routing or altitude changes issued by ATC, like altitude crossing and airspeed restrictions.

The FMC's VNAV capability also allows flight crews to make enroute performance computations such as optimum speeds, projected fuel burn, and recommended cruise altitudes, thus aiding flight planning decisions. One particularly popular FMC feature among flight crews is the ability to make altitude and airspeed computations quickly and accurately.

Upon reaching the destination, FMS can fly the entire transition from cruise to landing, including profile descents, instrument approaches, and in many cases, touchdown and even braking.

As you can see, flight crews depend on FMS through all phases of flight. Using high-powered computers to manage navigation and atmospheric and fuel flow data, the FMS allows pilots to fly their aircraft as efficiently as possible.

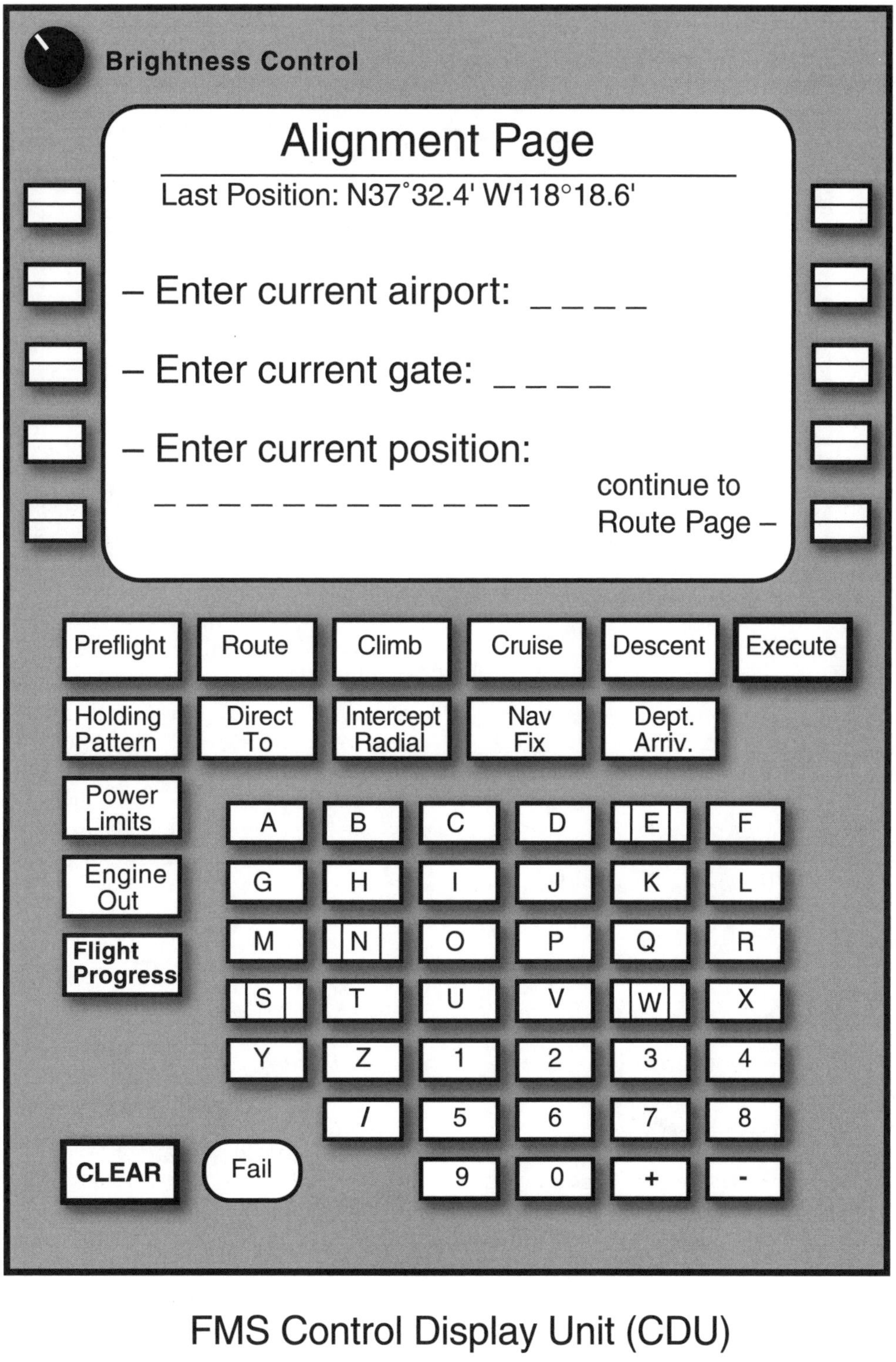

13.20. FMS alignment page.

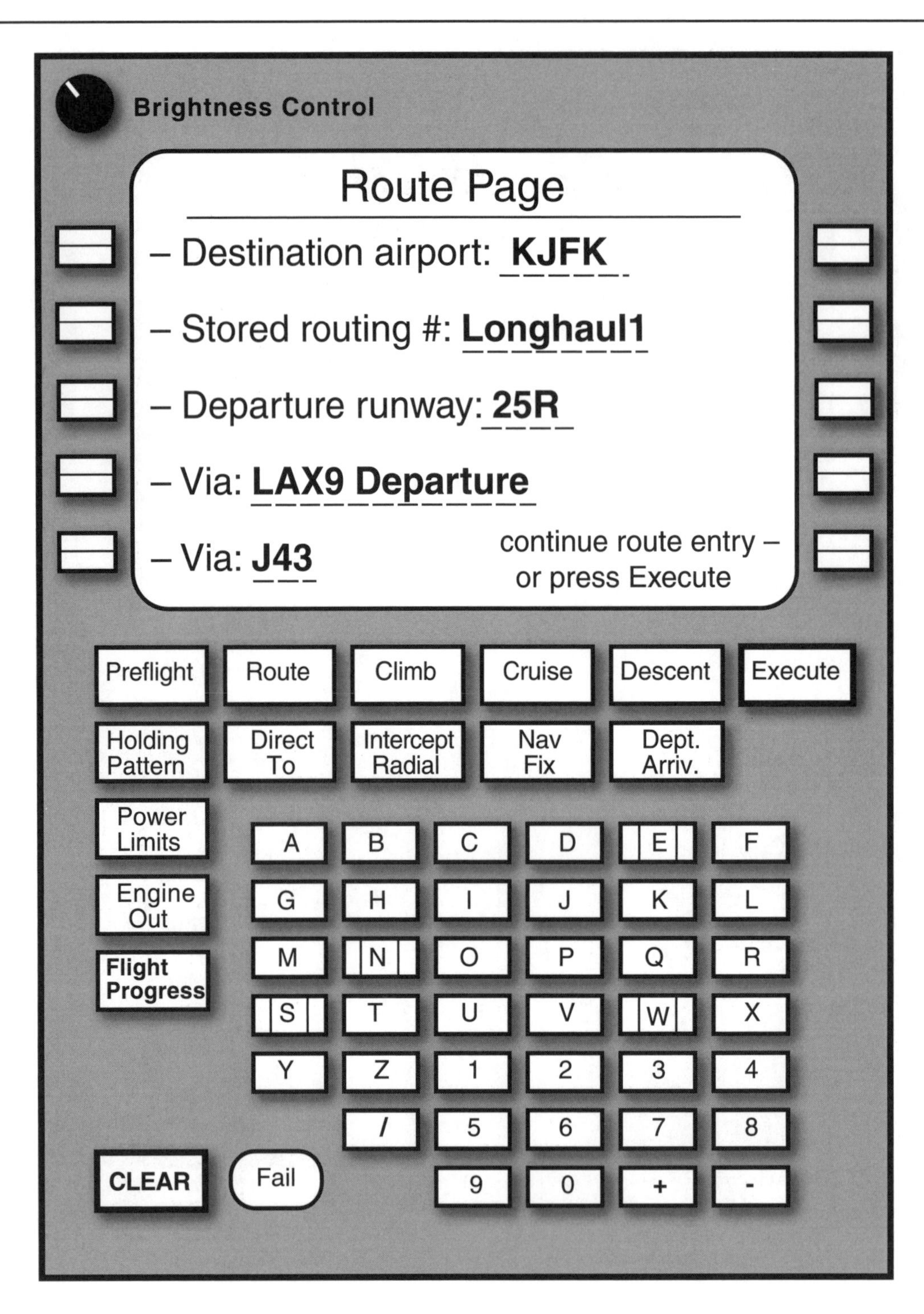

13.21. FMS route page.

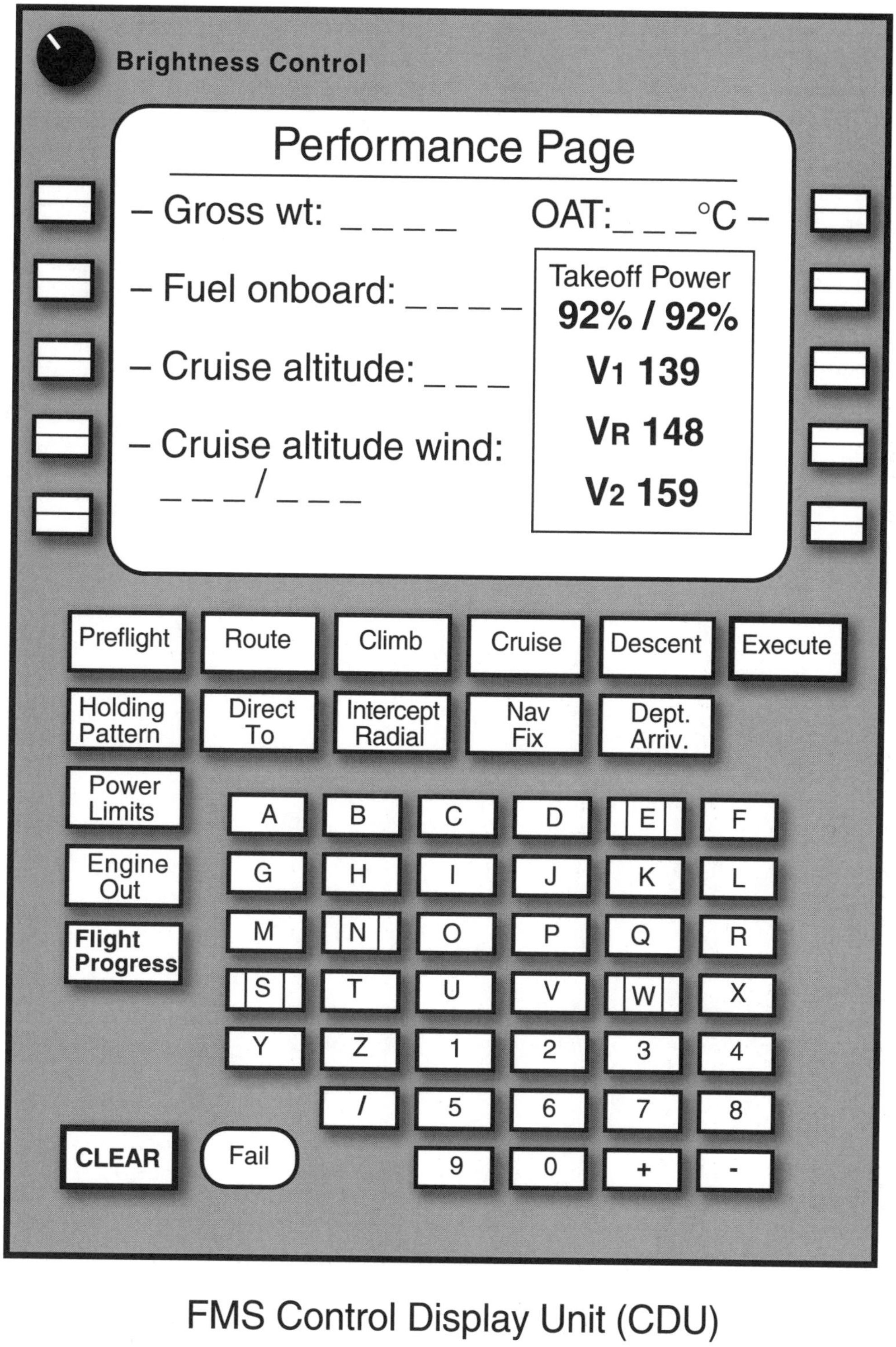

Brightness Control

Performance Page

– Gross wt: _ _ _ _ OAT:_ _ _°C –

– Fuel onboard: _ _ _ _ _

– Cruise altitude: _ _ _

– Cruise altitude wind:
_ _ _/_ _ _

Takeoff Power
92% / 92%
V_1 139
V_R 148
V_2 159

Preflight | Route | Climb | Cruise | Descent | Execute
Holding Pattern | Direct To | Intercept Radial | Nav Fix | Dept. Arriv.
Power Limits
Engine Out
Flight Progress

A B C D E F
G H I J K L
M N O P Q R
S T U V W X
Y Z 1 2 3 4
/ 5 6 7 8
9 0 + -

CLEAR Fail

FMS Control Display Unit (CDU)

13.22. FMS performance page.

Hopefully, this description has given you a useful taste of the basic logic for programming flight management systems. In training you can look forward to learning the FMS specific to your aircraft, using ground-based FMS simulators or software.

It's worth mentioning that excellent training materials for many FMS systems are now commercially available on CD-ROM, allowing pilots to learn and practice programming of specific systems on their home computers. If hired or transitioning into an FMS-equipped aircraft, contact your training department ahead of time and see if such materials are available for preliminary study. Learning these systems is interesting, and you'll be that much further ahead on FMS procedures when moving into the aircraft.

Pilot Operations in the Glass Cockpit

You should be getting the impression that having FMS and EFIS in the cockpit is neat stuff. It is. In fact, one of the biggest challenges for crews operating such systems is to avoid becoming totally engrossed. It takes a good deal of concentration and knowledge to program and operate these devices, but today's airspace system still depends on the old see-and-avoid principle. Newly checked out crew members, in particular, must work harder than ever to keep their eyes outside the cockpit, in spite of all the exciting new technology.

CHAPTER 14

Hazard Avoidance Systems

Weather Avoidance Systems

The FAA requires aircraft seating ten or more passengers to carry some form of airborne thunderstorm detection equipment. This may be a lightning detector or weather radar. Large transport category aircraft registered in the United States and engaged in passenger-carrying operations are specifically required by FAA regulations to be equipped with airborne weather radar. Many aircraft carry both types of devices, combining the strengths of each system for more accurate real-time weather depiction and analysis.

Keep in mind that thunderstorms are among the most deadly hazards faced by pilots, and given the quirks of both weather avoidance equipment and the storms themselves, no pilot should attempt flying anywhere near them without proper training.

Airborne Weather Radar

Radar is used in civil aircraft primarily for detection and avoidance of weather associated with thunderstorms. To a lesser degree it is also used to supplement other cockpit information for terrain avoidance and navigational support.

Radar is a tremendous tool that has greatly enhanced thunderstorm avoidance in aircraft. At the same time, it's probably the single most challenging cockpit device to properly (and safely) interpret. Effective use of airborne weather radar requires specialized pilot training and experience. Equipment-specific training is often necessary, since there's a lot of variation by application and manufacturer. If your outfit doesn't provide thorough radar training, invest in one of the excellent seminars offered on the subject.

How Radar Works

Radar is an acronym for *ra*dio *d*etection *a*nd *r*anging. A transmitter and receiver are used in combination to measure distance and direction to targets through the use of radio "echoes." A high-frequency RF (radio frequency) energy pulse is transmitted from the radar unit's highly directional antenna. It travels at the speed of light until it strikes something, such as rain droplets, which reflect some energy back to the antenna. The radar receiver detects and measures the intensity of the "return" (or echo), which is then displayed as a range of bright or colored areas on the aircraft's radar screen. The intensity of the radar return gives an indication of the severity of the weather. If a lot of RF energy is reflected back, heavy precipitation is indicated at that location, implying thunderstorms (Fig. 14.1).

Radar calculates distance, or *range,* by timing the signal's travel to the target and back. Since the velocity of the RF signal is known, it's relatively simple for the unit to determine how far the target lies ahead of the aircraft.

Radar antennas transmit a cone-shaped beam of energy (similar in shape to that of a flashlight beam), its diameter increasing with distance. Since the beam is relatively narrow, the area it illuminates depends upon antenna elevation, and the distance between the aircraft and the target weather.

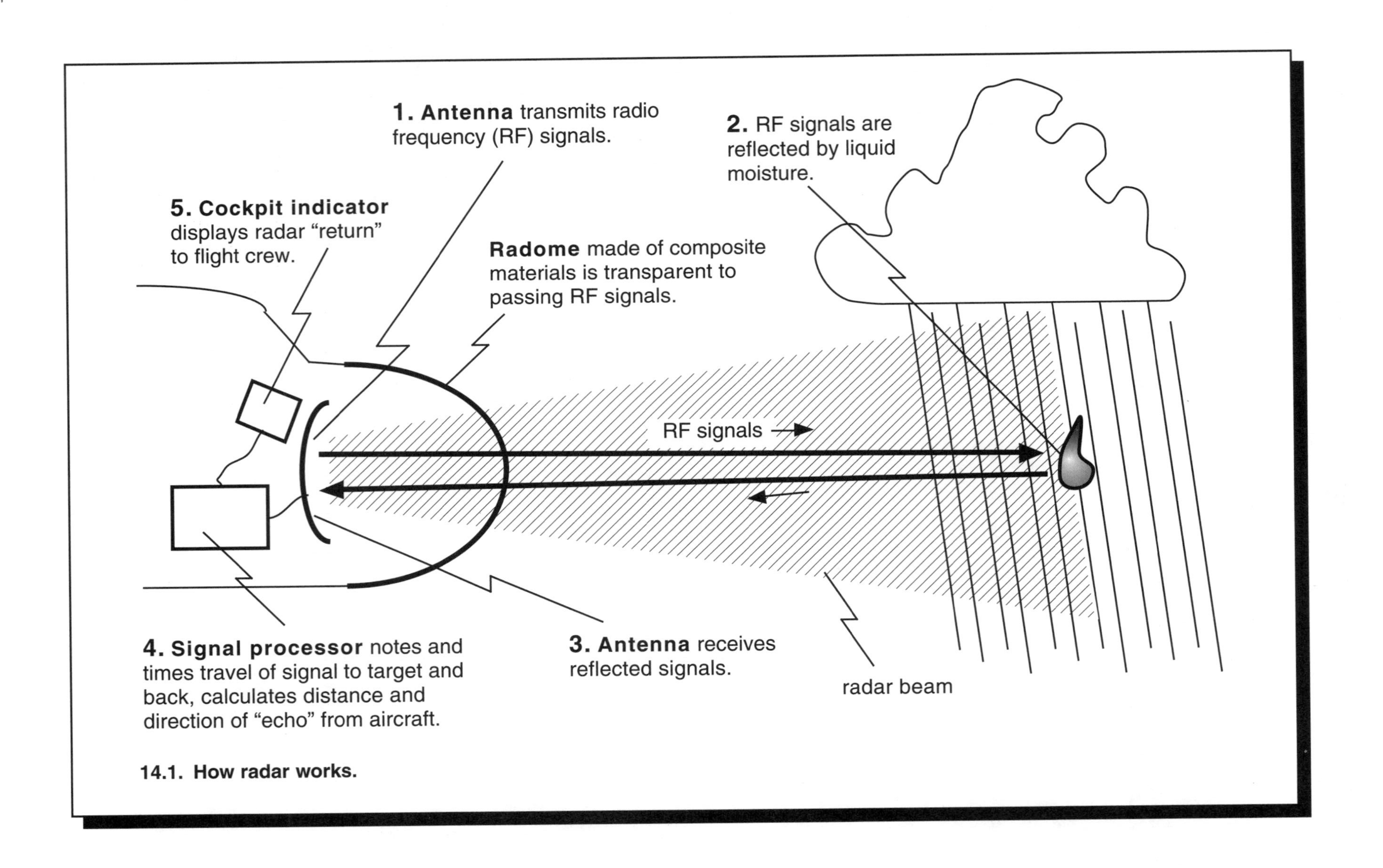

14.1. How radar works.

Azimuth Scan

In order to cover more area with the radar beam, an electric motor automatically and continuously "scans" the antenna back and forth in the horizontal plane, or *azimuth.* In most radar units, azimuth scan is set by the manufacturer to cover about 60 degrees either side of the aircraft's nose and cannot be controlled by the pilot. In some installations, however, the pilot may have a second "narrow scan" azimuth selection available, covering ±30 degrees of the aircraft's nose (see Fig. 14.2).

Elevation

The *elevation,* or vertical angle addressed by the radar antenna, can be selected by the pilot. This adjustment is made using the radar's *tilt control,* normally through a vertical range of ±10 degrees from the longitudinal axis of the aircraft. (Azimuth scan continues at its normal rate, regardless of tilt angle.) Some newer radar installations offer the pilot alternate selection of a motor-driven vertical scan. This is designed to determine the vertical cross section of a given weather buildup (see Fig. 14.3).

Radar Antennas

Radar capability is determined by several factors, including transmitting power, receiver sensitivity, electronic signal-processing capability, and antenna size and technology. There are two common types of radar antenna in use, *parabolic* and *flat plate.*

The radar beam patterns of the two different antenna types vary slightly, as you can see in Figure 14.4. Parabolic antennas project RF "side lobes" that are of sufficient energy to generate returns from targets, thereby providing broader coverage for the same antenna size. However, side lobe energy is drawn from the main lobe, so the range of a parabolic antenna isn't as great as that of the flat-plate type. A flat-plate antenna generally scans a smaller area than a parabolic model of the same size, but has greater range.

As with a flashlight reflector, diameter greatly impacts the effectiveness of radar antennas. Accordingly, radar performance depends somewhat on aircraft size. Large transport aircraft can accommodate the large-diameter flat-plate antennas best suited for radar performance. Smaller aircraft normally are equipped with parabolic antennas that provide broader coverage at smaller sizes. At the far end of the scale are the tiny wing-mounted parabolic antennas found on some single-engine planes; while useful, they can't possibly perform as well as the "big guys."

Many radar systems incorporate antenna gyrostabilization features to compensate for changes in airplane attitude. These use gyro-corrected inputs to keep azimuth radar sweeps parallel with the earth's horizon. (Non-gyrostabilized antennas always sweep parallel to the aircraft's wings, causing the radar beam to be reflected off the ground whenever the aircraft banks. The pilot must learn to ignore "monster" ground returns appearing at the sides of the display during each banking turn.)

Using Radar

To a casual observer, operation of an airborne weather radar system appears deceptively simple. Set the mode control to "weather" (Wx). Then adjust antenna tilt control down until a solid band of ground targets is displayed. Slowly increase tilt until ground returns appear only at the outer edge. Then monitor for returns (see Fig. 14.5).

Modern radar indicators normally display returns in three or four colors to differentiate between intensity levels. Green indicates areas of light rainfall, yellow shows areas of moderate rainfall rate, and red depicts heavy precipitation. Magenta or flashing red may be used to display areas of very heavy precipitation. Black indicates the absence of an echo and may be interpreted as areas with no detectable rainfall. Target range from the aircraft is depicted by a series of arcs superimposed over the screen.

Radar's not as simple to use as it sounds, however. It does not actually detect clouds or thunderstorms—just targets that reflect energy back to the antenna. This could be precipitation (which we associate with thunderstorms), mountains, or even level ground. In fact, radar is very selective in what it can "see." Water vapor, for example, does not reflect radar energy at appreciable levels. Ice does reflect it, but only minimally. From a weather standpoint, the only thing that radar does identify effectively is liquid water, or precipitation. This means that clouds without liquid moisture are invisible to radar. Ice crystals found, say, in the higher levels of thunderstorms, generate weak returns at best. A turbulent thunderstorm may lie ahead in its developing stage with no precipitation and be invisible to weather radar. For these reasons, a good deal of interpretation is required by the pilot to identify hazardous weather.

Precipitation Gradient

Generally, the greatest turbulence in storm cells occurs where precipitation increases or decreases sharply over a small horizontal distance. This is known as a "steep precipitation gradient" and is often identifiable on radar. (*Precipitation gradient* refers to the distance from the outer edge of a precipitation area to its core.) Much like the temperature and pressure gradients that you've studied in the past, steep precipitation gradients imply turbulence and vertical wind shear. Stay away from areas where precipitation returns go from light to heavy over a very short distance (see Fig. 14.6).

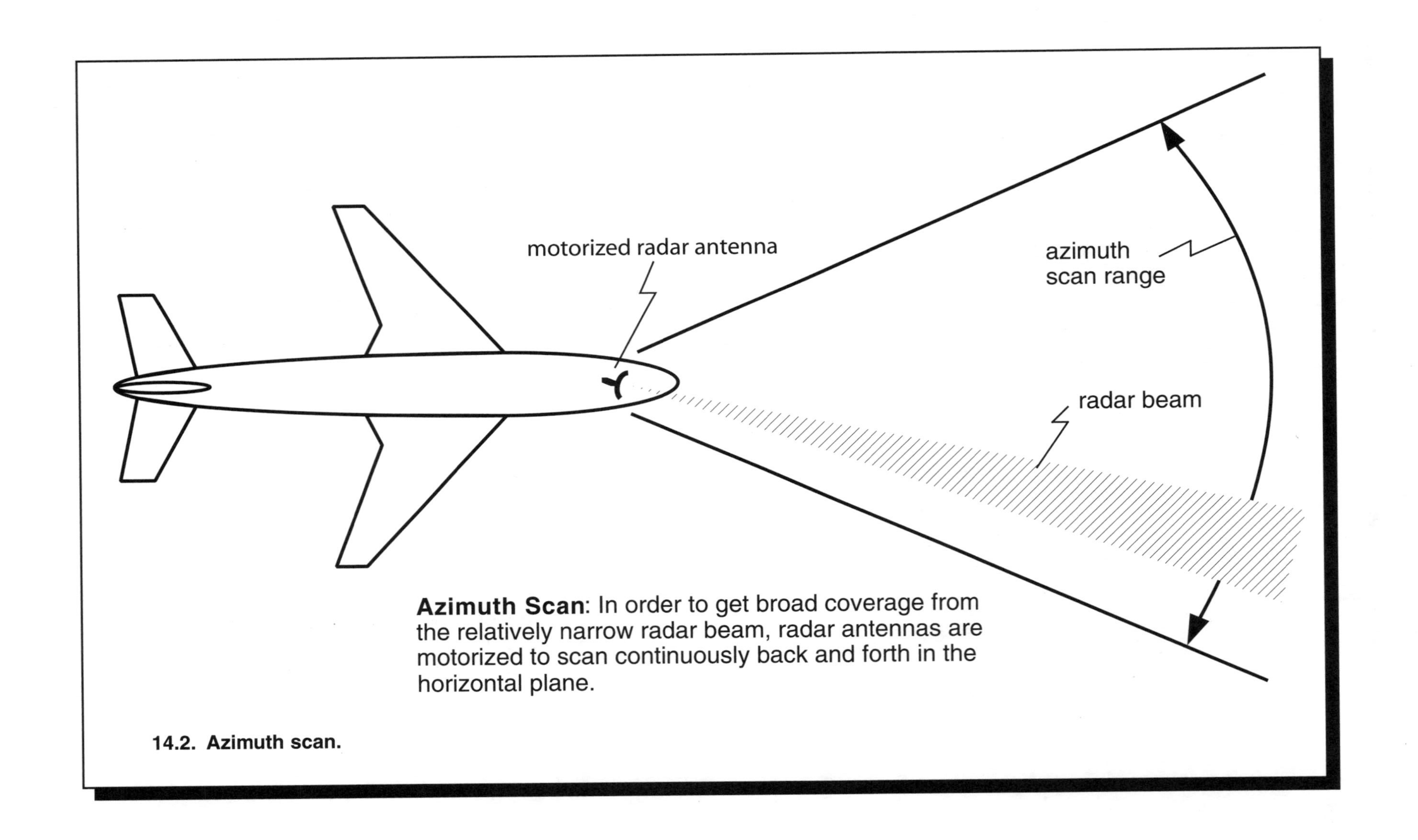

14.2. Azimuth scan.

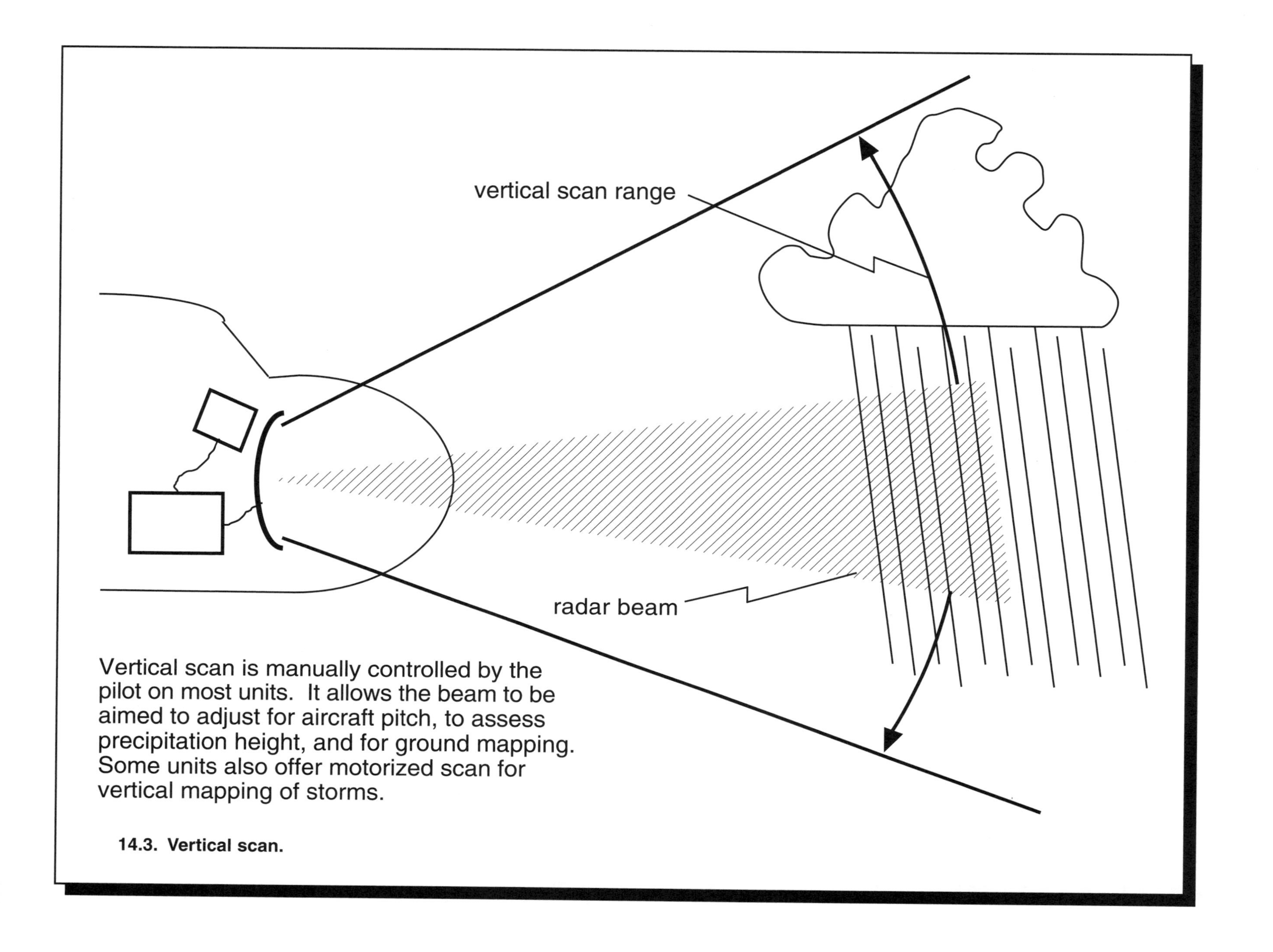

Vertical scan is manually controlled by the pilot on most units. It allows the beam to be aimed to adjust for aircraft pitch, to assess precipitation height, and for ground mapping. Some units also offer motorized scan for vertical mapping of storms.

14.3. Vertical scan.

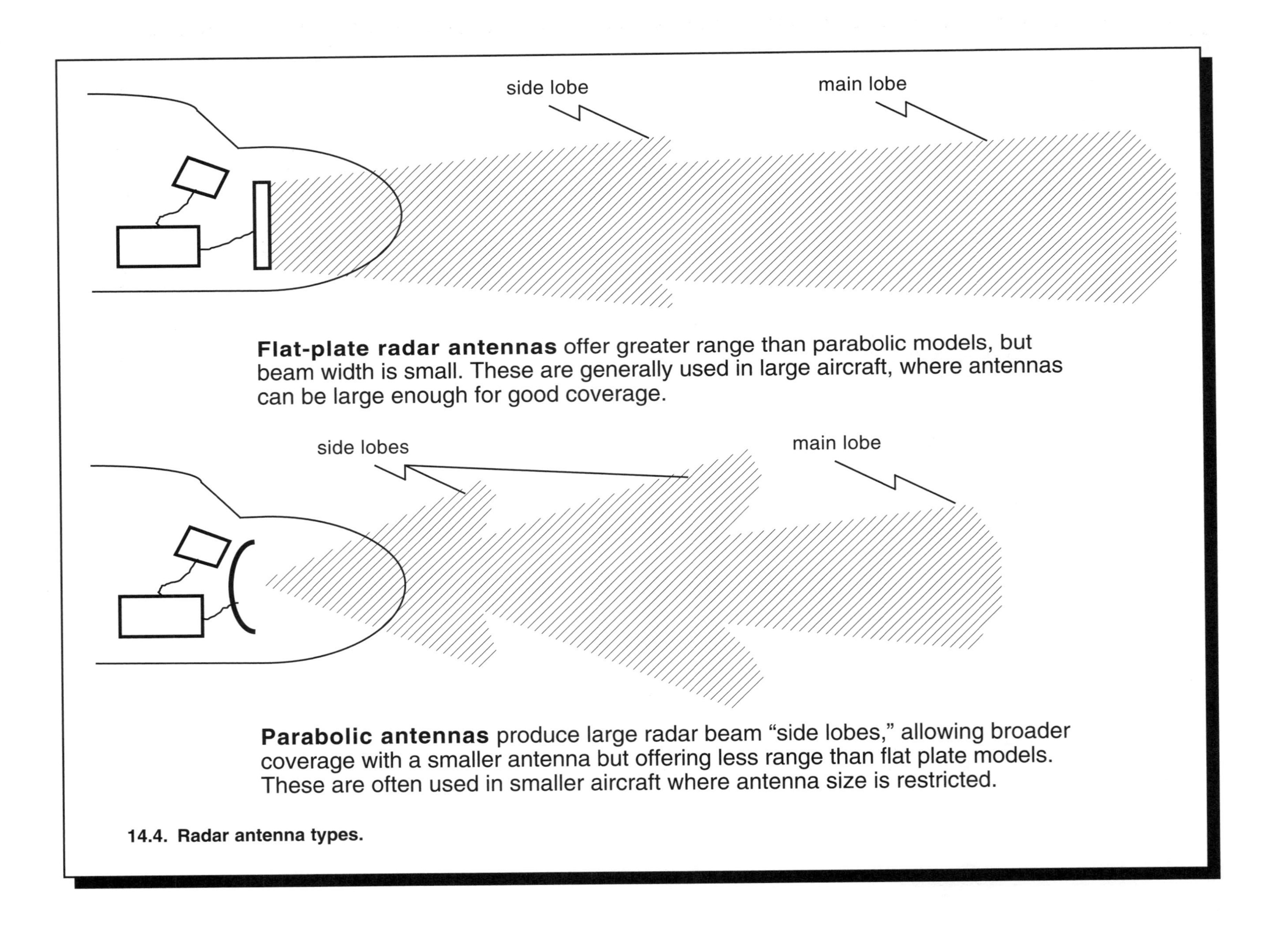

14.4. Radar antenna types.

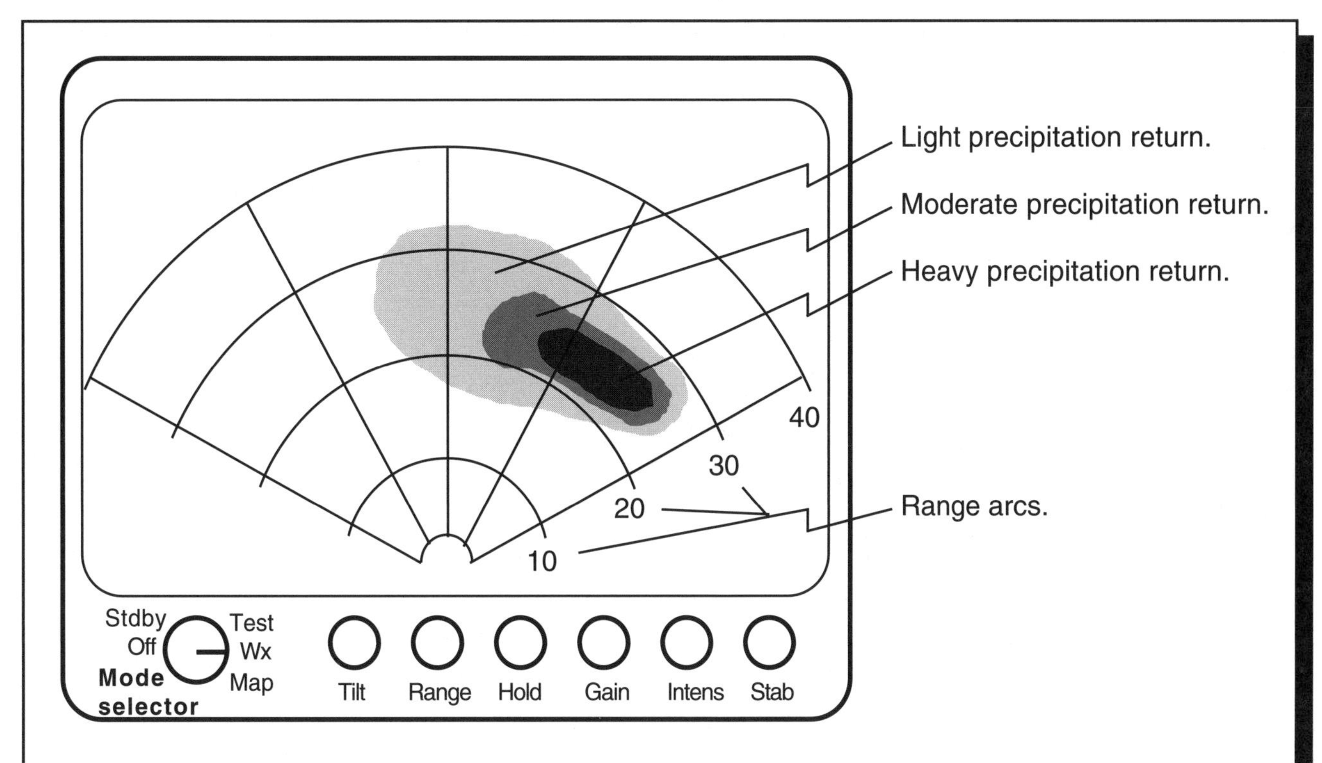

Test: tests radar system. Normally, a test pattern is displayed on the indicator, and antenna tilt and azimuth functions are automatically cycled to check integrity of the antenna pedestal. While some test cycles are fully automated, others require pilot interaction to test all functions.

Standby (Stdby): turns radar unit on for warm-up, but unit does not transmit.

Weather (Wx): normal weather avoidance radar operation. System is activated and screen displays weather returns.

Ground mapping (Map): used for surveillance of terrain. In this mode, the radar beam is depressed to scan ground targets, and the image is processed differently.

Tilt: directs radar scanning above or below horizontal by vertically tilting antenna.

Range: selects displayed distance scale. Arc-shaped range markings depict distance from aircraft to any displayed echoes.

Hold: freezes current screen presentation. Data is no longer updated while this position is selected.

Gain: adjusts receiver sensitivity. With gain properly set, the scope should not show too much light precipitation. Lower gain reduces display of light precipitation and thereby better defines areas of heavier precipitation.

Intensity: controls display brightness.

Stabilization (Stab): selects antenna gyro stabilization, if installed. This feature keeps antenna azimuth sweep parallel to the ground, independent of aircraft attitude.

14.5. Typical radar display and control functions.

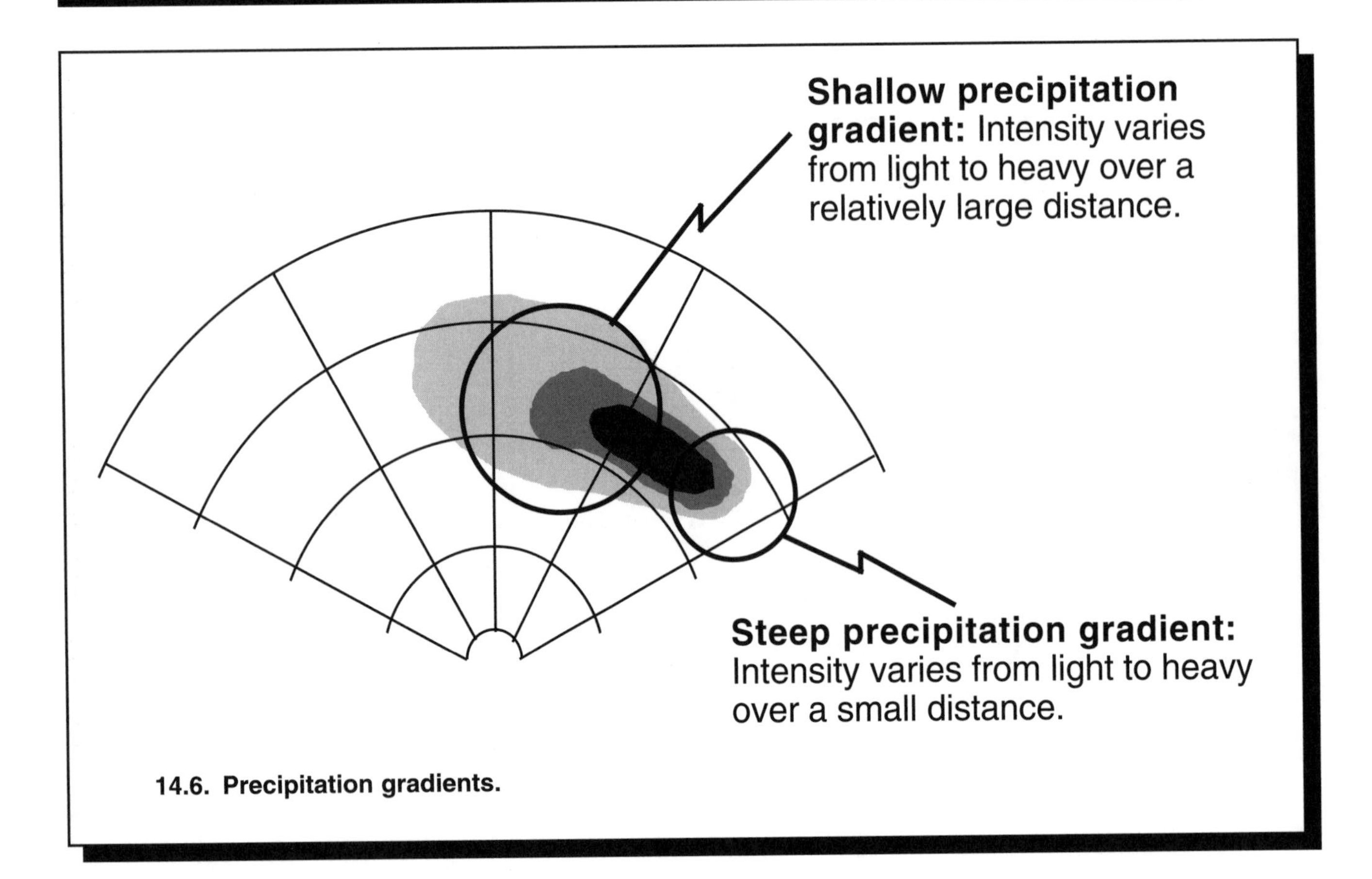

14.6. Precipitation gradients.

Attenuation and Radar Shadows

If there is no return in an area, it's safe to fly through, right? Not necessarily. While precipitation reflects radar well, it also absorbs RF signals very effectively. Therefore, weather behind areas of heavy precipitation may not be defined accurately, or even appear at all. This absorption of RF signals by heavy precipitation is called *radar attenuation.* (The word "attenuation" means to weaken or lessen in volume.)

A *radar shadow* is an area of severe attenuation, generally identified by a complete blackout of data behind a weather cell. Such an area appears to have no returns but actually results from precipitation so heavy that radar is prevented from penetrating to the airspace behind it. There could be a monster storm hiding behind that wall of precipitation. Never fly toward a radar shadow (see Fig. 14.7).

In some instances, radar echoes may be severely attenuated when passing through areas of moderate rainfall or even through large areas of light rainfall. This may mask some targets completely or cause distant targets to appear much less intense than they actually are. Flight through steady light rain can often reduce effective radar range to 10 or 15 miles or less. (That's only a minute or two, at jet speeds.)

It's also important to regularly check your aircraft *radome* (the fiberglass dome that covers the radar antenna) for scratches and pits. Water can puddle in such depressions and paint false returns on your screen that obscure real weather.

Echo Height

Two other ways of evaluating returns are height and shape. As you know, the height of a weather buildup is a major indicator of its severity—thunderstorm heights are directly associated with the ferocity of the storms. Accordingly, *echo tops* are always reported with ground-based radar information. With proper training and some clever work with the tilt control, it's possible to estimate tops of the precipitation you're "painting" on airborne radar. (You'll have to brush up on your trigonometry skills to do it, though.)

Echo Shapes

Echo shapes are also indicators of danger areas. Basically, any unusual precipitation shapes other than simple circles or ovals are suspect. Think of it this way. What would it take to modify the shape of an area of rain? Wind shear

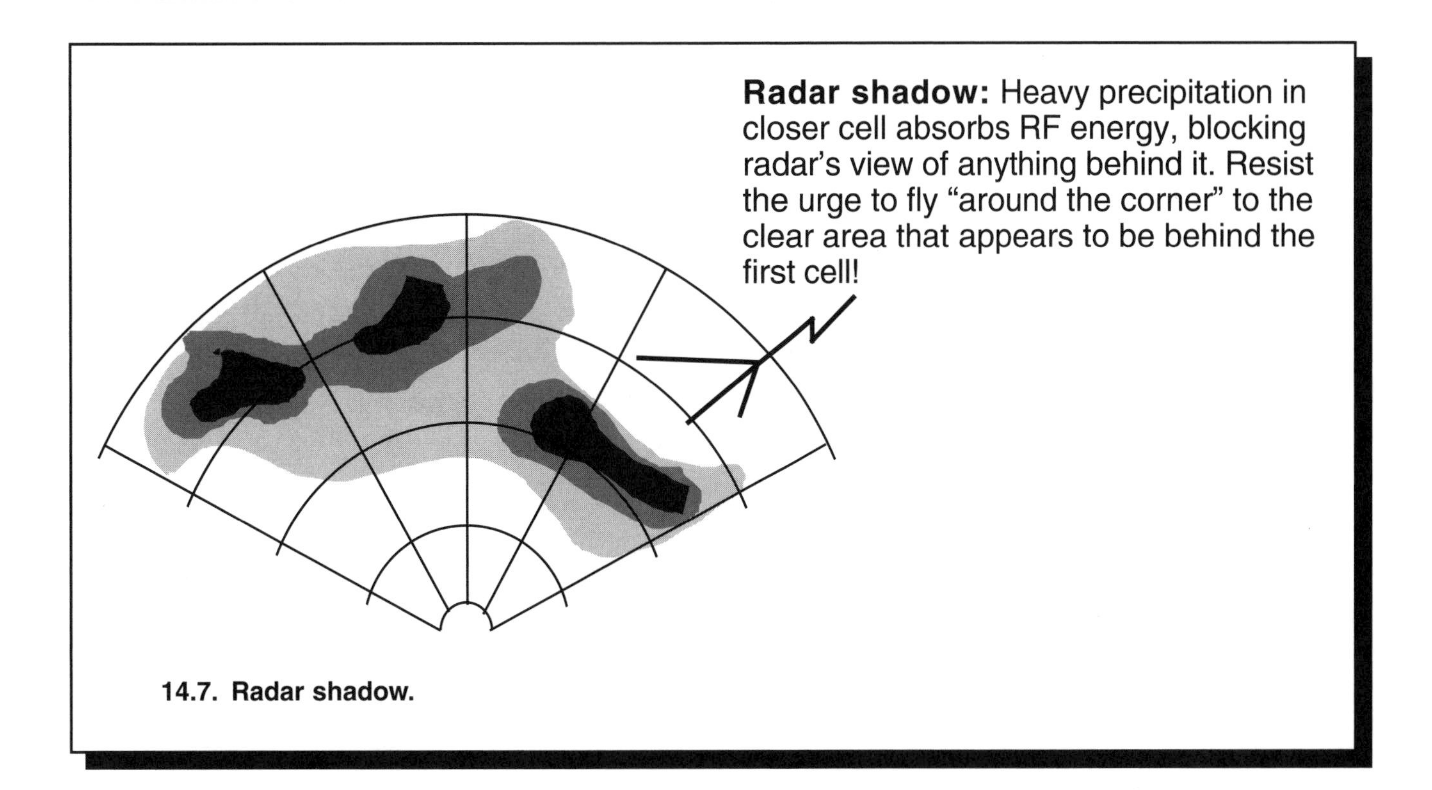

14.7. Radar shadow.

"scooping out" areas of rainfall or updrafts and downdrafts changing rainfall characteristics in certain areas.

Accordingly, odd shapes like "hooks" and "fingers" indicate large amounts of vertical wind shear. Remain well clear of any such cells; they may contain hail or even tornadoes. An hourglass-shaped echo indicates two closely spaced cells, one of which may be energizing the other. Ragged edges around a cell often indicate areas of hail. Square corners suggest extreme hail and very strong vertical updrafts or downdrafts, possibly exceeding 6000 feet per minute vertical movement. Radar echo "bulges" and "dips" indicate that the radar beam is penetrating farther in some areas than others, suggesting the presence of significant turbulence (see Fig. 14.8).

Avoiding Weather

To avoid indicated weather, first determine the heading change necessary to detour around the storm safely. Although there is no definite rule as to the minimum distance required to bypass a storm, there are some generally accepted guidelines. The best plan is to avoid storm cells by at least 20 nm, preferably to the upwind side. Begin by monitoring detected weather at long ranges. This allows time to assess weather movement and development along your intended flight path and then to determine the best course for avoiding the storm cells. When using short radar ranges, periodically switch back to longer ranges so you can estimate the extent of the detected weather activity. Always assume the presence of turbulence above any storm cell. Never attempt to fly through or just over a cell.

The Big Picture

It takes knowledge and quite a bit of experience to correctly interpret a radar display. A smart pilot uses radar to back up information from other sources such as ground-based radar, pilot reports, other weather avoidance equipment, and the old-fashioned eyeball. Intelligent weather avoidance decisions are made by understanding the "big" weather picture. Radar must be considered as one small, though valuable, part.

Ground Operations

A few words of caution are also in order regarding ground use of aircraft radar systems. Radar is a very powerful transmitter. *Never operate radar on the ground,* except in standby (Stdby) or test modes. The radar beam can injure people, damage electronic equipment in other aircraft, and present a hazard during refueling. Operating radar in congested ramp areas or hangars is also dangerous because the beam can be reflected off objects or surfaces back toward your own aircraft. Most operators include "radar off" in the After Landing Checklist and don't activate it again until takeoff. For use on takeoff, the radar antenna tilt should be raised so that the beam doesn't strike the ground.

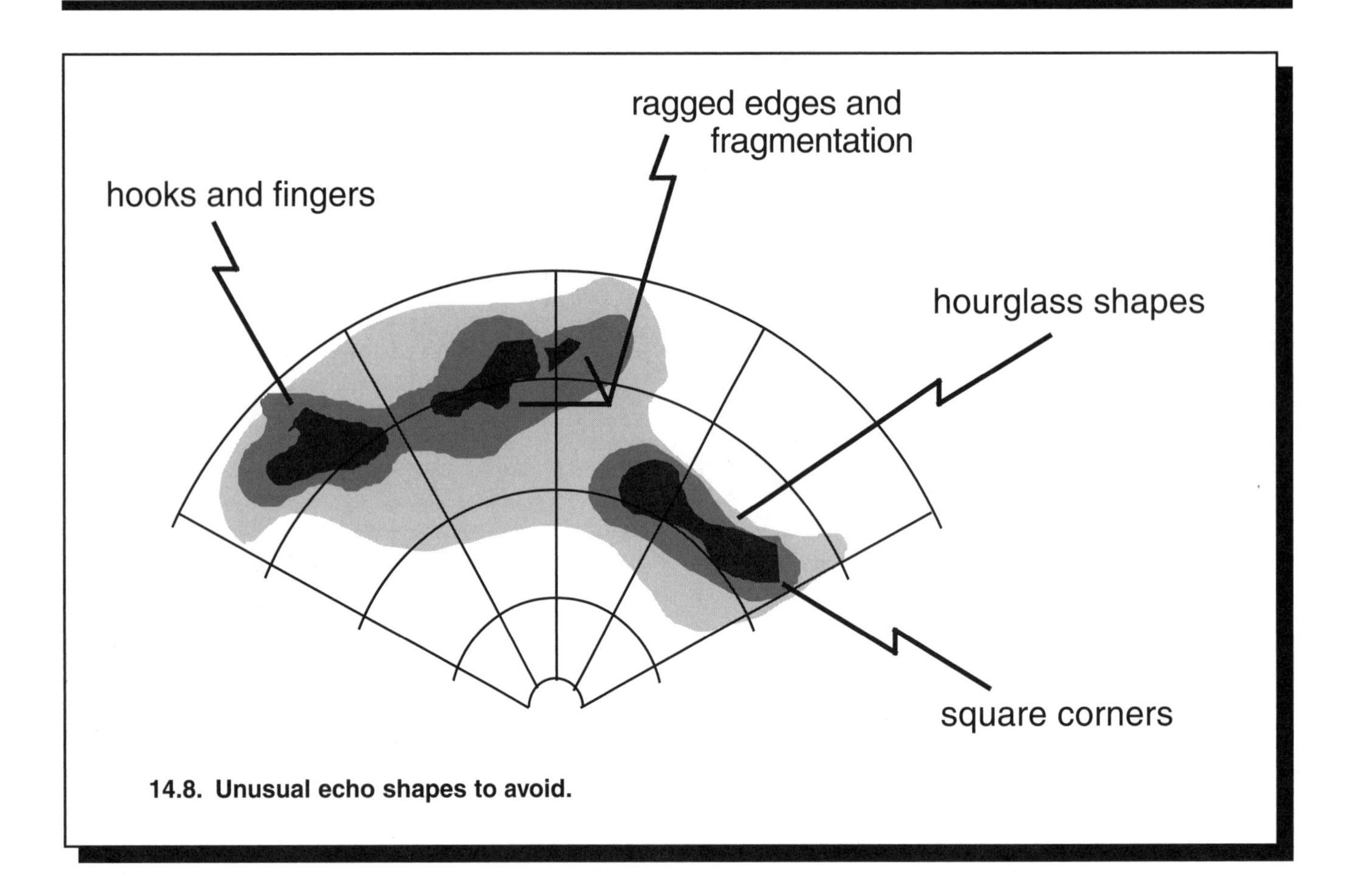

14.8. Unusual echo shapes to avoid.

Doppler Radar

Doppler radar is an added capability of newer generation radar systems used in large corporate and transport aircraft. Doppler units contain a computer that identifies frequency changes in reflected echoes. Such changes in frequency are caused by velocity differences within a given target and indicate turbulence.

To understand how this works, consider the classic "Doppler effect" example. A stationary train engine puts out a constant sound in the form of sound waves moving in all directions from the train. Now consider what happens when the train is moving at 30 miles per hour. The sound waves in front of the train approach the waiting listener at their normal speed *plus* the train's velocity of 30 miles per hour. Once passed by the train, the listener hears the sound waves passing at their normal speed *minus* 30 mph. To the listener in front of the train, the sound appears to be higher since the faster waves are passing at a higher frequency than those behind the train. That's why the sound of a train changes when it passes. Based on the relative frequency of passing sound waves, sound metering equipment can tell you if the train is stationary or if it's moving toward or away from you.

Doppler radar works on the same principle. A sophisticated computer measures the RF frequencies of the reflected echoes within a given target. If there's a variety of reflected frequencies, indications are that droplets are moving in a variety of directions, some "toward" the radar receiver and some "away" from it. This implies that turbulence is present since, if all the rain were falling straight down, the aircraft would be approaching it at a constant velocity (Fig. 14.9).

Therefore, radar units equipped with Doppler processors are capable of detecting and displaying storm-related turbulence as well as precipitation rate. The Doppler radar system determines areas of hazardous turbulence by measuring velocity (and therefore directional) differences within the target precipitation. The turbulence processor "looks for" large changes in the horizontal velocity of precipitation droplets to predict areas of severe turbulence. However, because the radar identifies turbulent areas by measuring velocities of water droplets, it can still function only when precipitation is present. Therefore, Doppler radar can only detect turbulence associated with precipitation. Clear air turbulence (CAT), for example, is not detectable by current Doppler radar systems. (Though development of CAT-detection systems is ongoing.)

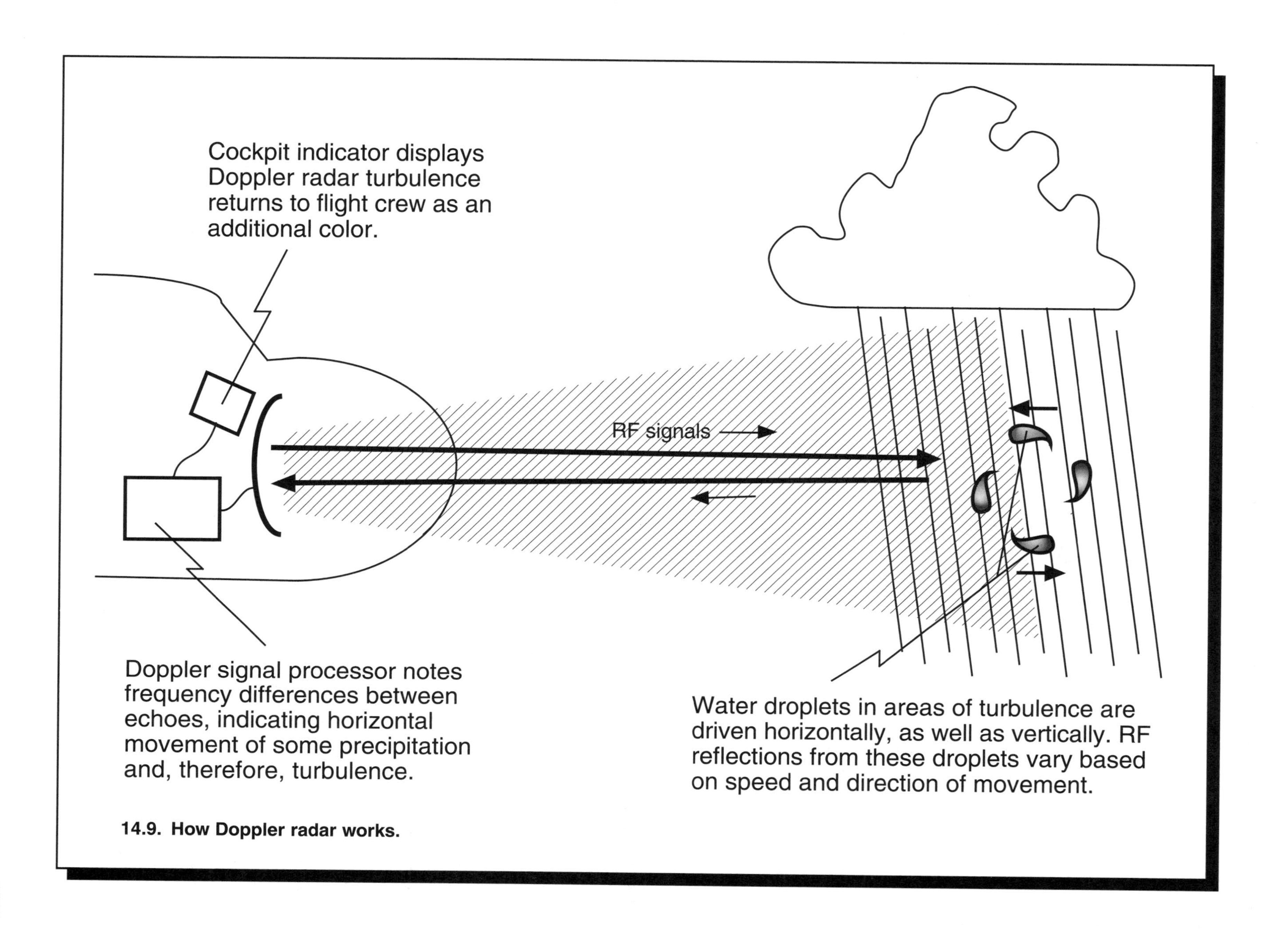

14.9. How Doppler radar works.

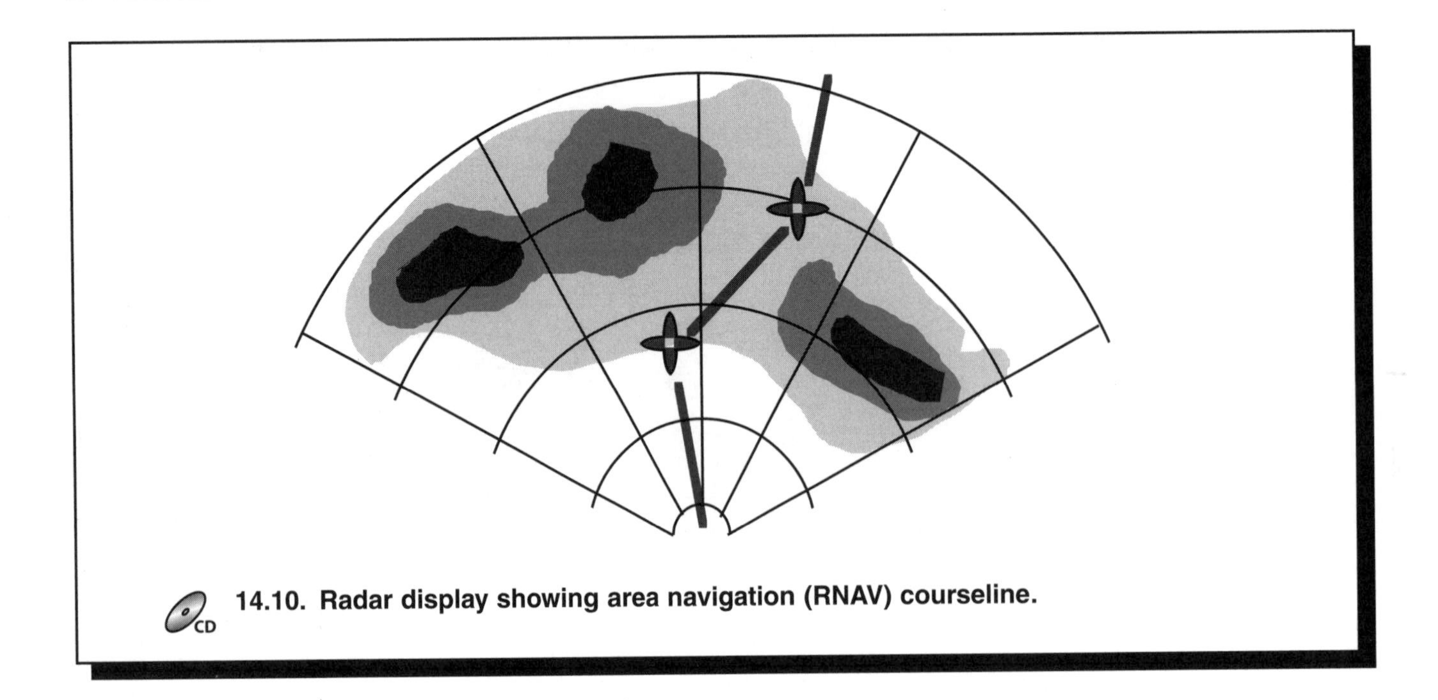

14.10. Radar display showing area navigation (RNAV) courseline.

Pilot operation of Doppler weather radar is similar to that of the basic systems we have already discussed. The only equipment differences noticeable to the operator are a couple of added mode controls. The flight crew may select "Wx" (weather radar only), "TURB" (Doppler radar only), or "Wx + T" (Doppler radar superimposed over the normal weather radar display). Magenta shading on the Doppler radar, or "TURB" mode, indicates areas of possible severe turbulence.

Doppler radar is valuable because it sometimes identifies turbulence in areas where you would otherwise expect smooth conditions, based on conventional weather radar returns. Turbulence can certainly occur in areas of light or moderate rainfall rates, especially during the developing stages of a storm. Doppler radar allows avoidance of these otherwise undetectable and potentially dangerous areas.

Incidentally, Doppler technology is not limited to airborne radar. Ground-based Doppler radar systems are now being used to collect winds aloft data and wind shear information. This is again accomplished by identifying differential movements of minute water droplets in the atmosphere.

Combined Weather Radar and Navigation Displays

Current generation weather radar displays now have the capability to be wired to a compatible navigation computer, allowing weather and navigational information to be displayed simultaneously on the same screen. This display actually shows the aircraft's courseline map and displays various navigation information on the screen (Fig. 14.10). Such data as ground speed, distance to waypoint, heading, and bearing to storm cells can be selectively depicted. Some systems also display a set of preprogrammed checklists.

Future advances in "glass cockpit" technology promise to make the processing of weather and navigation information much easier, increasing the efficiency and safety of flight crews in operating their aircraft.

Electrical Discharge or Lightning Detectors

Electrical discharge detectors, or *lightning detectors,* are airborne units designed to detect lightning. (Pilots often refer to this type of device as a "Stormscope," after the popular brand.) Lightning detectors, like weather radar, are thunderstorm avoidance systems. There are, however, fundamental differences. An electrical discharge detector is a true thunderstorm detector because it detects only lightning. Weather radar, on the other hand, detects only precipitation.

How Lightning Detectors Work

Radar is an "active" weather detection system, since it actively transmits a signal. Electrical discharge detectors, on the other hand, are considered passive since they transmit no signals to detect weather. They use only the equivalent of a radio receiver to sense electrical discharges of lightning. A computer is used to process the information and display it on a screen (Fig. 14.11).

The operating principle of an electrical discharge detector is similar to that of the ADF (automatic direction finder)

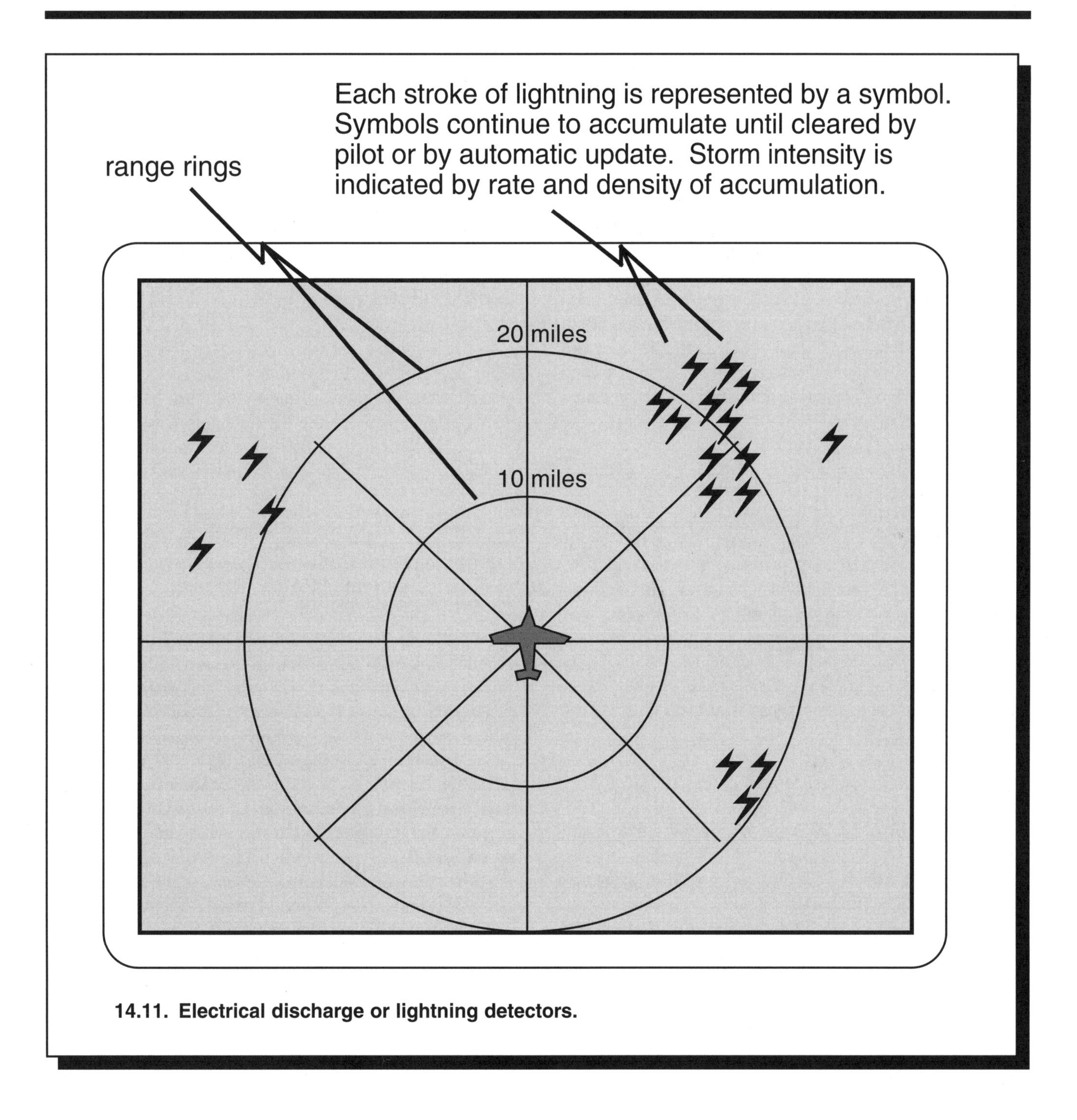

14.11. Electrical discharge or lightning detectors.

used for aircraft navigation. An ADF senses a radio transmission from a nondirectional beacon (NDB), and its bearing pointer indicates direction to the station. You may have noticed that navigating by ADF during a thunderstorm is difficult. When there's a bolt of lightning, the bearing pointer often points to the lightning instead of the NDB. Lightning detection systems take advantage of this phenomenon to identify and to display locations of electrical discharges.

Compared with weather radar, operation of a lightning detector is quite simple. Most units depict each "stroke of lightning" with a symbol plotted at the proper relative location on the indicator. Discharges accumulate on the screen so that rate and frequency of lightning discharges in different areas can be seen. The screen can be manually cleared when it becomes cluttered, or it can be set for continuous update where "oldest" discharges are replaced with the latest ones. Advanced lightning detectors are gyro-stabilized relative to the heading of the aircraft, while simple models must be cleared every time there's a heading change.

Interpreting Electrical Discharge Detector Displays

Like radar, lightning detectors are not foolproof. It takes experience and some weather knowledge to properly interpret their displays.

Electrical discharge detectors are very effective in displaying the correct bearings to areas with lightning. However, due to differences in strength of discharges they are less accurate in displaying distance from the aircraft to each cell. Unusually strong lightning discharges may be depicted as closer than they actually are, while weak ones may be shown as farther away. Therefore, it's important to avoid electrical discharge areas by large margins, unless you have visual or radar backup.

Lightning is always associated with thunderstorms and is largely caused by wind shear, so areas depicted by lightning detectors are definitely places to avoid. With these devices you can be confident that clear areas are not thunderstorms. However, you could still be in for a wet and bumpy ride, since lightning detectors do not sense precipitation or nonelectrical turbulence. Areas depicted as clear may still have heavy rain. As with weather radar, when flying in the soup always be a little suspicious of clear areas on the indicator.

Characteristics of Lightning Detectors

Electrical discharge detectors have a few other handy characteristics worth mentioning. Most models can be selected to show 360 degrees of weather, relative to the aircraft. While radar is limited in sensing to ±60 degrees either side of the airplane's nose, lightning detectors can keep you aware of activity to either side and even behind the aircraft. This is especially valuable when you're considering a "180" to get out of weather.

Also, electrical discharge detectors do not experience attenuation in the sense that radar does. Therefore, they provide some insight into what's behind that first line of weather.

Finally, since electrical discharge detectors are passive, they can readily be used on the ground. On days of embedded thunderstorm cells, pilots are often faced with takeoff into relatively unknown conditions before their airborne radars can be used effectively. Lightning detectors allow preliminary thunderstorm assessment on the ground so that a safe departure can be confidently planned before takeoff.

Lightning Detectors versus Radar

Historically, self-contained electrical discharge detectors have been installed primarily in lighter general aviation aircraft due to compact size and price tag. More sophisticated models integrate lightning displays with returns on the aircraft's radar screen.

Weather radar, on the other hand, has traditionally been installed primarily in larger aircraft due to antenna and weight considerations. Radar is also more expensive to install and maintain due to the numbers of components and moving parts.

It's easy to see that for successful weather avoidance it's really ideal to have both weather radar and electrical discharge detection onboard the aircraft. The weather radar warns of precipitation, and the lightning detector indicates areas of electrical discharge. The combination gives an excellent "big picture" of thunderstorm activity.

Traffic Alert and Collision Avoidance System (TCAS)

TCAS, or *traffic alert and collision avoidance system,* is installed in the aircraft to help aircrews avoid midair collisions. TCAS does this by receiving and interrogating the transponder signals of other nearby aircraft. Its computer processor analyzes those signals to determine altitude, range, and bearing of the traffic. If the TCAS computer processor determines that a traffic conflict exists, it automatically issues a traffic advisory to the crew. The crew then analyzes the situation and if necessary initiates evasive action.

Most TCAS systems include audio and visual traffic alert warnings. If conflicting traffic is identified, an audio voice warning system calls out, "Traffic, traffic." A visual display depicts the traffic in your vicinity and its altitude relative to your aircraft. The TCAS system may have its own display or be incorporated into the weather radar display (Fig. 14.12).

Some TCAS systems are coupled into special vertical speed indicators, incorporating a row of red and green lights around the perimeter. In the event of a traffic advisory, the flight crew pitches the aircraft in such a manner that the vertical speed indicator points toward the green lights.

It's important to remember that TCAS is unable to detect any aircraft that do not have operating transponders. So, while a TCAS system is a significant aid in avoiding midair collisions, it cannot replace the see-and-avoid concept of air traffic separation you've been using for years. Think of it as an extra set of eyes.

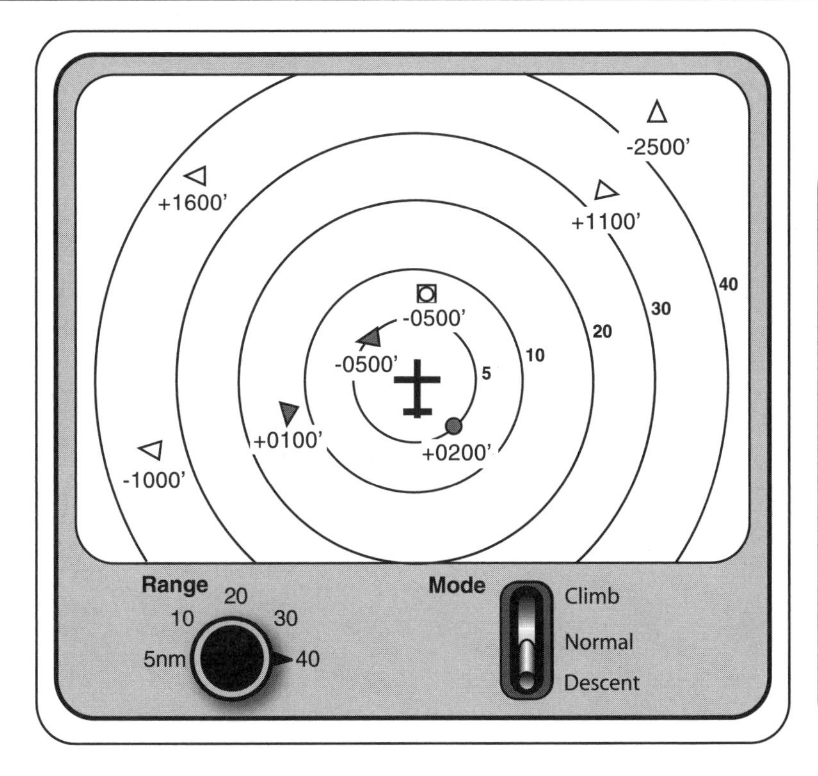

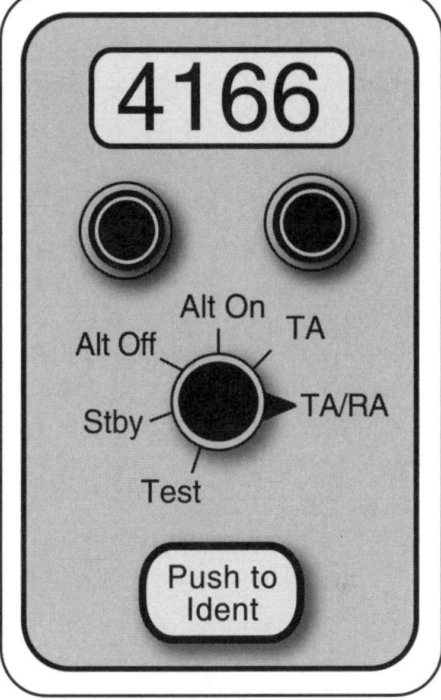

transponder and TCAS control panel

△ **Nonthreatening aircraft** (> 5 nm or > 1500 ft.)

▲ **Nonthreatening, close-proximity** (< 5 nm or < 1500 ft.)

● **Possible threat aircraft** (Traffic Advisory warning or "TA")
TCAS issues an oral warning, "Traffic, Traffic"

◘ **Immediate threat aircraft** (Resolution Advisory warning or "RA")
TCAS system issues a corrective action warning, "Climb," "Descend," "Monitor Vertical Speed," or "Clear of Conflict."

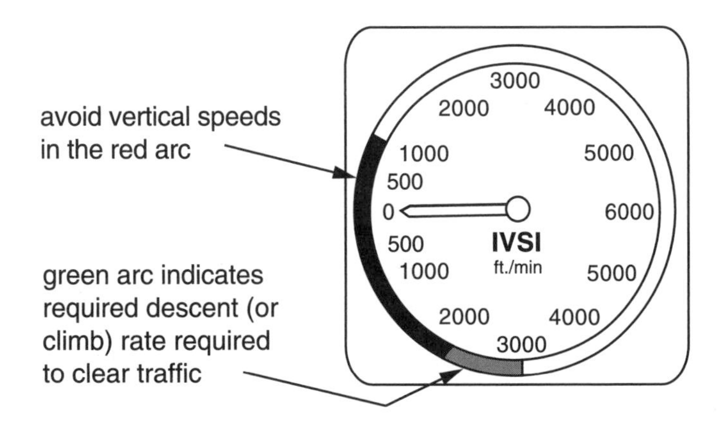

associated voice warning commands pilot action: "Descend! Descend!"

Many TCAS systems generate colored arcs on the vertical speed indicator (VSI), directing pilots to climb or descend at rates that will maintain adequate traffic separation.

14.12. Traffic alert and collision avoidance system (TCAS).

Ground Proximity Warning Systems and Enhanced Ground Proximity Warning Systems

The *ground proximity warning system* (*GPWS*) is designed to warn flight crews of excessive descent or closure rate with terrain. A GPWS typically consists of an electronic control unit, warning light, and audio warning located in the cockpit. The control unit monitors the aircraft's radar altimeter when the aircraft is within 2500 feet of the ground. When an excessive descent or closure rate with terrain is sensed, the warning light illuminates, and a synthesized voice calls out one of several warnings, depending upon circumstances.

GPWS units can be programmed to provide a variety of warnings. Among them (and their voice warnings): inadequate rate of climb after takeoff ("Don't sink!"); excessive descent rate and closure with the ground ("Pull up!"); and closure with rising terrain ("Terrain!"). Improper aircraft configuration for landing may also be covered by monitoring gear and flap positions. On MD-80s, for example, if the aircraft is not properly configured for landing, there are GPWS warnings at 500 feet above ground ("Too low gear!") and at 200 feet ("Too low flaps!").

In a further effort to reduce *CFIT* accidents (*controlled flight into terrain*), *the enhanced ground proximity warning system* (EGPWS) is a new advanced system specifically designed for "glass cockpit" aircraft. EGPWS operates by comparing surface terrain data contained in an onboard computer database with the aircraft's exact position as determined by the navigation computer. EGPWS systems display threatening terrain on the EFIS primary flight display's map mode, giving the flight crew a visual indication and therefore situational awareness of terrain around them.

CHAPTER 15

Operational Information

Aerodynamics of High-Speed/ High-Altitude Aircraft

This section introduces some basic concepts relating to high-speed aerodynamics. Even if you won't be flying high-speed jet aircraft any time soon, it's important to be familiar with some of the basic aerodynamic principles of aircraft you'll be sharing airspace with.

High-Speed Flight and the Sound Barrier

For our purposes, we'll define high-speed flight as cruise flight at near, but less than, the speed of sound. The speed of sound varies with air temperature. At standard temperature, sound travels in air through a range from about 740 mph at sea level to 660 mph or so up at 40,000 to 50,000 feet. Since the few civilian supersonic jobs are pretty well locked up by senior pilots on the Concorde, there's no need for us to delve too deeply into supersonic flight. At the same time, it's important to understand what happens to an airplane as it approaches supersonic speeds, speeds at which it encounters some of the major aerodynamic limitations of high-speed flight.

Mach Number

Airspeeds of high-speed aircraft are measured relative to the speed of sound using an index known as *Mach number.* Mach number is the ratio of an aircraft's airspeed to the speed sound travels under the same atmospheric conditions. If an airplane is flying at the speed of sound, it is said to be traveling at Mach 1. At 0.8 Mach, the aircraft is traveling at 80 percent of the speed of sound. An aircraft at 1.2 Mach is traveling at 120 percent of the speed of sound.

Three categories or regions of flight, based on the speed of sound, are commonly used to describe high-speed flight. Flight is grouped into one of the following categories, not by the speed of the aircraft per se but by the speed of the airflow over the aircraft's surface.

Subsonic flight includes Mach 0 up to approximately 0.75 Mach. The upper limit of subsonic flight for a given aircraft is the maximum airspeed at which none of the airflow over its surfaces reaches the speed of sound.

Transonic flight is generally considered to be between 0.75 Mach and Mach 1.2. (Again, the parameters of this region vary with the design of the aircraft.) In transonic flight airflow over aircraft surfaces is mixed between airflow below the speed of sound and airflow above the speed of sound.

In *supersonic flight* all airflow traveling over the surfaces of the aircraft is above the speed of sound. This is generally considered to be aircraft speeds above Mach 1.2.

Compressibility

Except for military pilot readers, chances are that during primary training your studies in aerodynamics concentrated on aircraft with relatively slow airspeeds. Because airflow at slow airspeeds undergoes only small density changes, it was simplest to treat air as incompressible, making airflow appear analogous to the flow of water.

Air is compressible, however, and at high airspeeds large air density changes occur. Therefore, high-speed aircraft must be designed to compensate for these compressibility effects. The concept of *compressibility* is relatively simple. When an aircraft flies at slow airspeeds, the surrounding air molecules have time to "get out of the aircraft's way." As the vehicle's speed increases, air molecules begin to "pile up" ahead of the aircraft, increasing air pressure, density, and temperature in that region. As the aircraft enters

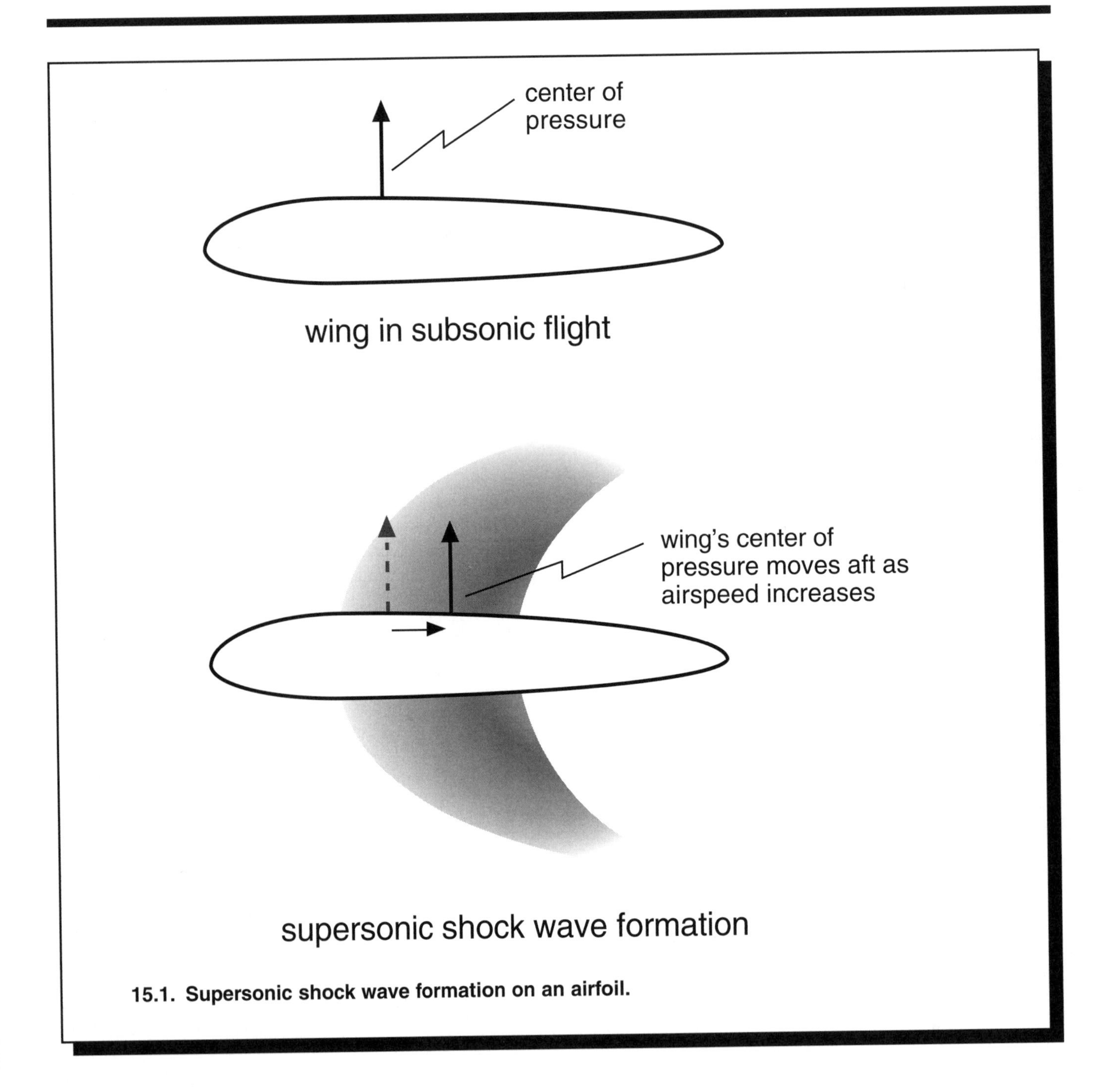

15.1. Supersonic shock wave formation on an airfoil.

transonic speeds, the air going over parts of the wing becomes supersonic, resulting in sudden tremendous increase in air density as air piles up in front of the vehicle.

Think of it this way. The maximum speed of a pressure wave through air is the speed of sound. As long as an airplane is traveling below that speed, the waves caused by its movement through the air radiate out in front of the vehicle. However, once the speed of sound is reached, the vehicle travels faster than the air can "get out of the way." A powerful compression wave, or *shock wave,* forms at the boundary between the disturbed air and undisturbed air around the now transonic aircraft (Fig. 15.1).

High-Speed Buffet

As an aircraft continues to accelerate into the transonic region, the shock wave generated on the wing travels farther aft, with increasing intensity. Much like a stall, the shock wave causes airflow separation toward the trailing edge of the wing. This disturbed air ultimately begins to affect the horizontal stabilizer, causing a *high-speed buffet* similar in feel to a stall buffet. A high-speed buffet may be encountered at different Mach numbers depending on altitude, load factor, and the weight of the aircraft.

Mach Tuck

Most corporate and airline jets are designed to cruise from around 0.7 Mach to 0.9 Mach. If these speeds are exceeded, a phenomenon known as *Mach tuck* may occur. You may remember that, on conventional aircraft, the wing's center of pressure lies aft of the center of gravity. Mach tuck develops because a wing's net center of pressure moves aft as its speed approaches the sound barrier. The farther aft the center of pressure moves the greater the pitch-down moment generated by the wing. The problem may then compound; as the aircraft pitches down, its airspeed continues to climb, increasing the effects of the shock wave and thereby worsening the Mach tuck effect. Left uncorrected, the aircraft may increase its pitch-down tendency and continue to accelerate until airframe failure.

Recovering from Mach tuck requires slowing the aircraft, either by throttle reduction, pitch change, or increasing drag. This is not necessarily as easy as it sounds. Airload on the tail under such circumstances can reach the point where there's not enough pitch authority for recovery. Accordingly, many aircraft are equipped with protective *Mach trim compensators.* These devices track airspeed and, if the onset of Mach tuck is projected, automatically apply up-elevator to slow the aircraft and avoid the encounter.

Critical and Limiting Mach Numbers

An airfoil's *critical Mach number* is the speed at which the airflow over any portion of the upper wing surface becomes supersonic in level flight. It is not an airspeed limitation but simply the speed at which a shock wave develops over the wing.

Limiting Mach number is the highest speed the aircraft can travel before it becomes noticeably uncontrollable. M_{MO}, *maximum operating speed* relative to the speed of sound, is automatically displayed on a jet's airspeed indicator by a self-adjusting indicating needle known as the "barber pole." (See Chapter 7, Limitations.) There is also an aural overspeed warning that warns of the aircraft's maximum operating speed. Due to its sound, this warning device is often called the *clacker.*

Coffin Corner

It might seem that the easiest way to avoid bumping up against the sound barrier would be to fly well below M_{MO}. However, there's another factor to be considered (besides the fact that nobody normal wants to fly slowly in a jet!). As you remember, stall speed for any given configuration increases (in terms of true airspeed) as a function of altitude. The speed of sound, however, decreases with altitude.

At high operating altitudes, the stall speed for a given aircraft may approach its M_{MO}, especially at high aircraft weights where stall speed is highest. This tight spot between stall speed and Mach limit is known as the *coffin corner* and merits close watching in some aircraft. Pilots must be aware of *buffet margins*—the airspeed range between stall buffet and high-speed buffet—and maintain airspeed comfortably between them. As the name implies, in several cases poor airspeed control or pilot confusion as to which limit was being encountered has led to fatal consequences. One way of improving buffet margins is to restrict aircraft altitude by weight (see Fig. 15.2).

Laminar-Flow or Super-Critical Wings

Laminar-flow wings are designed to give an aircraft a high cruise speed by decreasing, as much as possible, the drag created by the airfoil. To do this, a laminar-flow wing must be as thin and flat as possible, thereby delaying upper wing airflow from accelerating to transonic speeds until higher Mach numbers. Also characteristic of laminar-flow wings is that the point of maximum thickness of the wing is located as far aft as possible, allowing air to flow over the wing smoothly with a minimum of turbulence. Many corporate jets are equipped with this kind of airfoil.

Recently, aerodynamic researchers have developed super-critical wings that, like laminar-flow wings, are designed to yield high cruise speeds with minimal drag. The most noticeable characteristic of super-critical wings is that the lower side of the forward portion of the wing, instead of being flat, is more curved than the wing's upper surface. The top of the wing is nearly flat with only a gradual slope downward at its trailing edge. Again, the thickest part of the wing is as far aft as possible, allowing air to flow smoothly over the entire surface with minimal separation. A downward curve near the wing's trailing edge helps slow airflow in that vicinity to subsonic speeds, resisting the formation of a shock wave and reducing the amount of drag.

Swept Wing Aerodynamics

Modern airline and corporate jet aircraft are designed for high-speed, high-altitude, long-range performance. For these reasons, most of these aircraft are constructed with a swept wing configuration. Typically, a sweep angle of around 35 degrees allows jet aircraft to operate at high speeds without a large drag penalty.

The main advantage of a swept wing is that it produces less drag than a straight wing. It does this by delaying the

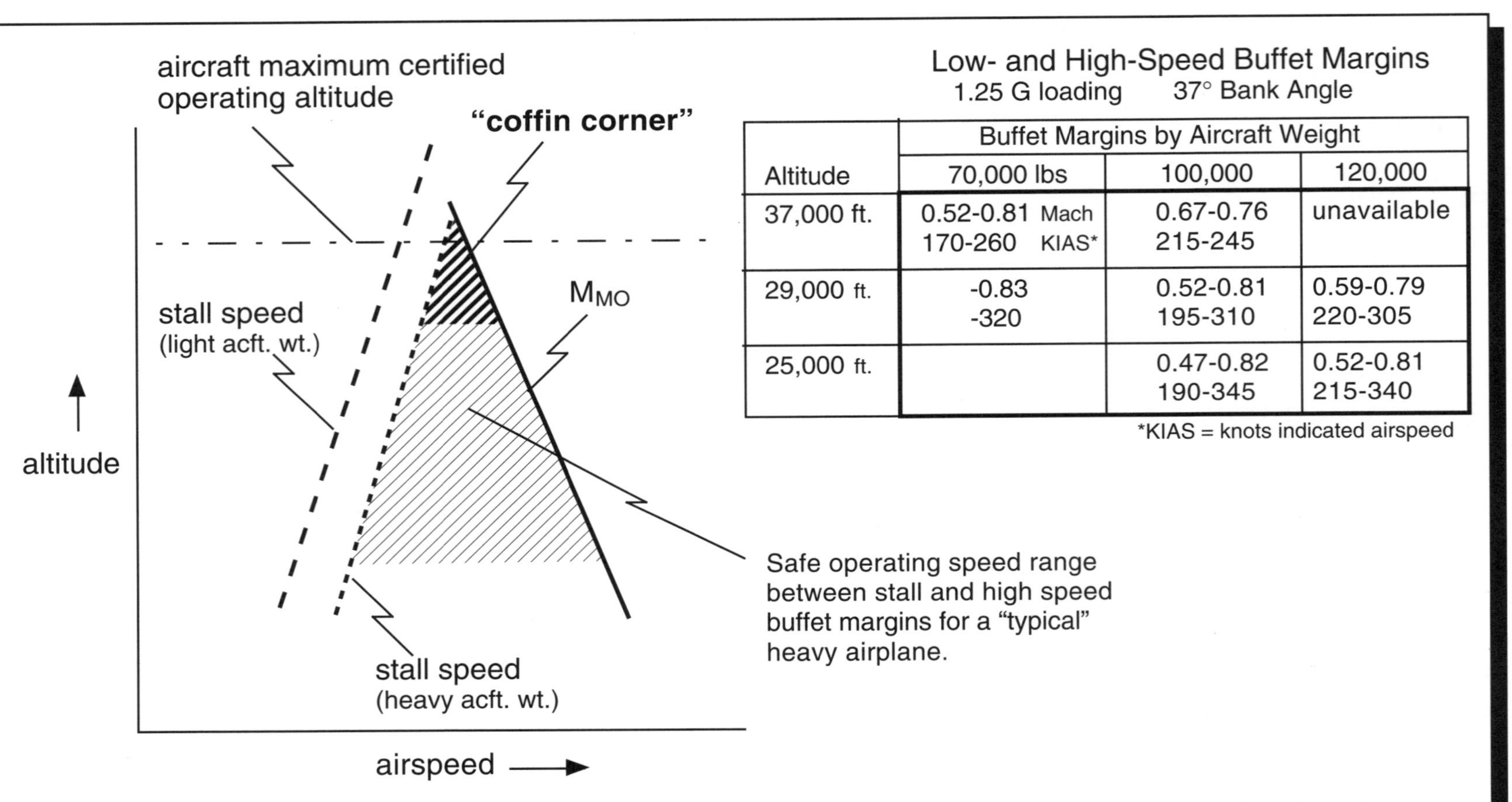

Low- and High-Speed Buffet Margins
1.25 G loading 37° Bank Angle

Altitude	Buffet Margins by Aircraft Weight		
	70,000 lbs	100,000	120,000
37,000 ft.	0.52-0.81 Mach 170-260 KIAS*	0.67-0.76 215-245	unavailable
29,000 ft.	-0.83 -320	0.52-0.81 195-310	0.59-0.79 220-305
25,000 ft.		0.47-0.82 190-345	0.52-0.81 215-340

*KIAS = knots indicated airspeed

Stall speed increases with altitude, while the speed of sound decreases. Therefore, an airplane's safe airspeed range between stall and M_{MO} narrows with altitude. Anything that increases stall speed, such as heavier aircraft weights and steep turns, further reduces safe airspeed range. **Buffet margin charts** depict safe operating airspeeds between stall and high-speed buffets as functions of altitude, angle of bank, and aircraft gross weight. At high gross weights, buffet margin considerations often prevent aircraft from operating at maximum certified operating altitudes.

15.2. "Coffin corner" and buffet margins.

peak effects of drag to higher airspeeds, thereby allowing the aircraft to go faster. This is accomplished because only a component of the airflow velocity flows directly aft or "chordwise" over the swept wing; the balance moves outward or "spanwise" along the wing. Spanwise flow along a swept wing has a lower velocity than the chordwise flow, increasing the wing's critical Mach number and thereby allowing the aircraft to fly faster.

Ordinarily, swept wings produce less lift than straight wings. Therefore, an airplane with swept wings must fly faster than one with straight wings in order to generate the same amount of lift. This can cause problems at the slow airspeeds desirable for takeoff and landing. The issue is addressed through installation of retractable lift-enhancing devices for slower airspeeds, such as leading edge devices (LEDs) and trailing edge flap systems. (See "Flight Controls" in Chapter 5.) The reduced lift of a swept wing does have one other benefit in addition to speed: swept wing aircraft are less sensitive to turbulence than most straight-winged models.

Since swept wings have a tendency to stall at the tip first, most swept-winged aircraft are designed with a slight aerodynamic twist in the wings. The thin outer portion of the wing is designed with a smaller angle of attack, at a given pitch, than the thicker inner portion near the root. This causes the wing to stall first in the middle, rather than at the tips, making the aircraft more controllable at slower airspeeds and/or higher angles of attack. The leading edge devices of larger aircraft are designed to do the same thing at low airspeeds. The outboard "slats" are effective at a higher angle of attack than the inboard "flaps." (See Chapter 5, Figure 5.1.)

Sweeping an aircraft's wings also increases its lateral stability. For further improvement in this area, aircraft designers normally incorporate some positive wing dihedral.

Dutch Roll

One annoying characteristic of swept wing aircraft is a predisposition toward *Dutch roll,* which refers to the tendency of an aircraft to roll whenever it yaws. On many swept wing aircraft any oscillation in yaw develops almost immediately into a rolling movement back and forth. Normally, Dutch roll is caused by turbulence, but it can also be caused by poor pilot technique. In any case, the combined yawing and rolling oscillation is undesirable and dangerous, if let out of hand, from standpoints of aircraft control and structure. At the very least, once those passengers in the back get to "rocking and rolling," someone's gonna hear about it!

Dutch roll is caused by an aircraft's tendency to sideslip slightly when the aircraft yaws. One wing yawing forward in this situation changes the effective span between left and right wings. The wing yawed forward momentarily creates more lift than the one on the other side. The result is that the forward wing rises and starts a rolling movement. The problem is aggravated by the fact that the forward wing, due to its increased lift, also generates more drag, pulling that wing back once again and starting an oscillation in the other direction (see Fig. 15.3).

A *yaw damper* is installed in aircraft with a tendency toward Dutch roll. This automatic device (often part of the autopilot) senses yaw and counters it with control inputs before Dutch roll can develop. In many swept wing aircraft an out-of-service yaw damper is a "no-go" item.

Fixed Aerodynamic Surfaces

In recent years, a number of aerodynamic design details have begun to appear on turbine aircraft in the form of *fixed aerodynamic surfaces.* These additional surfaces serve to enhance or correct aircraft performance in various ways without the need to re-engineer existing major structures of the aircraft. One driving parameter for these add-on surfaces has been fuel efficiency, especially in airline aircraft where fuel costs account for as much as 50 percent of the operating costs. Let's look at some of the most common fixed aerodynamic surfaces.

Many newer high-speed aircraft are equipped with *winglets.* These devices look like small vertical stabilizers mounted on the wing tips. On a normal wing, air tends to flow from the underwing high-pressure area around the wing tip to the wing-top low-pressure area. The result is reduced lift. By restricting airflow around the wing tips, winglets improve the effective span of a wing and therefore its lift. Aircraft performance is improved, especially at altitude (see Fig. 15.4).

Tailets are small vertical surfaces mounted on the horizontal stabilizer. They are designed to provide additional directional stability, especially on T-tail aircraft at low airspeeds and high angles of attack when airflow to the tail may be partially blocked by the wing (see Fig. 15.5).

Vortex generators are small, fixed aerodynamic surfaces that may be mounted on wings, tail, elevator, engine pylon, or fuselage. Normally installed in multisurface arrangements, vortex generators create small vortices that direct airflow into the boundary layer to prevent airflow separation from the wing at high airspeeds (see Fig. 15.5).

Boundary-layer wing fences are fin-like vertical surfaces mounted to the upper surface of a wing and used to control airflow. On swept wing aircraft, boundary-layer wing fences are located approximately two-thirds of the way toward the wing tips to prevent airflow from drifting toward the tip of the wing. Straight-wing turbine aircraft employ wing fences to direct airflow around the trailing edge flaps.

Vortilons may be mounted on the underside of an aircraft's wing, near the ailerons. They are similar in appearance

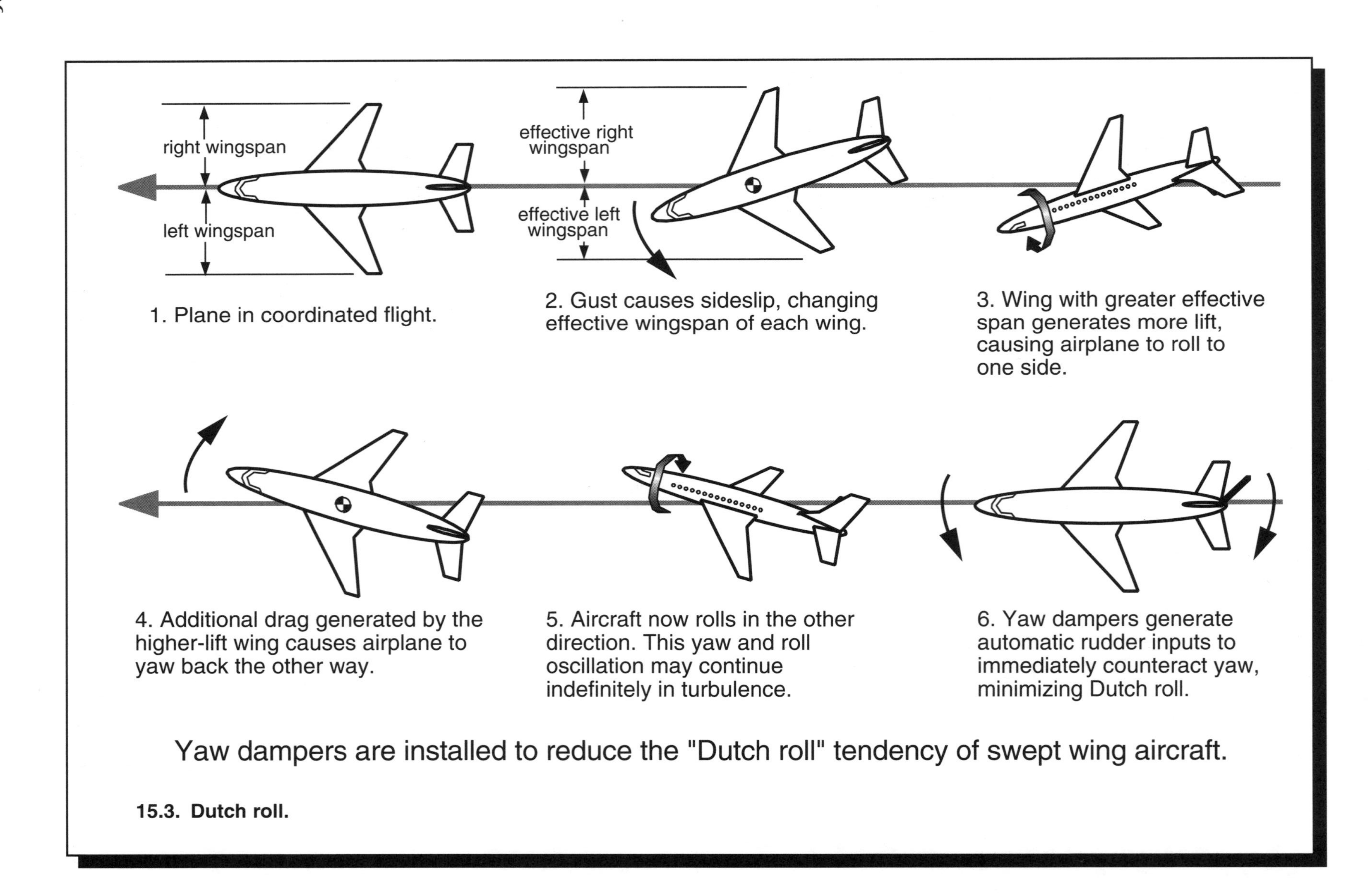

15.3. Dutch roll.

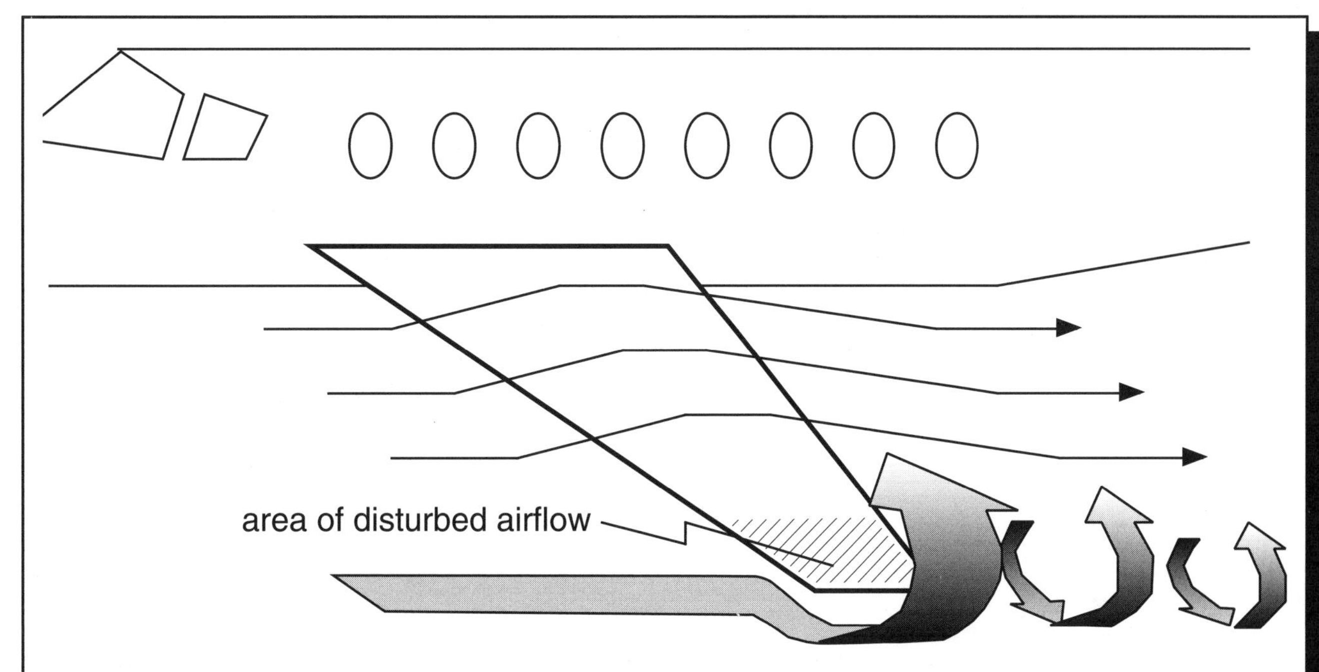

On a normal wing, underwing airflow rolls around the wing tips to the low-pressure area on top. Effective wingspan is reduced and drag increased due to the disturbed airflow atop outboard portions of the wings.

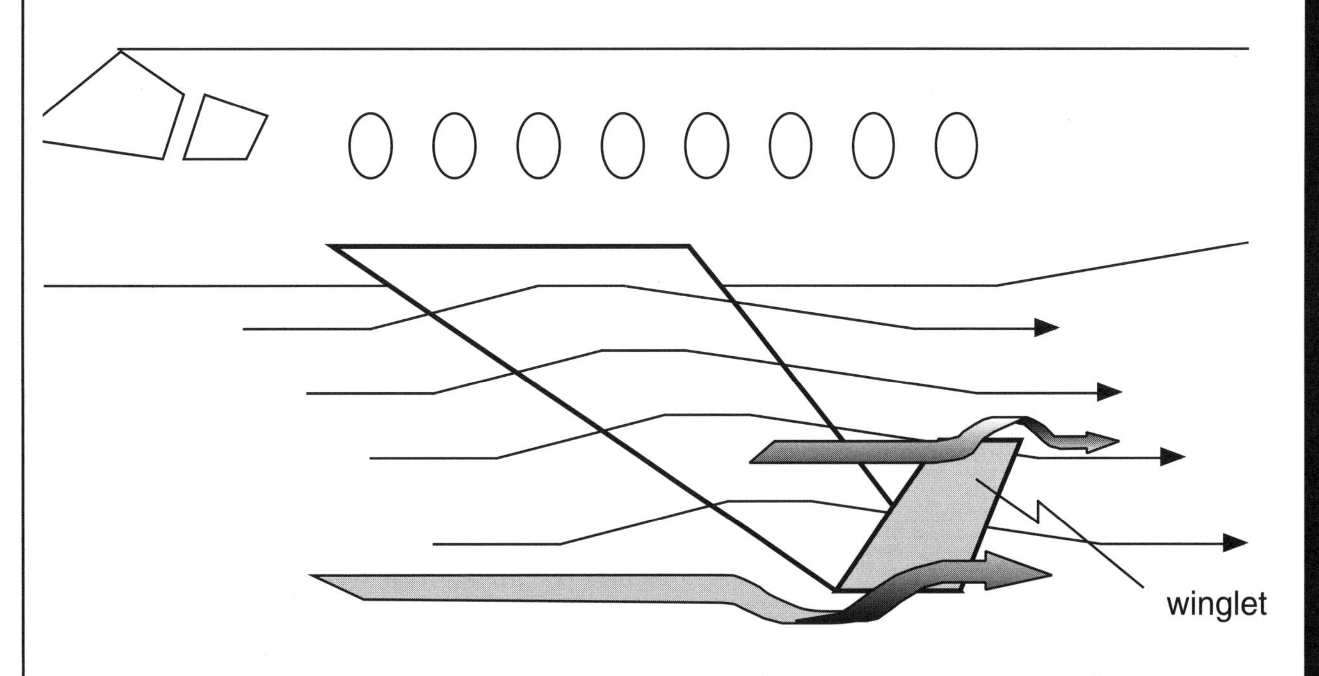

By reducing airflow around the end of a wing, winglets increase effective span and reduce drag.

15.4. Winglets.

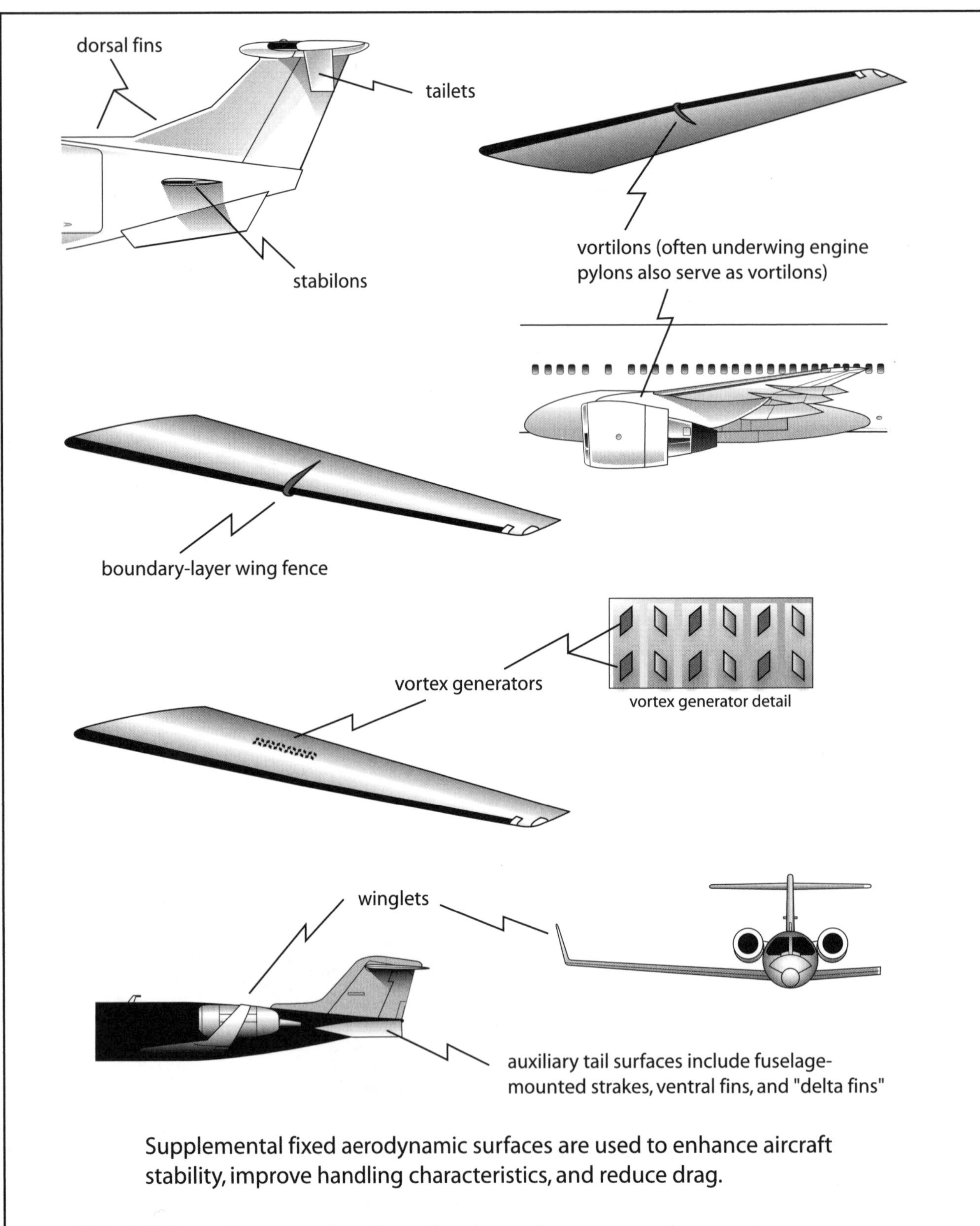

Supplemental fixed aerodynamic surfaces are used to enhance aircraft stability, improve handling characteristics, and reduce drag.

15.5. Tailets, vortex generators, boundary-layer wing fences, vortilons, stabilons, and ventral fins.

to a boundary-layer wing fence; however, they are really just large vortex generators designed to control turbulent airflow. Incidentally, many engine pylons serve the dual purpose of attaching the engine to the airframe and acting as large vortilons (Fig. 15.5).

Strakes are plates or fins that are mounted on the aircraft to direct airflow for specific purposes such as stall prevention or to assure airflow to the engine intakes at all flight altitudes (Fig. 15.5).

Stabilons are small horizontal surfaces sometimes installed near the tail of an aircraft. They help provide additional pitch stability and also improve stall recovery performance (Fig. 15.5).

Dorsal fins are sometimes mounted atop the fuselage and faired into the vertical stabilizer to increase directional stability (Fig. 15.5).

Ventral Fins are vertical surfaces mounted under the aft portion of the fuselage. They are used to increase both high-speed and low-speed directional stability. Related surfaces are often mounted diagonally near the tail cone to resist stalls at high angles of attack, in many cases eliminating the need for other stall-preventing devices (Fig. 15.5).

Stalls

At the low end of the speed range in turbine aircraft there is little difference in stalls to discuss, relative to the aircraft you've been flying. However, it is worth noting that turbine aircraft generally have *stick shakers* installed, rather than horns or lights, to indicate approach to stall. These devices literally shake the pilots' yokes as the airplane approaches stall.

Some aircraft come equipped with a system called a *stick pusher.* A stick pusher is designed to automatically lower the pitch of the aircraft when a stall is imminent (e.g., 1.07 Vs). The stick pusher may operate on the control wheel or the elevator control surface, depending on the particular airplane.

IFR Operations in Turbine Aircraft

In most respects, operations under IFR (Instrument Flight Rules) are similar for all civil aircraft. However, there are a few IFR procedures specific to flight at high altitudes (the realm of most turbine aircraft) and to turbine aircraft themselves. Note that in some cases, regulations and procedures specify turbine-powered aircraft; in other cases, turboprops are differentiated from turbojet aircraft. For purposes of IFR operating regulations, the term "turbojet" includes both turbojet and turbofan-powered aircraft and means any "pure jet" or nonpropeller-driven turbine aircraft.

Profile Descents

A *profile descent* is simply a published transition procedure from cruise flight to an instrument approach. Courses, headings, and altitudes are prescribed, usually from a high-altitude cruise structure intersection or navaid down to interception of an ILS (instrument landing system) or nonprecision approach. Profile descents may begin as far out as 100 miles or more from the destination airport, depending on the aircraft's cruising altitude. A profile descent normally allows a continuous descent from cruise, interrupted only by a brief level-off to slow to 250 kias (knots indicated airspeed) by 10,000 feet MSL (mean sea level).

Jet Routes

Turbine aircraft routinely travel at high altitudes above 18,000 feet. Altitudes at and above 18,000 feet are known as *flight levels.* Twenty-one thousand feet, for example, is known as "FL 210" and pronounced as "flight level two-one-zero." At these altitudes, aircraft no longer travel on victor airways but instead use *jet routes,* or J-routes. The jet route system is a VOR-based route structure, similar to the victor airway system. J-routes are established from FL 180 up to and including FL 450.

On high-altitude enroute charts jet routes are depicted in black on NOS charts and blue on Jeppesen charts. They are identified on the charts by the letter "J" and the J-route number. When examining a high-altitude chart, the first thing you'll notice is the length of airway legs between VOR stations. Most legs are over 100 nm in length, due to the high speeds at which turbine aircraft travel. If J-routes had short legs like victor airways, flight crews would be continually changing frequencies and adjusting the course knob. Because of the high altitudes involved with J-routes, VOR reception at long ranges is possible.

Above FL 450, aircraft typically navigate directly from point to point, often with some type of RNAV system. (See Chapter 13, Navigation, Communication, and Electronic Flight Control Systems.) ATC approval and radar monitoring are required on these types of routings.

Altimetry and IFR Cruising Altitudes at Flight Levels

A few more words are in order regarding operations up in the flight levels. All operations at or above 18,000 feet must be under IFR flight rules, and all aircraft must set their altimeters to 29.92 inches Hg. One reason for the standard altimeter setting is that high-altitude weather, at any given location, doesn't necessarily correspond to the surface weather below where altimeter settings are established; it would be dangerous to project surface altimeter settings all the way up to very high altitudes. Another reason is that it would be impractical to require flight crews to reset altimeters to local reporting stations within 100 nm of the aircraft, as is required below 18,000 feet. At jet speeds, that would require an altimeter change roughly every ten minutes.

Requiring a standard altimeter setting above 18,000 feet does cause a few complications, however. If the surface altimeter setting is very low, aircraft flying up at 17,000 or 17,500 feet could start bumping into those flying at FL 180 with 29.92 altimeters. Therefore, the FAA has defined limitations as to what minimum flight levels may be used under given situations. For example, FL 180 may not be used when the local altimeter setting falls below 29.92. As the altimeter continues to fall, more flight levels are restricted. (See FAR 91.121 for a full breakdown of minimum flight levels.)

IFR cruising altitudes between FL 180 and below FL 290 are set up similarly to IFR cruising altitudes below FL 180. That is, when magnetic course is from zero degrees through 179 degrees, any odd-numbered flight level may be used. When operating with a magnetic course of 180 degrees through 359 degrees, even-numbered flight levels may be used. However, at 29,000 feet there's a significant change in allowable IFR altitudes. From FL 290 up to FL 450, altitude separation between opposite direction traffic increases to 2,000 feet. Therefore, on eastbound magnetic courses of zero degrees through 179 degrees, allowable flight levels beginning at FL 290 have 4,000-foot intervals (e.g., flight levels 330, 370, 410). On westbound magnetic courses of 180 degrees through 359 degrees, allowable flight levels above FL 280 begin at FL 310 and continue at 4,000-foot intervals (e.g., flight levels 350, 390, 430). Note that even-numbered flight levels are not used above FL 290.

Category I/II/III Approaches

As you instrument-rated pilots know, standard ILS approach procedures are certified by the FAA, based on specific installation, down to lowest minima of 200 feet DH (decision height) and 1/2-mile visibility. These *CAT I approaches* (Category I) may be flown by any IFR operator meeting basic IFR licensing, currency, and equipment requirements. However, the FAA has established that with proper crew training, installed equipment, and aircraft certification, lower precision approach minima may be individually approved for specific operators and airports.

Special low-minimum operations may be approved as (from highest minima to lowest) CAT II, CAT IIIa, CAT IIIb, and CAT IIIc. Each step to lower minima requires additional training and equipment. CAT II allows operations down to 100 feet decision altitude and 1200 feet runway visual range (RVR), while CAT III approaches have, in most cases, no decision altitudes, only visibility limits (e.g., CAT IIIa RVR must not be less than 700 feet, while CAT IIIb RVR must not be less than 150 feet). Although most CAT III approaches do not have a decision altitude per se, most operators use an "Alert Height" in its place. An *alert height* is typically 50 feet above the runway. If an aircraft or ground-based equipment failure occurs prior to this alert height, the approach is discontinued and a missed approach initiated.

Until recently, it was required that all approaches to lower than basic CAT I minima be flown with an autopilot with approach coupler or flight director system. However, with the latest-generation heads-up displays (HUDs) or heads-up guidance systems (HGSs), CAT IIIa approaches are now permitted with a DH of 50 feet and an RVR as low as 700 feet, *hand-flown!*

Under CAT IIIc, aircraft can land under "zero-zero" conditions without reference to ceiling or visibility. These operators must use aircraft with full "auto-land" capability, including redundant autopilots that cross-check each other throughout approach and landing. Procedures are highly refined and specialized and are not approved in some countries for zero-visibility landings. (Paris has a DH of 20 feet.) Most low-minima operators are major airlines, due to equipment, certification, and training costs. Only a few aircraft and operators are certified down to CAT IIIc; most of those fly extremely long international routes where weather diversion at the destination is highly undesirable.

Holding

In 1999 the FAA standardized holding speeds for all aircraft. From the published minimum holding altitude to 6000 feet, maximum holding airspeed is 200 knots. From 6001 feet to 14,000 feet, the maximum is 230 knots, and above that altitude maximum holding airspeed is 265 knots. In some areas of heavy air traffic congestion, ATC may restrict holding airspeeds between 6001 and 14,000 feet to 210 knots. (When applicable this limit is published on instrument charts next to the published holding pattern icon.)

For both turboprops and turbojets, higher-than-normal holding speeds may be approved by ATC if necessary due to turbulence, icing, or other factors such as aircraft performance or configuration. (For more information on holding refer to the Aeronautical Information Manual.)

DME (distance measuring equipment) holds are more commonly assigned to turbine aircraft than to piston models. For DME holding patterns, leg lengths are determined by distance from a navigational fix, rather than by timing. (ATC may assign leg length as part of the holding clearance, or it may be depicted on the instrument chart with the holding pattern icon.)

Extended Range Twin-Engine Operations (ETOPS)

Extended range twin-engine operations (*ETOPS*) is an FAA certification that allows twin-engine airplanes to operate for up to 180 minutes from the nearest enroute alternate airport.

For many years the FAA required twin-engine airplanes to operate over routes that kept the aircraft within 60 minutes of a suitable alternate airport at all times. When this "60-minute rule" came into effect more than fifty years ago, aircraft and engine reliability were poor enough that it was not uncommon for aircraft to experience engine failures during long-range flights.

But as you can imagine, on many overseas routes the "60-minute rule" required twin-engine aircraft to fly longer and less desirable routes to remain within range of alternate airports. Therefore, most extended range flying prior to the mid-1980s was done by three- or four-engine aircraft, which were allowed to fly more direct routes. However, as airframe and powerplant technology improved with each generation of turbine powered aircraft, so did reliability, systems redundancy, and aircraft range.

With introduction of the Boeing 757/767 in the early 1980s, major airlines began petitioning for relief from the "60-minute rule." Eventually, the FAA developed a comprehensive certification process for extended range twin-engine operations (ETOPS). This process dictates various operational performance requirements, dispatch reliability rates, operator qualifications, and maintenance requirements. ETOPS certification originally granted airlines the authority to operate on routings that were 120 minutes from a suitable enroute alternate airport. Authority can now be granted to carriers for twin-engine aircraft to fly up to 180 minutes from an alternate.

International Flight Operations

While international flight operations are beyond the scope of this book, the simple topic of IFR holding procedures offers a good opportunity to raise the issue. Turbine aircraft routinely fly to all corners of the earth. While aircraft aerodynamics and systems are similar anywhere in the world, flight rules are not. Accordingly, international flight operations require a good deal of specialized pilot expertise and planning. Over the time period when this book was written, FAA holding speeds for turbine aircraft changed at least three times. Seasoned international pilots take a few such changes in stride.

There are at least five different sets of holding procedures in international use, as of the time of this writing. It's not uncommon in some areas of the world to have different holding rules for approaches at two sides of the same airport! Similar variations apply to everything from cruising altitudes to altimetry.

Probably the most noticeable difference between international flight operations and domestic flight operations is radio communication, with phraseology and procedures varying in many respects. In most parts of the world outside the United States, strict adherence to proper terminology is required ("diagonal" versus "slash," "decimal" versus "point," "zero" versus "oh", etc.); however, other areas of the globe operate in a mix of aviation English, local language, and local slang. American slang is generally not understood and local phraseology may not be familiar to many U.S. pilots, so be careful!

International operations should be studied every time you fly out of your home country, not just at recurrent training sessions.

Wake Turbulence

When stepping up to turbine-powered aircraft, you will often be operating at airports with high traffic densities and where the majority of traffic is composed of large aircraft. Most high-density airports have closely spaced parallel or intersecting runways, with approach and departure aircraft strung out 15–75 miles in several directions.

Operating under these conditions you'll constantly be confronted with situations where understanding of *wake turbulence* is required. You must know its causes, effects, and how to avoid it in order to give your passengers a smooth and safe ride. This section reviews some common procedures for avoiding wake turbulence hazards.

Wing Tip Vortices

As you know, an aircraft wing creates lift by developing a pressure differential: low pressure over the wing's upper surface and higher pressure underneath the wing. For most of the wingspan, this pressure differential translates directly into lift. However, something a bit different happens out at the wing tip.

As lift is created, some of the high-pressure air from under the wing rolls up around the wing tip toward the lower-pressure area on top. The result is sort of a mini "horizontal tornado" called a *wing tip vortex.* These wing tip vortices trail from the aircraft's wing tips and are the source of the hazard known as "wake turbulence." Since vortices are byproducts of lift, they are created from the time an aircraft begins to fly until it touches down on landing (Fig. 15.6).

Wing tip vortices tend to be strongest when associated with a heavy aircraft that's flying slowly, with its gear and

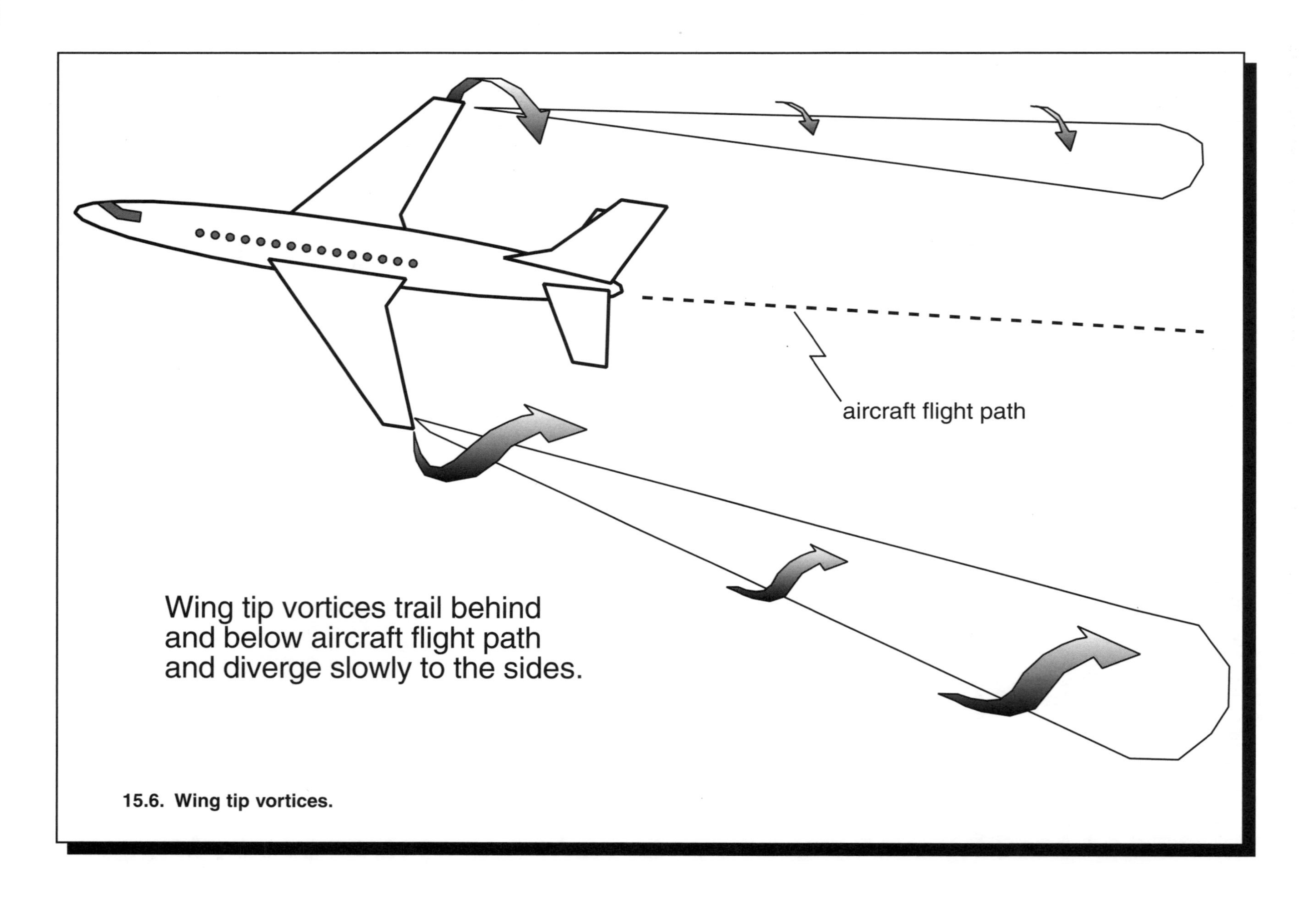

15.6. Wing tip vortices.

flaps retracted. (Extended gear and flaps tend to break up the vortices, to some degree.) Vortices are most dangerous to a trailing aircraft when the wingspan of that aircraft is less than the diameter of the vortex's rotational flow. This is especially true if the encountering aircraft flies into the wake vortex of another aircraft on the same heading. In that case the vortex may induce a violent roll, with potentially disastrous consequences if the trailing aircraft is near the ground.

When the wingspan of an aircraft flying into a wake vortex is greater than the diameter of the vortex, the resulting induced roll may be countered effectively by the ailerons. However, the resulting turbulence may be serious or even severe. Either way, pilots must learn to visualize the locations of wing tip vortices and then alter flight paths to avoid them.

Identifying Likely Areas of Wake Turbulence

In order to visualize wake turbulence locations, remember that wing tip vortices are created when an aircraft wing is creating lift. On its takeoff roll an aircraft begins to create wing tip vortices as it rotates and starts to fly. Upon landing, an aircraft ceases to create wing tip vortices around the touchdown point (see Fig. 15.7).

Flight tests have shown that wing tip vortices near the generating airplane are spaced approximately a wingspan's length apart and diverge gradually as distance from the aircraft increases. They tend to sink at a rate of 500–800 feet per minute, slowing and weakening as they descend.

Under no-wind conditions, large aircraft flying near the ground produce vortices that drift outward away from the flight path along the ground at 2–3 knots. Keep in mind, however, that wing tip vortices drift with the wind (see Fig. 15.8). Therefore, a few knots of crosswind may cause upwind vortices to remain on the runway for an extended period of time. That same crosswind can move downwind vortices toward a parallel runway. A head wind or tail wind can easily move wake vortices into touchdown or departure zones, so be aware of the wind and be careful!

Unfortunately, wake turbulence is not so predictable that every pilot can avoid it all of the time. The best wake turbulence avoidance skills are to understand its causes, to be continually aware of traffic activity, and to consider the effects of wind. Following is a summary of some widely practiced wake turbulence avoidance recommendations.

Departure

On departure, note the rotation point of the preceding airplane. If possible, within normal procedures, plan to rotate before that point. After a missed approach by a large aircraft, allow two minutes to elapse before beginning takeoff roll so there's time for the wing tip vortices to dissipate. Keep in mind the effects an engine failure would have on your aircraft's performance after departure and what it would do to your projected avoidance strategy.

Approach and Landing

Listen on the radio for the "big picture" of surrounding traffic. Always be aware of the aircraft you are following. What's the type? Speed and altitude? Check your TCAS (if so equipped); how far behind are you in trail? To avoid wake turbulence, many pilots choose to fly "one dot high" when trailing a heavy aircraft on an ILS (instrument landing system) approach. When landing behind a large aircraft, stay above that aircraft's flight path and plan to land beyond its touchdown point. Try to spot the touchdown smoke from the aircraft's tires ahead of you as an indication of which way and how fast the wind is blowing. This will give you some idea of the location of its wake turbulence.

General Procedures for Avoiding Wake Turbulence

Always comply with ATC separation assignments to take full advantage of wake turbulence precautions built into the ATC system. When operating in a terminal area, pay attention to the flight paths of aircraft around you. Avoid flying in areas where wake turbulence might be expected, such as behind and below heavier aircraft. Always keep in mind the effect of wind on wing tip vortices, especially when operating from airports with closely spaced parallel or intersecting runways. Listen to wind reports from the tower, and keep an eye on the wind sock!

Finally, remember that while ATC often provides separation for purposes of wake turbulence protection, the pilot (as always) is ultimately responsible. Ask for special handling when you're concerned about potential wake turbulence. For instance, let's say you're departing in a heavily loaded turboprop behind a large jet. It may be clear that you can lift off well before the jet's rotation point to avoid its wake. But the jet can outclimb you, and given the same course, you could fly into its wake. "Tower, Jetstream 3456Z requests an immediate right turn after takeoff for wake turbulence avoidance." You'll get it, or at worst the tower will offer you the option to wait.

Most turbine pilots operate in environments where their wake turbulence knowledge and judgment is tested daily. Remember to visualize the wing tip vortices of the aircraft around you and to alter your flight path accordingly.

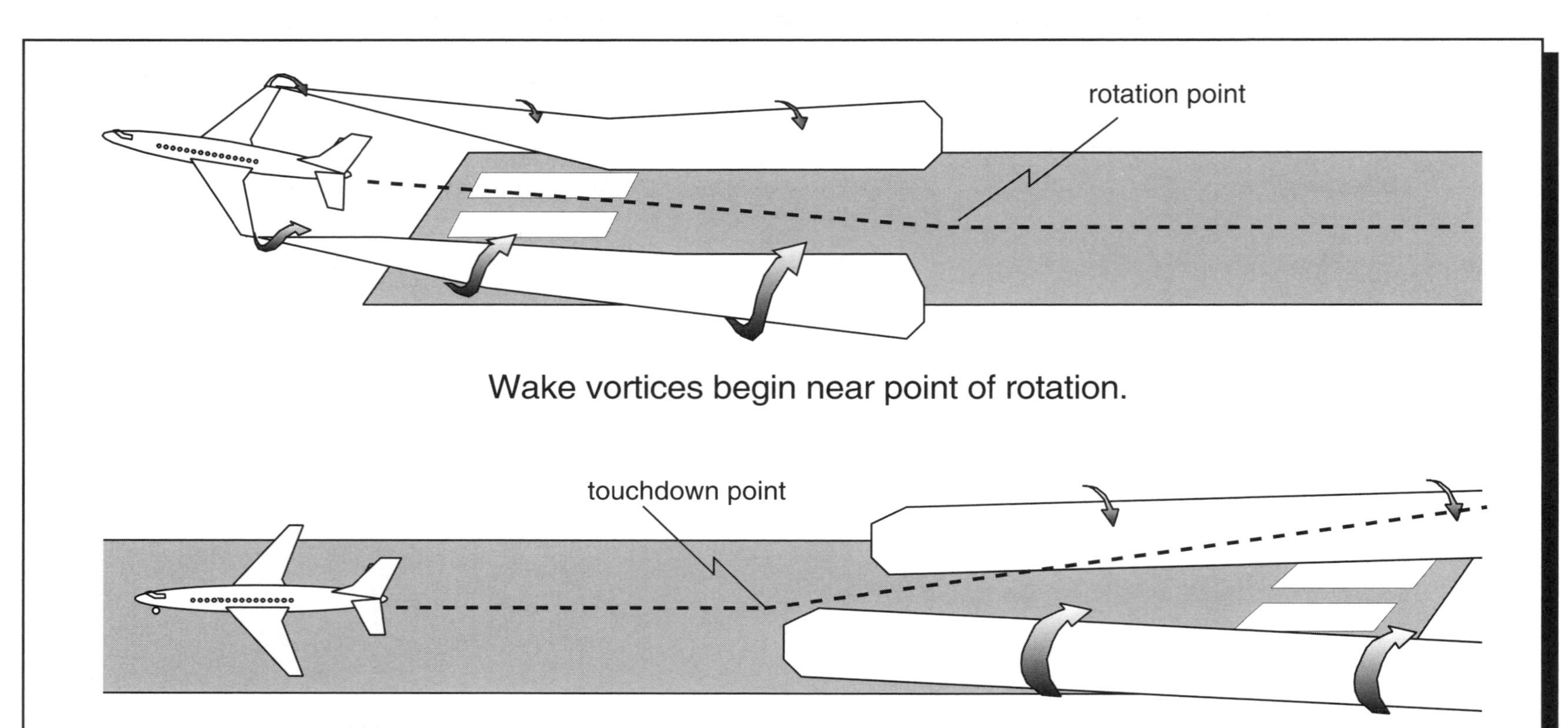

Most pilots plan rotation and touchdown so as to avoid flight segments of preceding aircraft over the runway. This often means rotating before the takeoff point of a preceding departure and landing beyond the touchdown point of a preceding arrival. Wind drift of wake turbulence must also be considered and, in any case, separation standards observed.

15.7. Wing tip vortices on takeoff and landing.

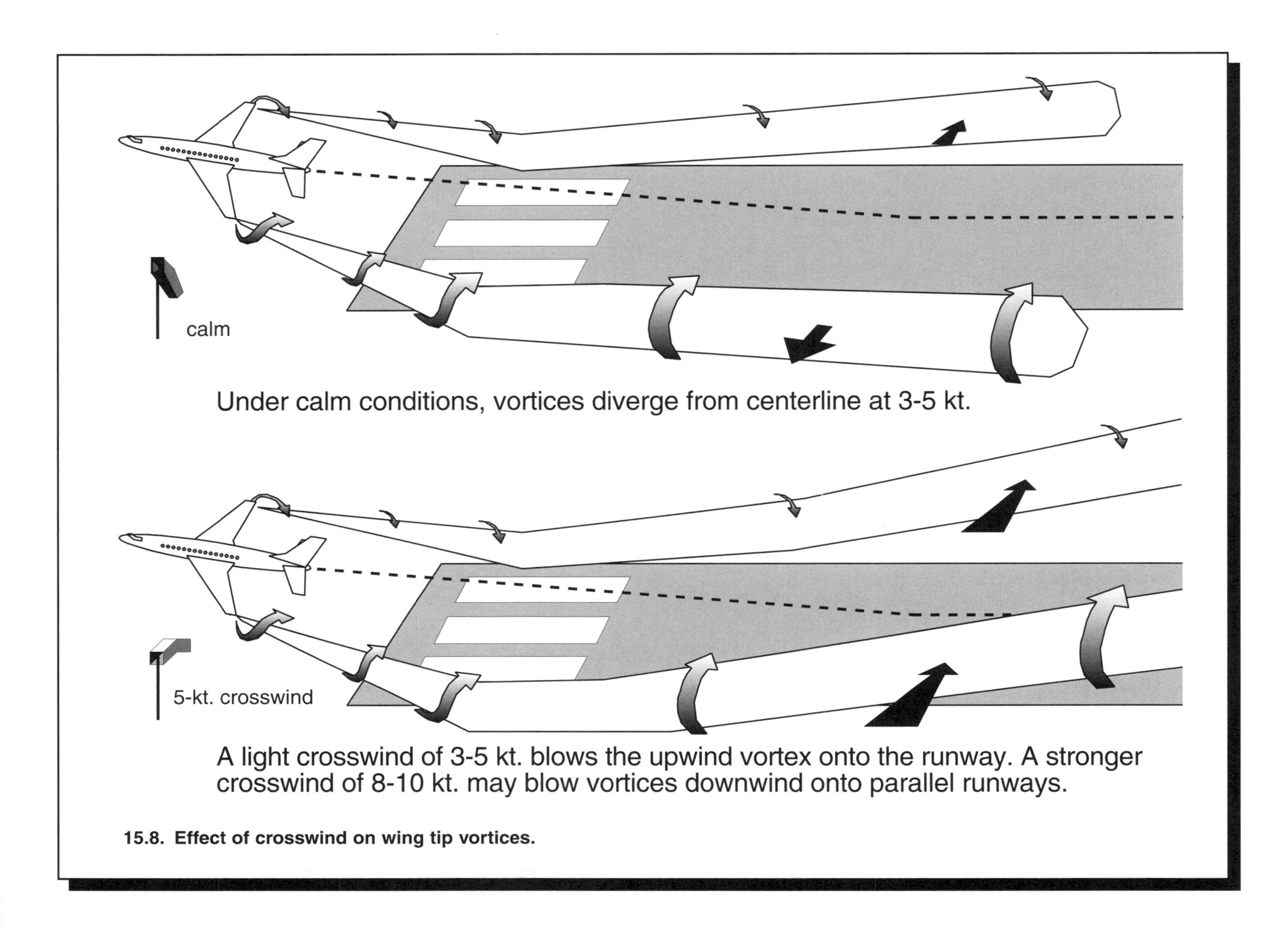

15.8. Effect of crosswind on wing tip vortices.

Air Rage

Unfortunately in these days of full flights and crowded airport terminals, long check-in lines and flight delays have become all too common. As a result, some emotionally unstable passengers, fed up with the whole travel process, misbehave to a point where they become a threat to air safety.

Air rage incidents are on the rise and occur at a variety of levels. The most common cases involve passengers being verbally abusive to flight attendants or other passengers, with extreme cases involving totally out-of-control passengers breaking down cockpit doors and attempting to wrest control of the airplane away from the pilots.

According to a recent NASA study of passenger misconduct and effects on flight crews, passenger disruptions may cause pilots to make serious errors such as altitude deviations, airspeed and navigational errors, and runway incursions. In almost half of air rage cases serious enough to warrant cockpit attention, flight crews experienced some level of distraction from flying duties, with half of those distractions leading to pilot deviations. In almost a quarter of study incidents, a flight crew member left the cockpit to assist flight attendants in dealing with an unruly passenger.

Given this alarming situation, it's no wonder that flight departments are increasingly training their pilots in methods to prevent passenger distractions from interfering with flight duties. Some airlines even simulate disruptive passenger situations during recurrent pilot simulator training. Clearly, air rage has become a serious threat to aviation safety, and pilots must be especially careful during such incidents to avoid making serious cockpit errors, not to mention the challenges of dealing with the unruly passengers themselves.

CHAPTER 16

Weather Considerations for Turbine Pilots

WEATHER is an especially important topic to turbine pilots because most fly true all-weather operations. Company CEOs authorize corporate airplanes because of their desire for flexible, dependable travel. Charter operators and regional and major airlines earn their money from passengers who expect arrival at their destinations on time, every time. As a result, the turbine aircraft operated by these companies are expected to fly in all but the most threatening weather. Pilots and equipment must be up to the task.

This chapter covers some of the weather phenomena that you're likely to experience in turbine pilot operations. Weather is a complex and interesting topic and one that deserves your attention in detail far beyond the scope of this book. Truly professional pilots are lifelong students of weather.

Turbine pilots must be expert on both low- and high-altitude weather. Low-altitude weather, due to the approaches and departures that must be made to meet the schedule; high-altitude weather, for enroute travel at turboprop and jet altitudes. Turbine aircraft are most fuel efficient at high altitudes; the ride there is generally smoother, and they are above most of the hazardous weather typically encountered at lower altitudes. High-altitude winds can be very strong and therefore tremendously helpful or problematic, depending on conditions and direction of flight.

Low-Altitude Weather: Wind Shear and Microbursts

By the time you transition to civilian turbine aircraft, you should be pretty knowledgeable about low-altitude weather and associated IFR operations. Therefore, very little of that is covered here. At the same time, you've probably not yet been flying in some of the really heavy weather encountered in corporate and scheduled flight operations.

Low-level wind shear and microbursts have been hot topics among turbine pilots in recent years. A string of major air carrier accidents has led to meteorological studies of wind shear associated with fast-moving fronts, downbursts, and microbursts. This section covers some of the characteristics and hazards of these phenomena, along with general procedures recommended by most training departments for successfully dealing with them.

Wind Shear

Wind shear occurs whenever two or more adjacent masses of air are moving in different directions, resulting in a "tearing action" where they meet. Wind shear can occur in horizontal or vertical planes. *Low-level wind shear* refers to occurrences within 1000 feet of the ground.

Wind shear can result from any source of shifting winds or vertical air movements. As such, wind shear is often associated with passage of fast-moving weather fronts. Strong winds and uneven heating in mountainous areas are also prime culprits. In each of these cases wind shear may occur in most any type of weather conditions, including clear skies with no associated visible weather.

Microbursts

Wind shear has long been associated with thunderstorms, where powerful updrafts and downdrafts exist side by side. These conditions can easily occupy areas exceeding 15 miles in diameter. Areas of extreme downdraft are sometimes called "downbursts." Exceptionally dangerous instances of wind shear have been documented with localized downbursts of air known as *microbursts* (Fig. 16.1).

Microbursts are typically associated with very strong thunderstorms. "Wet microbursts" typically occur in the portion of a thunderstorm cell containing the heaviest rain. "Dry microbursts" occasionally occur below virga, particularly in desert and high plains areas. ("Virga" is precipitation that evaporates before reaching the ground.) Look for blowing dust, in these cases, to indicate the presence of downdrafts.

While generally fairly small in size, microbursts are dangerous because of their intensities and the relative difficulty of predicting them. The downdraft of a microburst is typically concentrated in an area of less than a mile in diameter. Vertical velocities within and around them may reach several thousand feet per minute, with localized horizontal outbursts sometimes exceeding 120 knots. The force of a microburst can be incredible. In several recorded cases large circular patches of forest, including trees 2 feet in diameter, have been blown down.

Effects of Microbursts on Aircraft

Wherever a powerful downdraft exists, air striking the ground (within 1000 feet or more of it) blows radially outward from the core of the downdraft. Horizontal wind shear in the vicinity of a microburst may lead to airspeed changes of 40 knots or more. This can be a real problem for aircraft, especially when flying across a microburst within 1000 feet of the ground, due to the potential difficulties of recovering from the encounter. Large, heavy turbojet aircraft are particularly vulnerable to microbursts due to their large mass and the "spool up" time required for their engines to produce full power.

The classic scenario for a downburst encounter goes something like this. As an aircraft on an ILS approaches a microburst, it encounters a rapidly increasing head wind due to strong horizontal outflow. This causes the aircraft's indicated airspeed to suddenly increase, and it destabilizes approach relative to the glideslope. The crew chops power to regain its proper position and airspeed on the ILS but then enters the downdraft. The airplane is now in the unfortunate position of being at reduced power in a severe downdraft. If the crew is fortunate enough to exit the downdraft safely, the airplane immediately enters the area where wind is radiating away from the downburst, now as a tail wind. Indicated airspeed now drops. Losing 30 or 40 knots on approach can be deadly in any aircraft. You can see what a confusing and dangerous situation this combination represents (see Fig. 16.2).

Avoidance Procedures

In order to avoid microbursts and associated wind shear, it's important to analyze all available weather information prior to flight. Monitor frontal activity, since frontal passage indicates probable large changes in wind direction and possible low-level wind shear. Particular attention should be paid to thunderstorms in the departure area and to those forecast for arrival time at the destination airport. Warm temperatures and large temperature-dewpoint spreads are other signs that microbursts could occur. Pay close attention to PIREPS (pilot reports) or controller reports of local wind shear.

If conditions exist for wind shear in any form, monitor the reports of aircraft ahead of you on the approach. ATC may report that "The Lear ahead of you had no problems." Just remember that, especially when thunderstorms are around, the Lear's report doesn't ensure a safe ride.

Microbursts are particularly deadly due to their transitory nature. While typically lasting ten to fifteen minutes, a microburst goes through a definite life cycle. Descending air in the convective cloud mass rapidly develops intensity, plunges to the ground, spreads in a violent burst of outflowing winds, and ultimately dissipates. The danger to an aircraft is greatest when passing through at the moment of greatest intensity, when the downdraft reaches the surface and begins its curl outward.

Because microbursts can develop so quickly, your aircraft could be the one "in the wrong place at the wrong time," even if the airplane ahead did pass through uneventfully. The problem is especially difficult since there's little useful correlation between the appearance of a thunderstorm cell, its radar signature, and the presence of microbursts. Many companies prohibit their pilots from starting an approach when microbursts are reported. The safest course of action is to avoid takeoffs, landings, and approaches when thunderstorms are in the area.

Low-Level Wind Shear Alerting Systems

Some airports are particularly vulnerable to wind shear problems. Based on criteria such as severe weather climatology, topography, and the number of commercial operations, many airports have been equipped with *low-level wind shear alerting systems* (*LLWAS*). This type of computerized system incorporates a series of remote sensors, located around the perimeter of an airport, to measure wind speed and direction near the ground. The LLWAS computer compares remote sensor data with wind measurements at the center of the field. If the vector wind difference between these sensors is large enough, tower controllers are alerted to provide traffic with a wind shear advisory. Ground-based Doppler radar is also being installed at many airports to

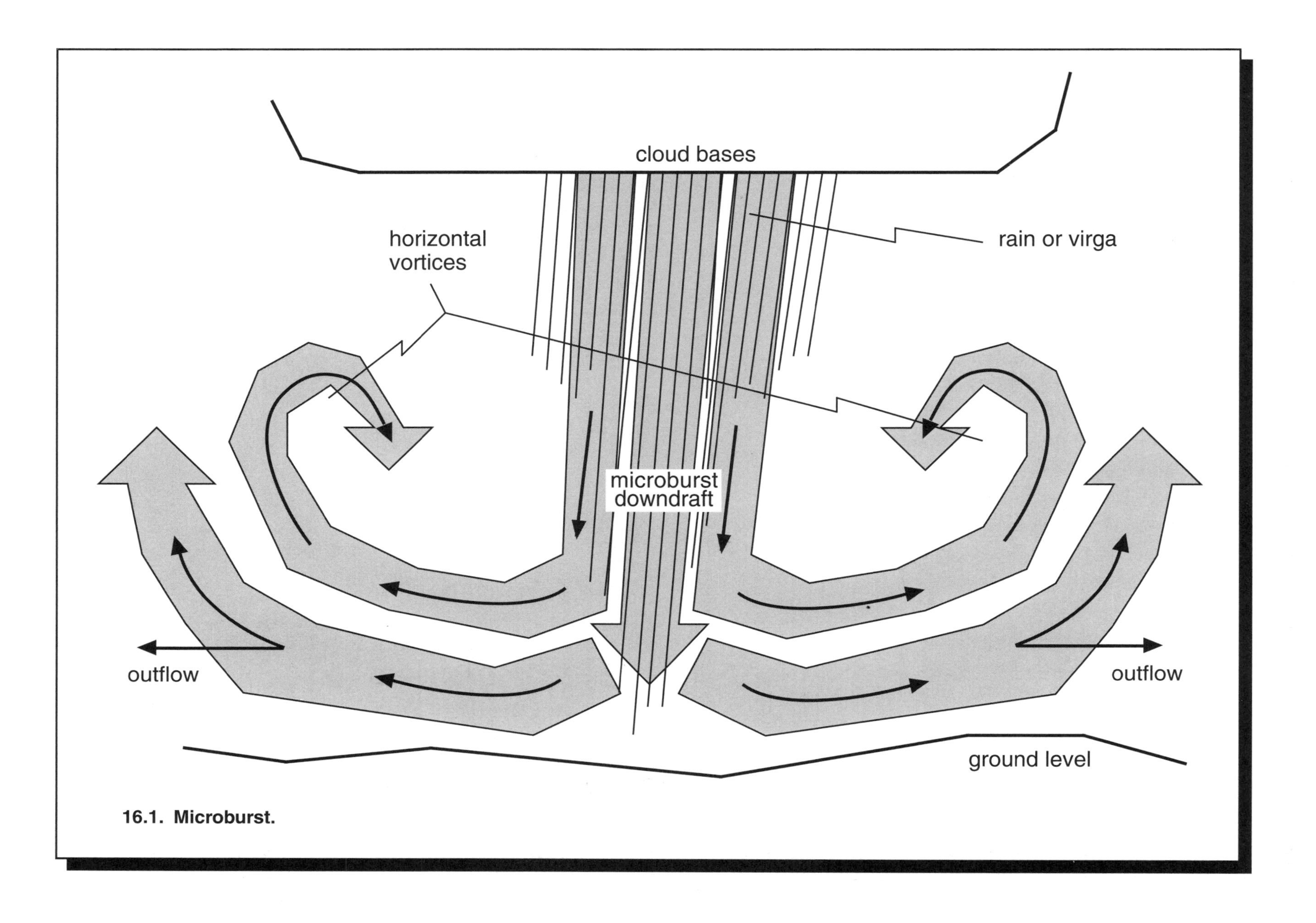

16.1. Microburst.

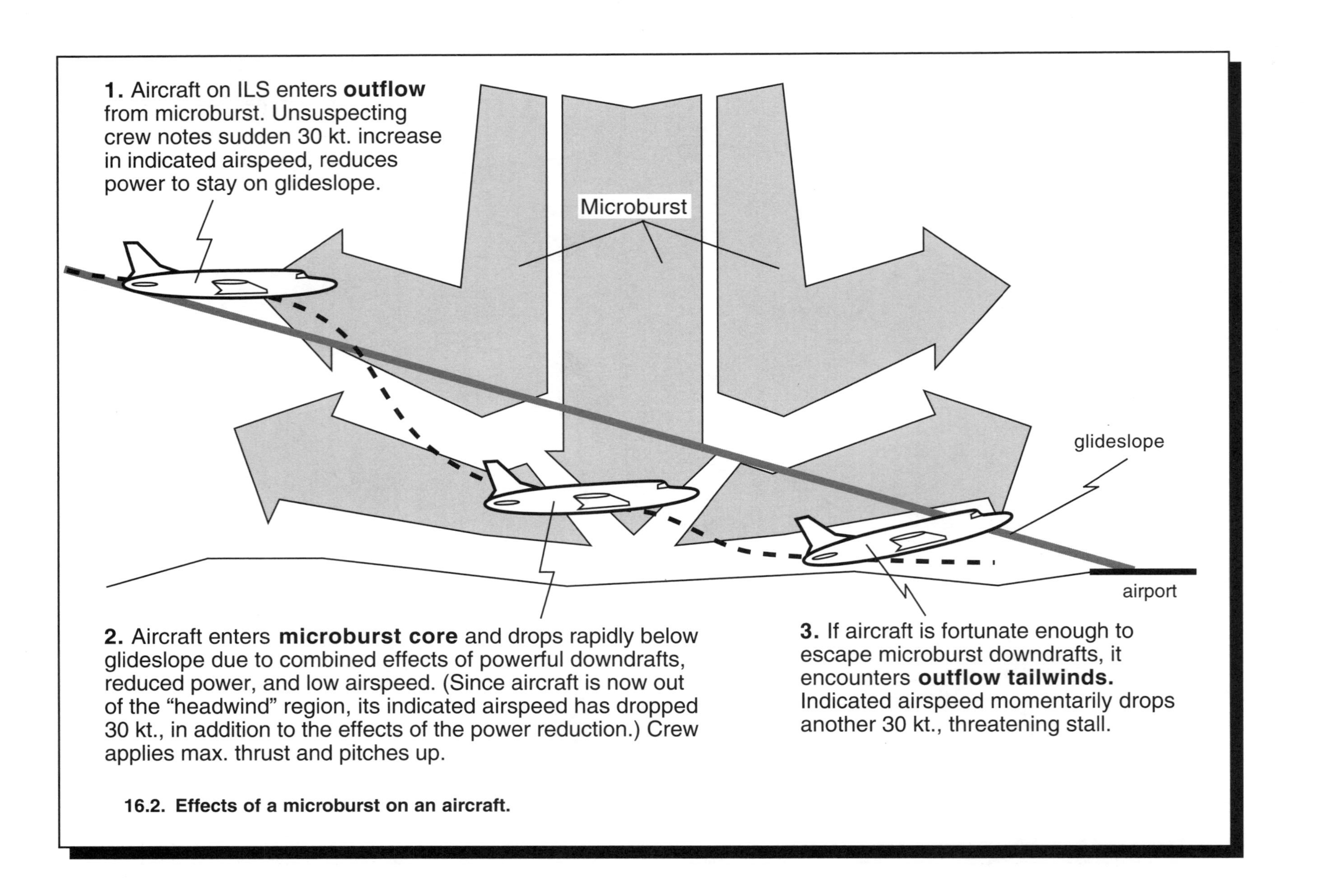

16.2. Effects of a microburst on an aircraft.

identify wind shear conditions as they develop. (See "Doppler Radar" in Chapter 14.)

Recognizing and Responding to Wind Shear

Along with monitoring the weather, it's important that pilots know their aircraft. Any deviations from expected aircraft performance are good cause for suspicion. It's common practice to increase approach speed when possible wind shear is anticipated, thereby providing some protection from a sudden airspeed loss. As you should know by now, a sudden airspeed gain is cause for alarm, too. In anticipation of a possible downdraft and subsequent momentary airspeed loss, you should accept the airspeed fluctuation when attempting to recenter on the glideslope. Missing the approach is always a good idea if things don't straighten out quickly. Better yet, when thunderstorms are around and wind shear alerts are in effect, it may be wisest to delay the approach or to proceed to an alternative airport.

If dangerous wind shear is encountered, most flight departments recommend procedures similar to the following:

1. Immediately apply maximum power. (In this case, maximum power means "to the stops," not just to normal power limitations. There's no point in babying the engines if every bit of available power may be required to escape.)
2. Increase pitch attitude in order to avoid contact with terrain. This may require pitching up to the stall warning, stick shaker, stall buffet, or other recommended pitch attitudes. (Many newer aircraft have wind shear alert systems installed that provide a visual pitch reference for recovery.) Hold pitch attitude until clear of the wind shear condition.
3. Maintain current flap and gear configuration until clear of wind shear.

Training for Wind Shear Encounters

Wind shear procedures are well addressed in most company ground schools, though the best place for wind shear training is in a flight simulator. Many operators have actually programmed in microburst wind shear profiles from past deadly accidents. The hazards can be safely reproduced there in detail. Once you've experienced wind shear and microbursts in the simulator, you'll have a healthy respect for these conditions and some valuable practice in dealing with them.

High-Altitude Weather

When pilots transition from piston-powered aircraft to turbine models, the study of high-altitude weather is often overlooked. Most airline and corporate ground schools concentrate almost entirely on aircraft systems, flight profiles, pertinent regulations, and basic operational considerations. Even flight examiners tend to focus more on those areas and bypass examination of an applicant's high-altitude weather knowledge.

This section introduces weather phenomena that influence flight planning and safety at the higher altitudes frequented by turbine aircraft. While turbine aircraft travel above much of the weather that dogs piston flyers, there are still some important high-altitude weather considerations that merit our close attention as pilots.

To begin with, some of the weather found "down low" does impact high-altitude flying. Thunderstorms, in their most severe incarnations, often extend above the ceilings of even the most high-flying aircraft. Therefore, avoidance, fuel, and route planning (both enroute and in the terminal area) are as important to turbine pilots as to anyone else.

Icing

Icing is another weather phenomenon that follows aircraft up to high altitudes. Most low-altitude pilots are familiar with icing as a winter problem—come spring they can finally forget about it. The problem exists year-round, however, in visible moisture at high altitudes due to perpetual low temperatures. Icing encountered at higher, subfreezing levels of thunderstorms is particularly hazardous (as is turbulence). Pilot awareness of potential icing conditions and timely use of anti-ice equipment is imperative for high-altitude flying. (See "Ice and Rain Protection" in Chapter 6.)

Wind

For flight and fuel planning, winds are also critically important weather phenomena, especially since turbine aircraft (particularly older ones) burn such large amounts of fuel. Turbine fuel consumption goes up dramatically at lower altitudes, while true airspeeds go down. (See Chapter 10, Performance.)

Consider a westbound jet flying against strong head winds. If the airplane operates at high altitudes where winds are strongest, fuel consumption is reasonable, but range is compromised due to reduced ground speed. However, if the same aircraft flies at lower altitudes where winds are less strong, it faces sharply increased fuel consumption, even if it picks up a few knots of ground speed. Things get really complicated when you factor in the need for IFR fuel reserves along with, say, a bad forecast.

You might think that such situations are rare. (It takes a lot of low-altitude flying to line up a few good 60-knot head wind stories.) However, there's a permanent high-altitude weather feature that makes long-distance turbine fuel planning interesting during much of the year: the jetstream.

The Jetstream

Pilots and nonpilots alike are familiar with the term "jetstream." For years now, we've all watched the evening weather personality plot jetstream location on TV weather maps. The announcer might briefly explain that the latest cold spell or dry spell was caused by a shift in the location of the jetstream. Unfortunately, that's the extent of most people's knowledge of the topic. The fact is that pilots benefit greatly by having a working knowledge of jetstream characteristics and their effects on weather.

The *jetstream* is a narrow band of high-speed winds that meanders around the globe at altitudes near the tropopause. It occurs near a "break" or "step" in the tropopause and is characterized by wind speeds from 50 knots to greater than 150 knots in an elliptical cross section. The jetstream core includes the area of strongest jetstream winds. It is generally surrounded by a larger area of somewhat lesser winds, with decreasing velocity farther from the core (Fig. 16.3).

Knowing where the jetstream is at any given time and how to use it to increase ground speed can be very important to a jet pilot. Consider a Boeing 727 flying from San Francisco to Chicago at standard cruise power setting for 0.80 Mach at FL 330. At a gross weight of 162,500 pounds, a 727 burns approximately 9800 pounds of fuel per hour (1470 gallons per hour). If the aircraft can hitch a ride on 130-knot jetstream winds, it could shave as much as an hour off the trip. That's almost 10,000 pounds of fuel savings! Imagine the fuel saved by a whole airline fleet taking proper advantage of the jetstream. Of course that same jetstream impacts the return trip, too. That same 727 may require a Denver fuel stop in order to make it back to San Francisco.

You can see that it's of utmost importance to correctly determine jetstream location and to plan cruise altitudes and routes accordingly. Eastbound, it may pay to fly well north or south of the direct course to your destination in order to remain in the jetstream. Westbound, a 100-mile deviation or relatively low cruise altitude may be justified in order to avoid jetstream head winds.

Even crosswind components for north and southbound flights must be considered. Jetstream winds can be so strong that, even as 90 degree crosswinds, they measurably reduce an aircraft's ground speed. Where the jetstream is concerned, there's almost always a head wind or tail wind component to consider during flight planning, regardless of wind direction.

Along with optimizing ground speed and fuel consumption, a working knowledge of the jetstream is useful for interpreting general development of weather systems. Jetstream location helps in predicting the weather to be encountered on a given trip and the best routes for avoiding adverse weather.

The jetstream develops in areas where large temperature differences exist at the tropopause level, creating large pressure differentials between air masses. Therefore, the jetstream occurs at the boundary between warm and cold air at high altitudes.

Anyone who's ever taken an FAA written exam knows that, during the winter, the jetstream moves farther south and picks up intensity. This corresponds with the more southerly extent of cold air in winter and the greater temperature differentials between northern and southern air masses. The jetstream also tends to be lower in altitude during the winter. To remember this, consider that tropopause height (and therefore jetstream altitude) is proportional to air mass temperature. The jetstream is higher and weaker in the summer, and lower and stronger in the winter.

Effects of the Jetstream

When upper-level winds move in a relatively straight line from west to east, a generally stable situation prevails. To the north relatively cold air may be found both at low altitudes and aloft, while to the south there's warm air both at low altitudes and aloft. This combination makes for a low temperature lapse rate and a stable air mass. The jetstream, however, rarely flows in a straight line. Rather, it meanders its way around the globe, dipping far to the south in some locations, while retreating far to the north in others. When the jetstream takes on a wavelike path, the weather gets more interesting.

Experienced pilots learn that low-pressure areas and fronts often follow the path of a jetstream. The jetstream may be thought of as a "steering current" for low-pressure areas and the weather systems that accompany them. For a surface low to develop, there must be someplace for the air converging into the low to go. The jetstream meets that need by drawing airflow up into its core. That's the main reason why the intensity of a surface low is generally associated with the wind speed of the jetstream. It is also the reason why the jetstream frequently adds to the instability of the air masses beneath it. Most adverse weather is found on the southern or eastern sides of the jetstream.

Finding the Jetstream

Jetstream location is best determined using constant pressure analysis charts. As you remember, these display wind direction and speed at specific pressure altitudes. For locating the jetstream, the 300-millibar chart is commonly used; it illustrates airflow at around 30,000 feet. Use the 500-millibar chart (18,000 feet) to check the relationship of jet core to surface weather. (The 500-millibar chart provides a good average view of what's happening in the tropopause,

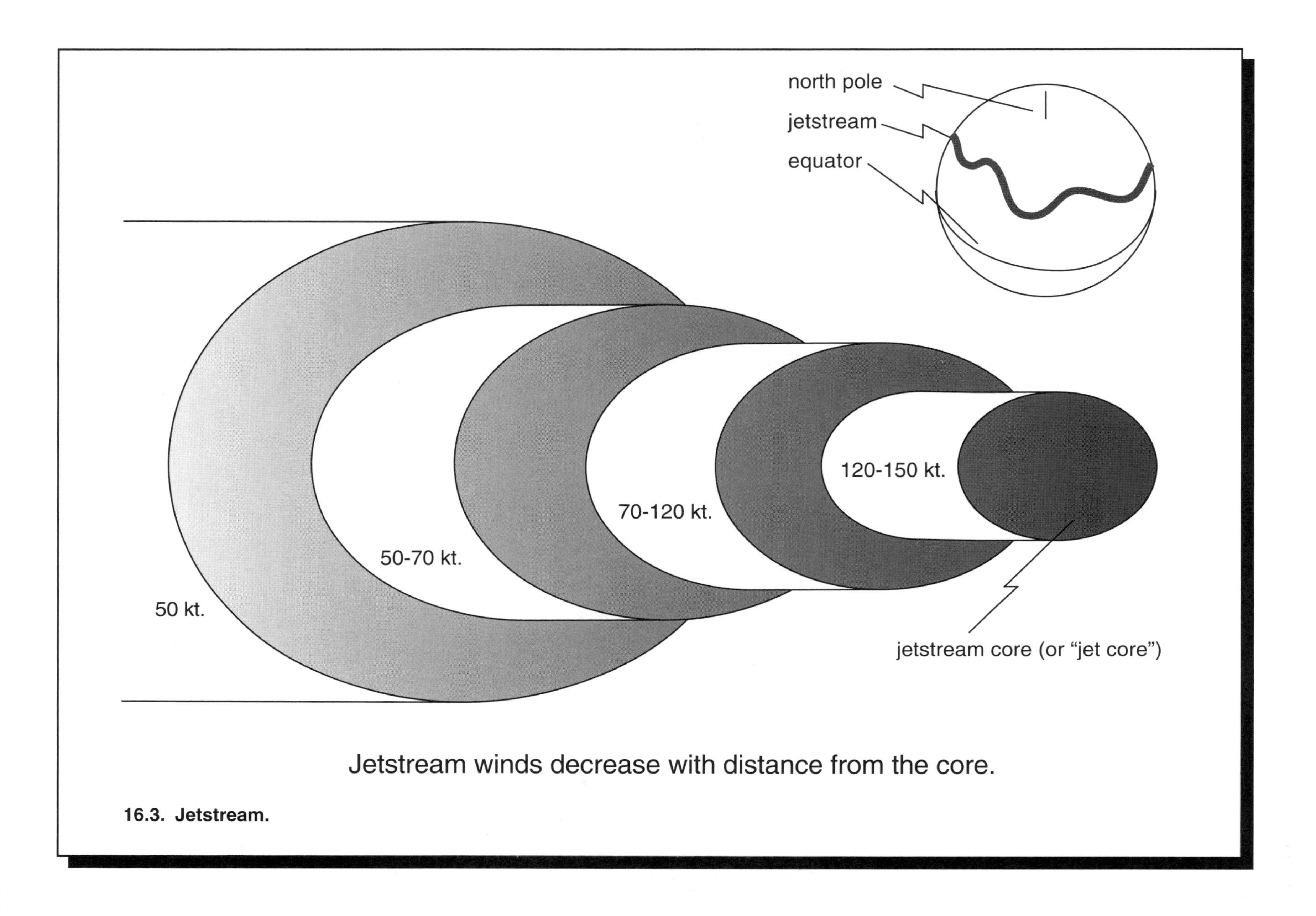

16.3. Jetstream.

since 18,000 feet falls about halfway through the vertical mass of the atmosphere.)

When constant pressure analysis charts are not available, look to the winds and temperatures aloft forecast (FD) for jetstream information. Remember that this forecast is computer-generated and may be less accurate than constant pressure charts, which show observed winds. (See the FAA's publication *Aviation Weather Services* for more information on constant pressure analysis charts and on winds and temperatures aloft forecasts.)

Clear Air Turbulence (CAT)

The term *clear air turbulence* (*CAT*) is commonly used to describe high-altitude wind shear turbulence. Clear air turbulence can range from light continuous chop to severe turbulence capable of causing aircraft structural failure. In most cases, CAT occurs in clear air, high above cloud tops. However, if there is enough moisture in the air mass, cirrus clouds may be associated with the turbulence. Because CAT is rarely connected with any visible weather, it is often encountered in areas where there seems to be no reason for its development.

Clear air turbulence is often caused by converging air masses of different temperatures: a cold blast of Arctic air mixing with northbound warm air, for example. This is because areas of converging warm and cold air masses normally mark the location of intensifying weather systems that create ascending and descending air currents. The larger the temperature difference between air masses, the more likely CAT is to be associated with the convergence.

Clear air turbulence may also be caused by large differences of wind speed or direction in a relatively small area. Conditions are suspect where strong winds converge or vary widely in direction over a small area. Either case often occurs when two different air masses come into contact with one another, as when a cold front underlies a warm air mass.

The jetstream is a prime CAT culprit since it consists of closely spaced airflow moving in a common direction but at widely varied speeds. The large velocity differences between a jetstream core and lesser surrounding winds can create significant wind shear. It should be no surprise that aircraft flying perpendicularly across the jetstream often experience serious turbulence. Sharp bends in the jetstream indicate areas particularly prone to turbulence.

Mountain waves can also create clear air turbulence. Mountain waves are created when a stable air mass is forced up the side of a mountain by strong winds. Because the air is stable, its flow is resistant to further lifting. Once the air mass has moved beyond the mountain, it descends. The result is wavelike motion of airflow downwind of mountain ridges. This situation is compounded when airflow crosses several mountain ridges in sequence. Mountain wave conditions can extend downwind of the ridges that caused them for 100 miles or more and reach altitudes above the tropopause. Aircraft flying through these areas can experience moderate to severe CAT (Fig. 16.4).

Significant, cyclic changes in airspeed every few minutes often indicate mountain wave. This is because the peaks and troughs of mountain waves are often miles apart. Pitch changes are required to maintain altitude as the aircraft crosses these large alternating areas of dramatically rising and sinking air, with resulting airspeed deviations. Mountain wave CAT is frequently found east of the Rocky Mountains and sometimes east of the Appalachians.

Hazards of CAT

Since turbine aircraft encounter clear air turbulence at cruise altitudes, and often without warning, the hazards are significant. One problem stems from the fact that cruise at the flight levels truly is high-speed flight. Aircraft encountering turbulence below 18,000 feet normally cruise at airspeeds where turbulence may be serious but is rarely life threatening. At high-speed cruise, however, an aircraft is particularly exposed to the possibility of structural damage or failure in turbulence.

Think back, for a moment, to your commercial pilot studies. The structural integrity of an aircraft, and its controllability without damage, are designed to accommodate a range of wind gusts and associated control movements within a specific speed envelope. Furthermore, the structural load imparted by a given vertical gust increases dramatically with increasing airspeed. Since the cruise speed of a turbine aircraft is far above its turbulent air penetration speed (V_B), sudden gusts can exceed design load limits and break the airplane.

This problem is particularly dangerous when a turbine aircraft is flying near its "coffin corner," the narrow range between stall speed and supersonic flight at high altitudes. (See also "Aerodynamics of High-Speed/High-Altitude Aircraft" in Chapter 15.) Gusts (or pilot control inputs) that slow the airplane threaten stall and loss of control. Gusts that effectively increase airspeed can lead to "high-speed upset" or to structural damage caused by encounter with the sound barrier.

Finally, flight attendants and passengers are often unbelted during cruise and moving about the cabin. Their belongings are spread about, along with refreshments and serving containers. The sudden onset of CAT can seriously injure people as they and their belongings are thrown about the cabin.

Avoiding CAT

For pilots, it's obviously best to avoid CAT by changing routing or altitude. If that's not possible, the next best thing

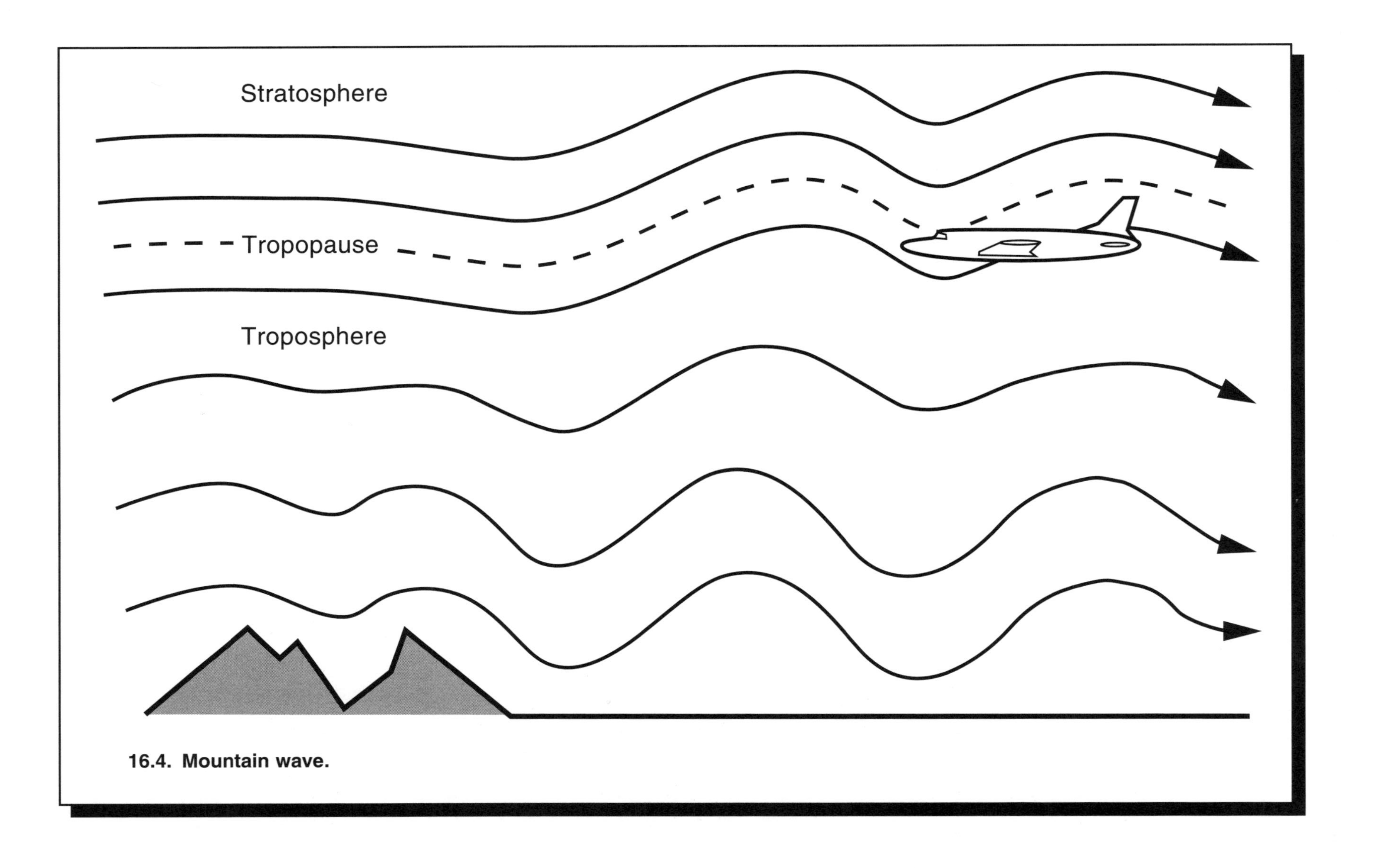

16.4. Mountain wave.

is to anticipate the turbulence, get the airplane slowed down before entering it, and then get out as soon as possible.

During preflight planning, several factors may aid in predicting the whereabouts of clear air turbulence. When checking weather before departure, take time to review locations of the jetstream and associated weather patterns. Areas of clear air turbulence commonly occur on the polar side of the jetstream. This is because of a larger temperature gradient on that side, resulting in a stronger wind speed gradient and therefore greater turbulence. Also expect to find CAT where the jetstream curves around surface lows; these are areas of intensifying temperature change.

Clear air turbulence associated with mountain waves is difficult to predict at higher altitudes and therefore hard to avoid. Check for strong winds near mountain peak levels along your flight route. Be suspicious when they exceed 30 knots. Mountain waves tend to be strongest from late morning to late afternoon. When winds aloft are strong, avoid lower altitudes near mountains or downwind of them.

Once aloft, monitor high-altitude "Flight Watch" frequencies (Enroute Weather Advisory Service) along with ATC. Because CAT is so hard to forecast, pilot reports (PIREPS) are the best source of enroute information. In many cases, areas and altitudes affected by clear air turbulence can be identified and avoided, based upon the reports of other aircraft.

For clear air turbulence associated with the jetstream, note changes in wind direction when entering the area of turbulence. If CAT is associated with strong head winds or tail winds, a change of course may help in avoiding the most turbulent areas. This is because areas of CAT lying along the jetstream are often long and narrow. If the turbulence is associated with a stiff crosswind, change altitude but maintain course, the objective being to quickly cross "widthwise" the narrow dimension of the turbulent area. An altitude change of 4000–8000 feet is often sufficient.

APPENDIX 1

Handy Rules of Thumb for Turbine Pilots

THERE ARE MANY, many rules of thumb used by turbine pilots for various situations and aircraft. Some are very useful, some are difficult, and a few are just neat. Here are a few basic techniques that are simple enough and broad enough to be particularly handy for pilots transitioning into turbine aircraft.

Operational Techniques

Turbine Fuel Planning

While the fuel gauges (and planning charts) of most turbine aircraft are calibrated in pounds, jet fuel is ordered from the fueler in gallons. Since the volume of a given weight of jet fuel varies with temperature, charts are required for precise calculations. However, the following rule of thumb is very handy for ordering fuel under normal noncritical situations. (All units are U.S. measure.)

Example: You wish to add 2000 pounds of fuel for your next leg. Divide 2000 pounds by 10 (= 200), and multiply by 1.5. Your order should be 300 gallons.

Another version of this rule of thumb, which works on aircraft fueling two tanks only, is to divide the total fuel load by two, drop the last digit, and add fifty. This will give you the fuel load in gallons per side.

Example: You have two wing tanks and wish to add 2000 pounds of fuel for your next leg.

$2000 \div 2 = 1000$;

drop a zero = 100; add 50 = 150 gallons of fuel per side.

Example: You have two wing tanks and wish to add 1500 pounds of fuel for your next leg.

$1500 \div 2 = 750$;

drop a zero = 75; add 50 = 125 gallons of fuel per side.

Common Practice for Altitude Level-off

Due to the high climb and descent rates sustainable in pressurized turbine aircraft, pilots must be especially careful not to overshoot target altitudes. It is common practice for many pilots of passenger-carrying aircraft to reduce rate of climb or descent to 1000 fpm for the last 1000 feet before target altitude and 500 fpm for the last 500 feet.

For turbine aircraft you will find this works better than the old instrument training method of beginning level-off at 10 percent of climb/descent rate before target altitude. Descent rates of 3000 fpm are not uncommon in turboprops, suggesting level-off 300 feet before target altitude. You may get leveled off in time, but will anyone ever ride with you again? This old saw works fine for mild climb/descent rates but is not always appropriate for turbine aircraft, especially with passengers onboard.

Most turbine aircraft are equipped with extremely capable autopilot systems. One thing you will notice when first using a turbine aircraft autopilot is the amazing job it can do leveling the aircraft. The autopilot will maintain, to within a few hundred feet of your target altitude, what seems to be an impossibly high rate of climb or descent, right up to a smoothly executed level-off. If you are flying an aircraft without auto-throttles, you must be prepared to smoothly and immediately adjust throttles to whatever power setting is necessary to maintain target airspeed once the autopilot has leveled off.

If you are fortunate enough to fly an aircraft equipped with an auto-throttle system, a word of caution is in order regarding use of the previously discussed level-off techniques. The flight director will command the same spectacular level-off maneuver while the aircraft is being hand-flown as it does when the autopilot is flying. If you are hand-flying the airplane with auto-throttles engaged (e.g., after takeoff up to 10,000 feet), you must follow the flight director commands precisely during level-off, or else turn off the auto-throttle system. If you instead try to reduce aircraft rate of climb to 1000 fpm for the last 1000 feet with the auto-throttle system active, the throttles will automatically advance in an attempt to achieve the climb rate commanded by the flight director. Airspeed may quickly increase beyond your assigned airspeed, or worse, exceed aircraft design limitations, so be careful! Whenever autopilot or auto-throttles are not doing what you want them to do, turn them off.

Maintaining Altitude in Turns

High-performance aircraft operate over broad ranges of airspeed, load, and configuration. Accordingly, in turns, pitch control forces required to maintain altitude can vary significantly. If a flight director is not available to help you maintain pitch for constant altitude in turns, be sure to include the IVSI (instantaneous vertical speed indicator) in your scan. As long as the IVSI needle remains zeroed "in the doughnut," altitude will be maintained. As with any aircraft, a smooth entry, constant control pressure, and constant airspeed simplify the turn.

Jet Aircraft Deceleration in Level Flight

As we have discussed, the clean, aerodynamic nature of most turbojet or turbofan aircraft sometimes makes it difficult to slow down, especially when attempting to "come down and slow down" at the same time (even with thrust levers at idle). A commonly used rule of thumb for decelerating jets in level flight is to allow 1 nm for every 10 KIAS (knots indicated airspeed) of airspeed you are trying to lose.

Example: The aircraft is indicating 320 knots, and you are given a speed restriction of 250 knots at a certain fix. How far before the fix must you begin slowing?

$320 - 250 = 70$

or 70 knots of airspeed to be lost. Using this handy rule of thumb you would bring the thrust levers to idle 7 nm before the fix.

If an airspeed restriction is combined with a clearance to cross a fix at a certain altitude, simply determine your anticipated indicated airspeed during the descent. This is usually somewhere near M_{MO}. Using the example above, let's say it's 340 KIAS.

$$340 - 250 = 90$$

or 90 knots of airspeed to be lost. The transition from 340 KIAS to 250 KIAS will take approximately 9 nm. Add 9 nm to whatever distance you came up with for a top-of-descent point.

Thunderstorm Avoidance at Altitude

Avoid thunderstorms by a distance, in miles, equal to the altitude of the aircraft in thousands of feet (minimum of 10 nm at any altitude).

Example: At an aircraft altitude of 10,000 feet, give thunderstorms a minimum of 10 nm clearance. At 25,000 feet, allow 25 nm between aircraft and thunderstorms.

Estimating Ground Speed Based on Mach Number

Here's a method for getting you "in the ballpark" when estimating ground speed for aircraft equipped with a Machmeter. Use each 0.1 Mach to represent 1 nm per minute.

Example: 0.6 Mach $\cong$ 6 nm per minute.
Example: 0.8 Mach $\cong$ 8 nm per minute.

Navigating by Automatic Direction Finder (ADF) or Radio Magnetic Indicator (RMI)

Virtually all turbine aircraft are equipped with a combined moving compass card ADF (*automatic direction finder*) and VOR *radio magnetic indicator* (RMI). While ADF and VOR/RMI navigation may not be common during normal flight operations, you can bet you'll be expected to perform either an ADF approach or VOR hold using only the RMI during some portion of your turbine flight training. Many pilots have stared blankly at an ADF or RMI bearing pointer, trying to understand what the bearing pointer was depicting. With this in mind, we've included a few handy rules of thumb to help get you headed in the correct direction.

- Rule #1: Always tune and I.D. the intended NDB or VOR station.
- Rule #2: Always picture the aircraft at the tail end of the bearing pointer and the NDB or VOR station at the tip (arrowhead).

- Rule #3: Determine if you want to fly toward the station or away from the station. If you are flying toward the station, use the tip to navigate and forget about the tail. If you are flying away from the station, use the tail to navigate and forget about the tip.
- Rule #4: Decide where and by how much (in degrees) you want either the tip or the tail to move on the compass card. Double that number for the amount of aircraft heading change required.
- Rule #5: When tracking inbound, turn opposite the direction you want the tip to move on the compass card. Outbound, turn in the same direction you want the tail to move on the compass card.
- Rule #6: As the bearing pointer moves toward your "target" bearing or radial, constantly fine-tune your heading using Rule #4 to keep the pointer where you want it.

Calculating Visual Descent Point (VDP)

A visual descent point (VDP) is a point on nonprecision instrument approaches where visual descent must begin to intercept a standard 3-degree descent angle toward the runway. While some nonprecision approaches include a published VDP, many others do not. There are a number of ways to determine VDP for an approach; here are two of the most common. Both assume that the missed approach point (MAP) is at the runway, not before.

1. The time method: To determine VDP as a function of time, take 10 percent of the height above touchdown (HAT). This is the amount of time, in seconds, that must be subtracted from the computed time for the final approach fix (FAF) to the MAP.

Example: If HAT = 480 feet, then $0.10 \times 480 = 48$. Therefore, your VDP may be found by subtracting 48 seconds from your computed FAF-to-MAP approach time.

2. The DME method: Divide HAT by 300 to determine VDP location in nautical miles from the runway.

Example: If HAT = 480 feet, then

$480 \div 300 = 1.6$ nm,

and VDP is then 1.6 nm from the end of the runway.

Adjusting Final Approach Speed for Wind

Standard approach speed for most turbine aircraft during calm no-wind conditions is V_{REF} plus 5 KIAS. This is the speed to target for the final 300 feet of altitude prior to aircraft flare. Most turbine pilots use a simple rule of thumb to adjust computed V-speeds to counteract the effects of wind.

The rule of thumb for approach speed adjustments is: add to approach speed, one-half of the steady headwind component plus all of the additional gust. (Typically, this value should not exceed a total addition of 20 knots.)

Example: You are landing Runway 27 with wind reported to be 290 degrees at 10 knots, gusting to 15 knots. A quick look at the crosswind component chart shows the headwind component is 9 knots. Half of the steady 9 knots is 4.5. Adding the additional gust of 5 knots equals 9.5. Round this value to a whole number and add 10 knots to the computed V_{REF} for your target approach speed.

Time, Speed, and Altitude Computations

Calculating Required Descent Rate to Stay on Glideslope, Based on Ground Speed

Every instrument pilot knows (or should know) that the rate of descent required to stay on glideslope varies with ground speed. Sometimes it's handy to know what that descent rate should be, say, when flying an unfamiliar aircraft type, when facing unusually strong head or tail winds, or when requested by ATC to shoot an approach at other than normal airspeeds.

To calculate rate of descent for a 3-degree glideslope based upon ground speed, halve the ground speed and add a zero.

Example: An aircraft is shooting an ILS at 180 knots DME. For rate of descent, divide ground speed of 180 knots by 2 (= 90), and multiply the result by 10 for 900 fpm.

Or putting it another way, add a zero to your ground speed and divide by two.

Example: $180(0) \div 2 = 900$ fpm.

Either way, target that 900 fpm on your IVSI, and you'll stay right on the glideslope.

Descent Planning

To achieve a standard 3-degree descent based upon distance from destination airport or any particular fix (say for a crossing altitude restriction), your altitude at any given time should be 300 multiplied by distance, in nm, from destination or fix.

Example: An airplane is 32 DME from its destination airport. For a 3-degree descent, its altitude at that distance should be

300 × 32 miles = 9600 feet.

In practice just multiply the altitude change (expressed in number of thousands) by three. This equals the required distance to make the descent. Once you know when to begin the descent, you must decide on the descent rate. To do this just use the same rule of thumb you used for calculating the required descent rate to maintain an ILS glideslope.

Example: An airplane is cruising at FL 330 and given an ATC clearance to cross TYLER intersection at 13,000 feet. When should a descent begin?

33,000 feet – 13,000 feet = 20,000 feet

of altitude to be lost; expressed in number of thousands, this = 20. Therefore

20 × 3 = 60,

or 60 nm before Tyler intersection is where you should begin your descent.

What should your descent rate be? Take your ground speed readout, add a zero, and divide by two. For example, let's say your ground speed is 525 knots, then

525(0) ÷ 2 = 2625 fpm

descent required.

Once the aircraft begins descending, the ground speed will change. Simply monitor the ground speed readout every thousand feet or so, add a zero on the end, and divide by two.

Example: An airplane, 150 miles from Conner City VOR and level at FL 370, is cleared by ATC to cross 40 nm from Conner City VOR at 10,000 feet. How far out should descent begin? (This is known as top-of-descent point.)

37,000 feet – 10,000 feet = 27,000 feet

of altitude to be lost; when expressed in thousands, this = 27. Therefore

27 × 3 = 81,

or descent should begin 81 nm prior to the 40 nm fix, or simply 121 nm west of Conner City VOR

(81 + 40 = 121).

What should your descent rate be in this example? Let's say your ground speed is 410 knots.

410(0) ÷ 2 = 2050 fpm

descent required.

Another handy application for calculating required descent rate, using the above methods, is when you determine that the aircraft is too close to the airport or crossing restriction to make a standard 3-degree descent rate.

Example: You receive an ATC clearance to cross Delta intersection at 6,000 feet. Your ground speed is 300 knots, you are level at FL 210, and you're 35 nm from Delta intersection.

First calculate your ground speed in nautical miles per minute, then divide miles to the fix by the miles per minute ground speed, and then divide altitude to be lost by time to the fix. This gives you the required descent rate to make a crossing restriction.

First:

300 knots ground speed ÷ 60 minutes = 5 nm per minute.

Second:

35 nm to Delta intersection ÷ 5 nm per minute = 7 minutes to the fix.

Third:

15,000 feet altitude to be lost ÷ 7 minutes to the fix = 2000 fpm required to make the fix.

Temperature Calculations

Determining Standard Air Temperature (SAT) at Cruising Altitude

This rule of thumb works for cruising altitudes below 36,000 feet. Double the altitude (in tens), subtract fifteen, and then change the positive number to a negative number.

Example: What is the standard air temperature at FL 310?

(31 × 2) – 15 = 47;

change to a negative number to get SAT = –47°C

Celsius to Fahrenheit Temperature Conversions

A commonly used rule of thumb for temperature conversions comes in handy when relaying weather information to your passengers. You may have become comfortable with temperatures expressed in Celsius, but your U.S. passengers expect to hear them expressed in Fahrenheit. This conversion isn't exact, but it's simple and it will get you within a couple of degrees of the true temperature.

To convert from Celsius to Fahrenheit, double the degrees Celsius and add thirty.

Example: Your destination ATIS says the temperature is 7°C. Let's try our handy rule of thumb.

$(7°C \times 2) + 30 = 44°F$

Compare the same example using the conversion formula you learned in school:

$(9/5 \times °C) + 32 = °F$

$7(1.8) + 32 = 44.6°F$

(Whew! a little more accurate but a lot more work.)

This rule of thumb works just as well for converting degrees Fahrenheit back to Celsius; just work the problem backward using

$(°F - 30) \div 2.$

Example: $(78°F - 30) \div 2 = 24°C$

APPENDIX 2

Airline, Regional, and Corporate Aircraft Spotter's Guide

MOST PILOTS can spot the differences between the well-known Boeing 747 and a McDonnell Douglas DC-10. However, distinguishing between today's many types of airline, corporate, and commuter aircraft can be difficult. Good aircraft identification skills are important in today's complex airport environment. A typical Air Traffic Control clearance at a major airport sounds something like this: "King Air 4GC, hold for the United 777, then turn left on the inner; hold at taxiway Lima; then turn left behind the Comair RJ on taxiway Mike; follow the American 737 to 25R." Following such instructions can be tough if you can't recognize aircraft named in the clearance.

For this reason, we have included the following brief guide to civilian turbine aircraft. Aircraft are grouped into categories by similar appearance (for example, T-tailed turboprops or twin-engine turbofan airliners). This guide is not intended to be detailed or complete but rather is intended to point out key identifying features of commonly seen turbine aircraft. Since seating and performance vary tremendously by model, engines installed, and operator, such information should be considered extremely general. It's worth your time to keep current on the latest aircraft models by reading industry publications.

Corporate Turboprops

Single-Engine Turboprops

1. Pilatus Aircraft LTD. PC-12

Powerplant: one Pratt & Whitney PT-6 Turboprop.
Seats: 8–11.
Cruise speed: 255 knots.
Distinctive features: large cabin for a single-engine plane, extended nose cowling, large dual exhaust pipes, one on either side of nose cowling, T-tail.

2. Socata-Groupe Aerospatiale Matra TBM700

Powerplant: one Pratt & Whitney PT-6 Turboprop.
Seats: 6–7.
Cruise speed: 255 knots.
Distinctive features: small, pressurized single with extended nose cowling, low tail.

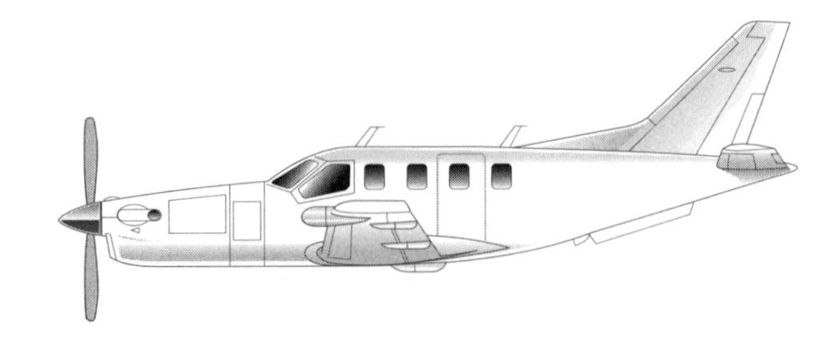

3. Cessna Caravan

Powerplant: one Pratt & Whitney PT6 Turboprop (or two in a common cowl sharing one propeller shaft).
Seats: up to 13.
Cruise speed: 160 knots.
Distinctive features: high-wing, long sloping windscreen, large fixed landing gear; usually has an externally mounted baggage compartment on the belly of the aircraft.

Low-Wing and Low-Horizontal Stabilizer

4. Cessna Conquest I/II

Powerplants: two Garrett TPE 331 or Pratt & Whitney PT6 turboprops.
Seats: up to 10 passengers.
Cruise speed: 270 knots.
Distinctive features: long rounded nose, flattened on top.

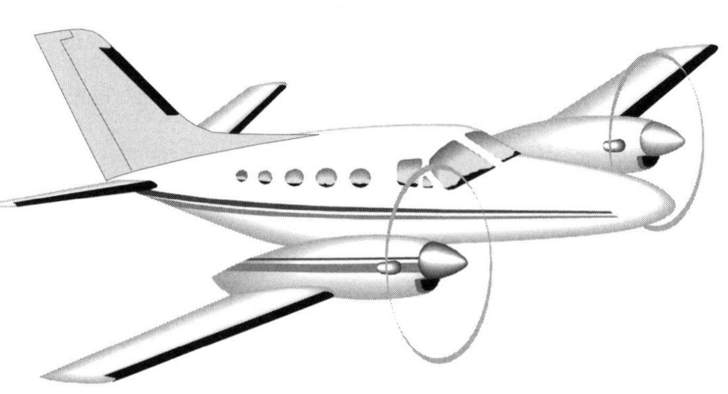

5. Beechcraft King Air Series 90

Powerplants: two Pratt & Whitney PT6 turboprops.
Seats: up to 7 passengers.
Cruise speed: 245 knots.
Distinctive features: short, stubby nose, fairly large fuselage for its length.

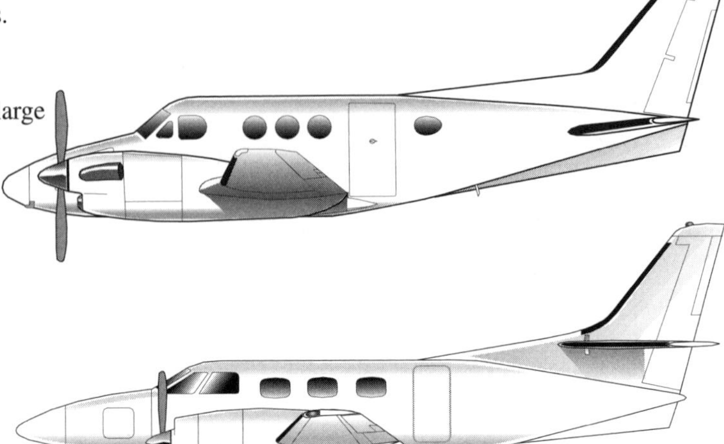

6. Fairchild Merlin

Powerplants: two Garrett TPE 331 turboprops.
Seats: up to 12 passengers.
Cruise speed: 280 knots.
Distinctive features: narrow tubular fuselage, long, narrow nose.

High-Wing Turboprops

7. Mitsubishi MU-2
Powerplants: two Garrett TPE 331 turboprops.
Seats: up to 8 passengers.
Cruise speed: 290 knots.
Distinctive features: small, high wings, tip tanks.

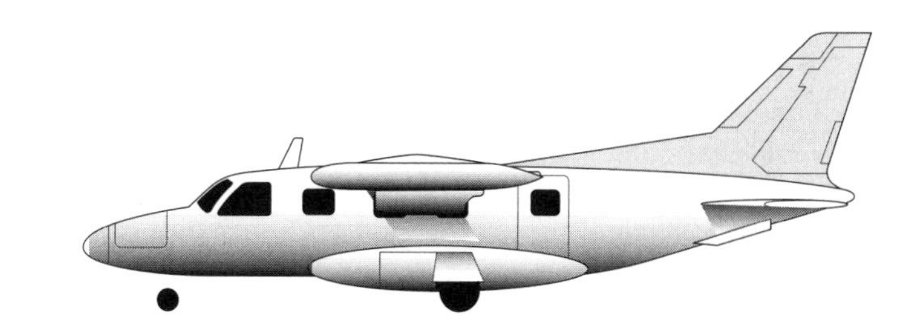

8. Commander 840/980/900/1000
Powerplants: two Garrett TPE 331 turboprops.
Seats: up to 10 passengers.
Cruise speed: 275 knots.
Distinctive features: high wings, pointed nose, fuselage sits low on the ramp.

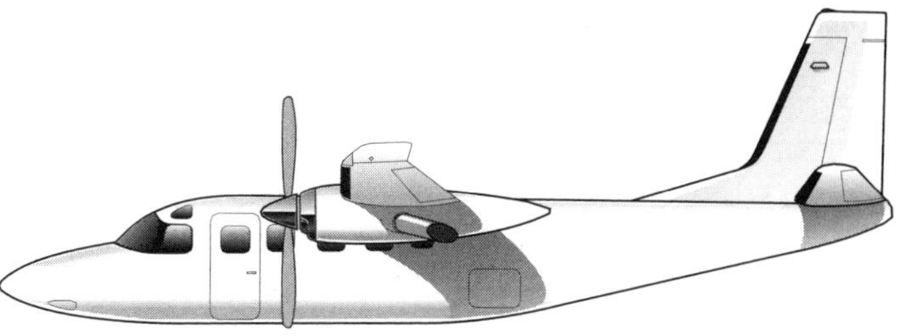

Low-Wing, T-Tail Turboprops

9. Piper Cheyenne
Powerplants: two Garrett TPE 331 or Pratt & Whitney PT6 turboprops.
Seats: up to 10 passengers.
Cruise speed: 260–350 knots, depending on model.
Distinctive features: relatively large T-tail, small wing tip tanks (early models have conventional tails).

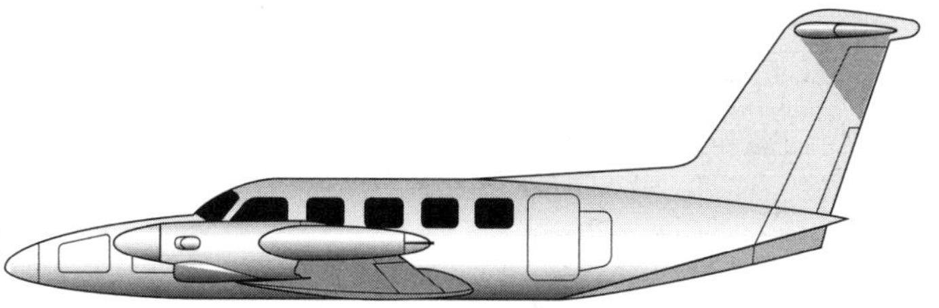

10. Beechcraft Super King Air 200/300/350
Powerplants: two Pratt & Whitney PT6 turboprops.
Seats: up to 15 passengers.
Cruise speed: 315 knots.
Distinctive features: T-tail, some models have winglets.

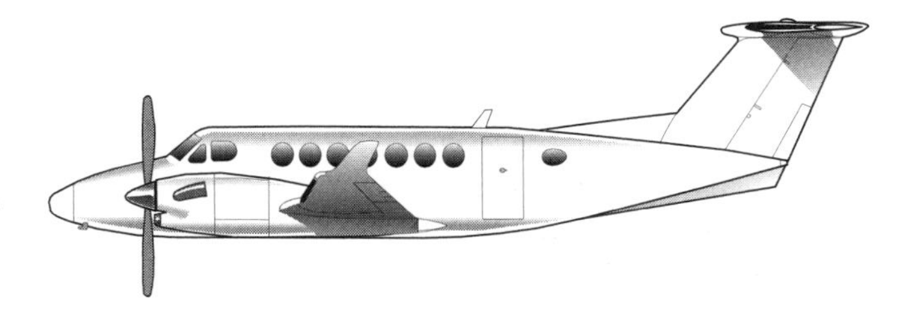

Pusher Propeller Turboprops

11. Beechcraft Starship
Powerplants: two Pratt & Whitney PT6 turboprops.
Seats: up to 7 passengers.
Cruise speed: 335 knots.
Distinctive features: composite construction, canard configuration, and winglets.

12. Piaggio Avanti

Powerplants: two Pratt & Whitney PT6 turboprops.
Seats: up to 9 passengers.
Cruise speed: 385 knots.
Distinctive features: three-surface design includes main and forward wings and T-tail, large wraparound cockpit windows.

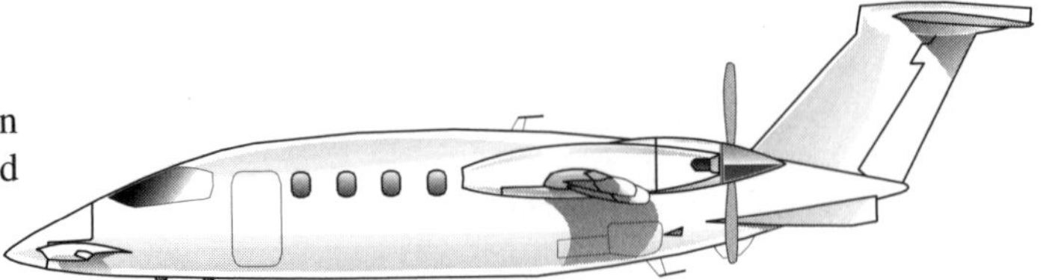

Corporate Jets

Mid-Tail and Low-Tail Jets

13. Lockheed Jetstar

Powerplants: four Garrett TFE 731 engines.
Seats: up to 10 passengers.
Cruise speed: 440 knots.
Distinctive features: four engines mounted on tail section, large external underwing fuel tanks.

14. Rockwell Sabreliner

Powerplants: two Pratt & Whitney JT12 turbojets, Garrett TFE 731 turbofans, or General Electric CF700 turbofans.
Seats: up to 8 passengers.
Cruise speed: 430 knots.
Distinctive features: cockpit "eyebrow windows," older models have triangular passenger windows.

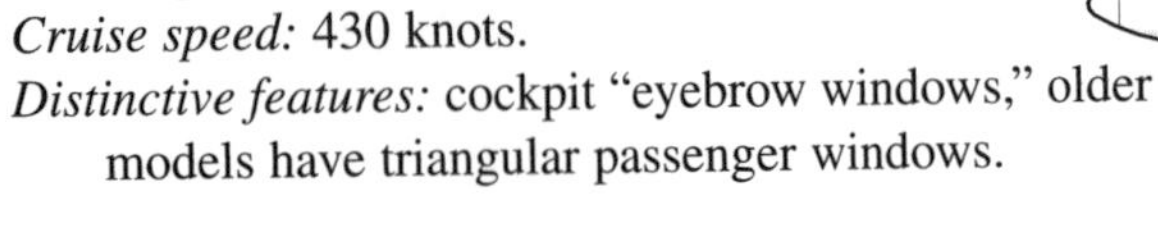

15. Dassault-Breguet Falcon 20

Powerplants: two General Electric CF700 or Garrett TFE 731 turbofans.
Seats: up to 10 passengers.
Cruise speed: 410 knots.
Distinctive features: large cockpit windows typical of most Falcon models.

16. Dassault-Brequet Falcon 2000

Powerplants: two General Electric/Allied Signal CFE738.
Seats: 8–19.
Cruise speed: 0.8 Mach.
Distinctive features: relatively large engines for a corporate jet; horizontal tail has noticeable anhedral.

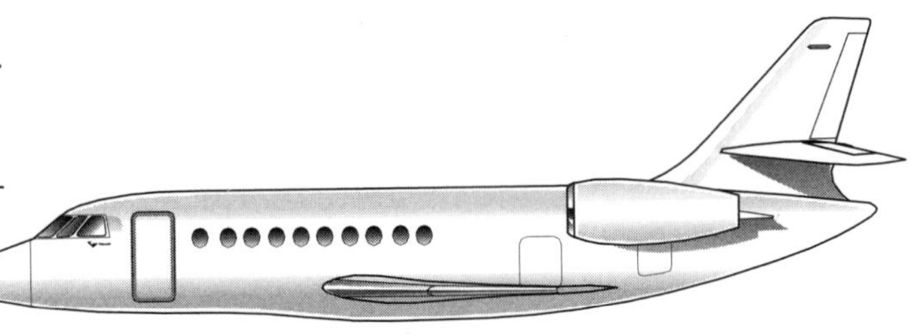

17. British Aerospace HS-125

Powerplants: two Garrett TFE 731 turbofans.
Seats: up to 14 passengers.
Cruise speed: 410 knots.
Distinctive features: pointed nose, small accessory intake at base of vertical stabilizer, four passenger windows.

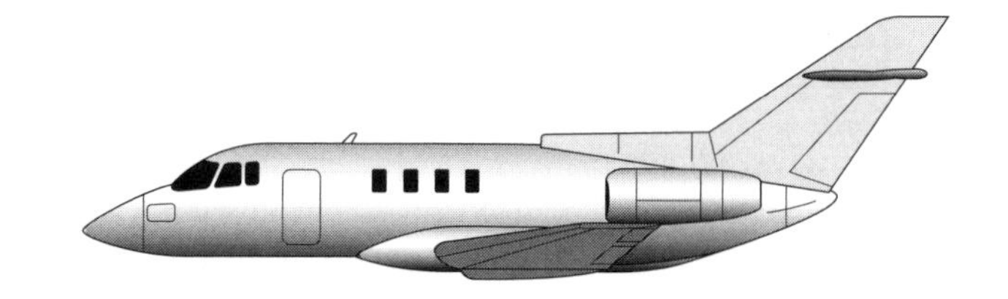

18. Hawker 800/1000

Powerplants: two Garrett TFE 731 or Pratt & Whitney PW305 turbofans.
Seats: up to 9 passengers.
Cruise speed: 450 knots.
Distinctive features: pointed nose, wraparound cockpit windows, accessory intake at base of vertical stabilizer, six passenger windows (800), seven passenger windows (1000).

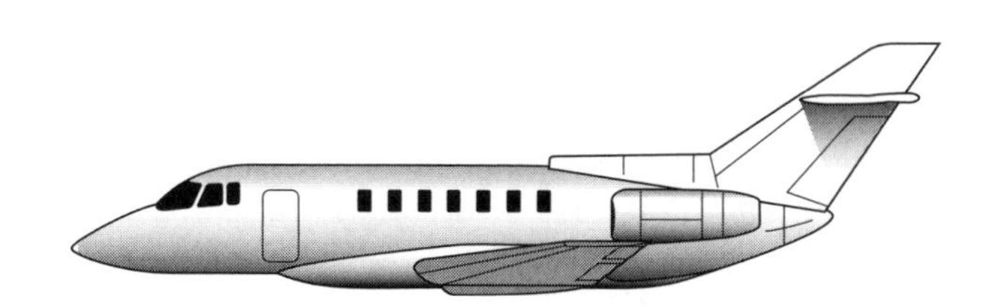

19. Cessna Citation I/II/V/Bravo/Encore/Excel/Sovereign

Powerplants: two Pratt & Whitney PW JT15D turbofans.
Seats: up to 12 passengers.
Cruise speed: 340–420 knots.
Distinctive features: large cockpit windows, straight wings (Excel and Sovereign have multipane windshields).

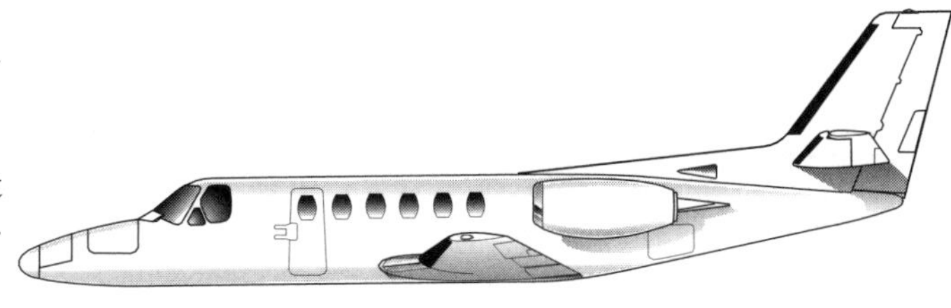

20. Israel Aircraft Industries Astra Jet

Powerplants: two Garrett TFE 731 turbofans.
Seats: up to 9 passengers.
Cruise speed: 460 knots.
Distinctive features: swept wing, small cockpit eyebrow windows above pilots' heads.

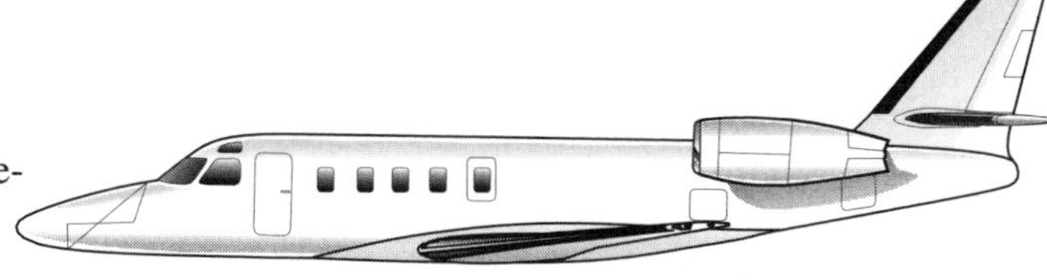

21. Israel Aircraft Industries Westwind 1124/1124A

Powerplants: two Garrett TFE 731 engines.
Seats: up to 10 passengers.
Cruise speed: 425 knots.
Distinctive features: straight wings, unusual wing tip tanks, with winglets on later models.

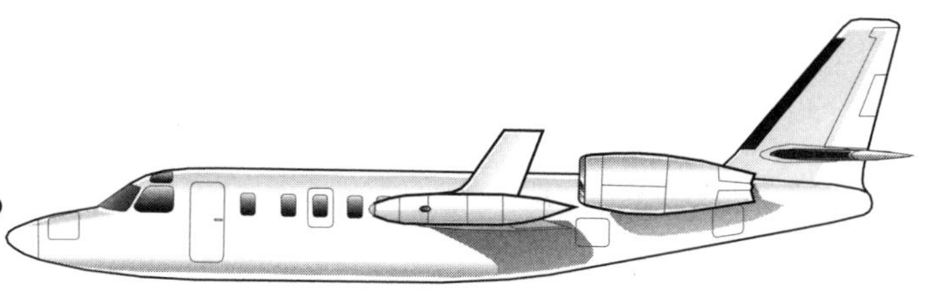

22. Dassault-Breguet Falcon 50/900

Powerplants: three Garrett TFE 731 turbofans.
Seats: up to 10 (Falcon 50), up to 19 (Falcon 900).
Cruise speed: 470 knots.
Distinctive features: typical Falcon cockpit windshield, three engines, horizontal stabilizer has a slight anhedral.

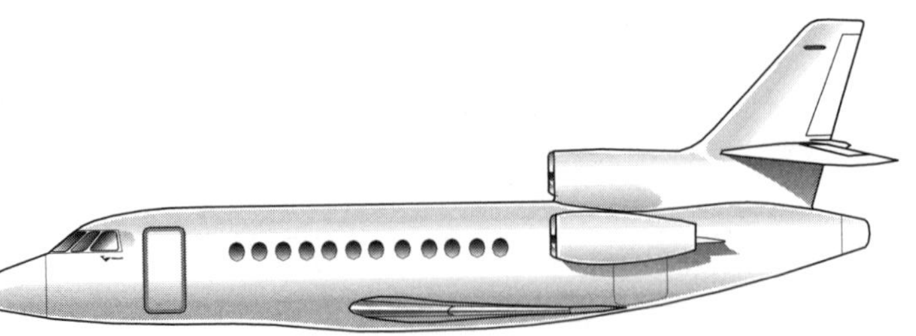

T-Tail Corporate Jets

23. Beechjet 400

Powerplants: two Pratt & Whitney JT15D turbofans.
Seats: up to 9 passengers.
Cruise speed: 450 knots.
Distinctive features: unique cockpit window arrangement with side window frame angled diagonally forward.

24. Cessna Citation Jet/Cessna Citation Jet2

Powerplants: two Williams-Rolls FJ44 turbofans.
Seats: 5–7 passengers.
Cruise speed: 370–400 knots.
Distinctive features: smallest Citation, straight wings, small engine nacelles.

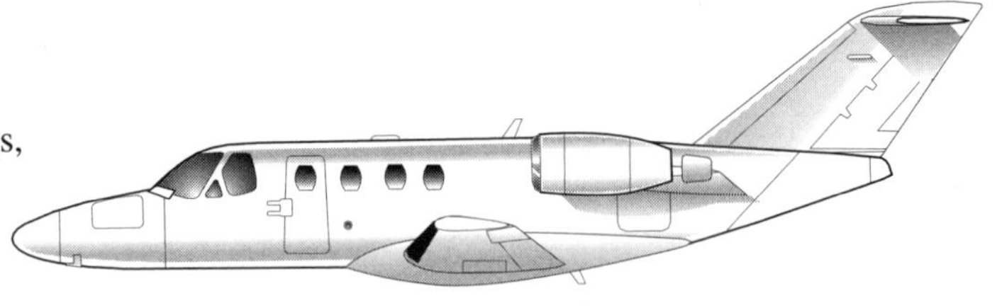

25. Cessna Citation III/VI/VII/X

Powerplants: two Garrett TFE 731 turbofans or Allison GMA 3007A turbofans (X).
Seats: up to 8 passengers.
Cruise speed: 450–510 knots, (0.9 Mach Citation X).
Distinctive features: swept wings, T-tail, unique shape of nose.

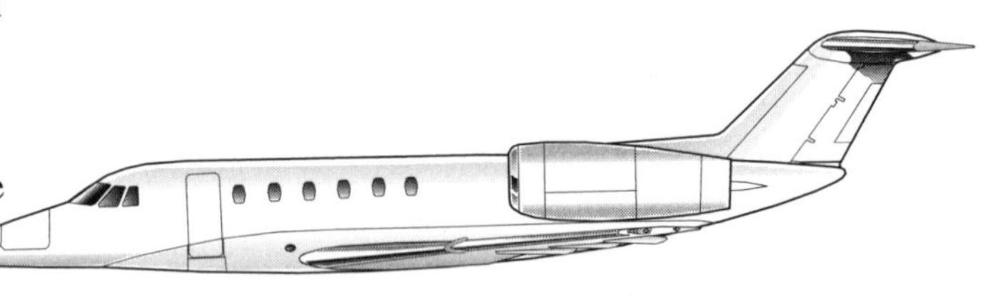

26. Gulfstream II/III/IV/V

Powerplants: two Rolls-Royce Spey or Rolls-Royce Tay turbofans or BR 710 turbofans (V).
Seats: up to 19 passengers.
Cruise speed: 480 knots.
Distinctive features: relatively large size for a corporate jet, flat oval passenger windows, G-III, G-IV, and G-V have winglets.

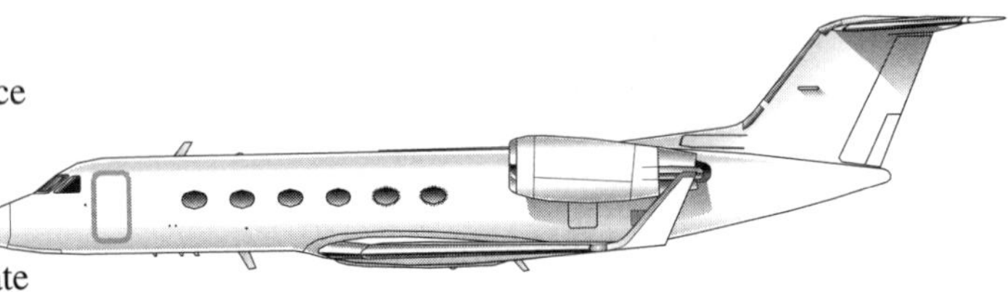

27. Lear 23/24/25/35/36

Powerplants: two CJ 610 turbojets (23, 24, 25) or two Garrett TFE 731 turbofans.
Seats: up to 10 passengers.
Cruise speed: 450 knots.
Distinctive features: distinctive sharp nose, narrow fuselage, wing tip tanks.

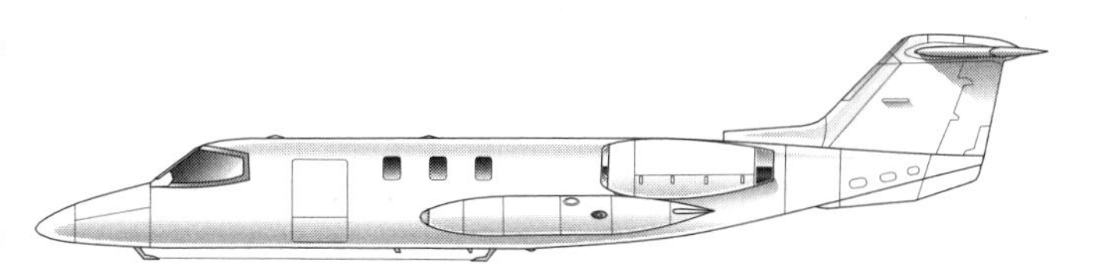

28. Lear 28/31/45/55/60

Powerplants: two Garrett TFE 731 or Pratt & Whitney PW305 turbofans.
Seats: up to 10 passengers.
Cruise speed: 440 knots.
Distinctive features: narrow fuselage, winglets.

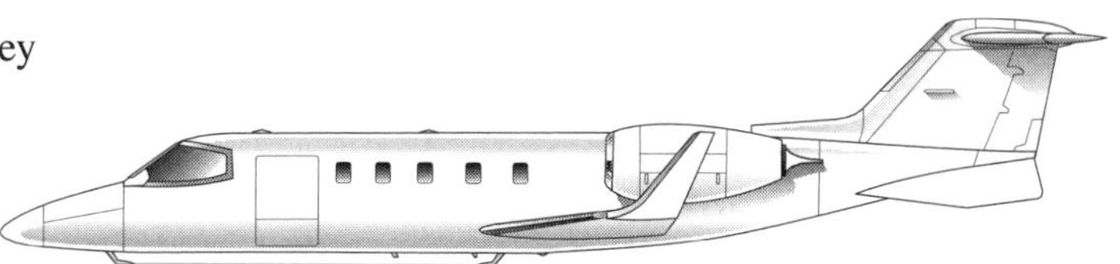

29. Canadair Challenger 600/601/604

Powerplants: two General Electric CF34 turbofans.
Seats: up to 30 passengers.
Cruise speed: 450 knots.
Distinctive features: wide fuselage, large turbofans, wraparound windshield, winglets (601).

30. Canadair Global Express

Powerplants: two BMW/Rolls-Royce BR700.
Seats: 2–19.
Cruise speed: 490 knots.
Distinctive features: winglets, oversized engines.

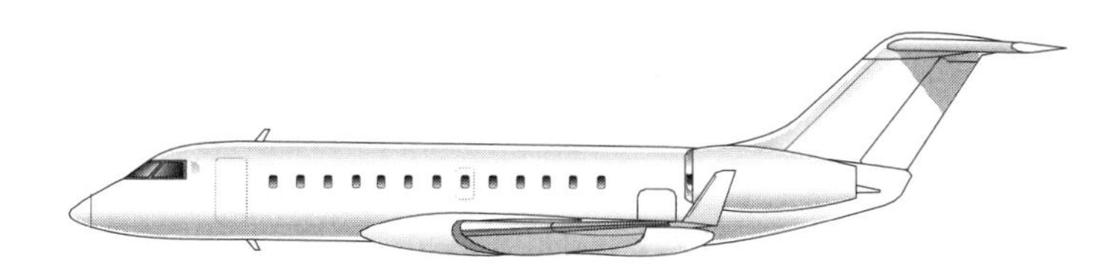

Regional Airline Turboprops

Low-Wing Turboprops

31. Beechcraft 1900C

Powerplants: two Pratt & Whitney PT6 turboprops.
Seats: up to 19 passengers.
Cruise speed: 240 knots.
Distinctive features: additional fixed horizontal tail surfaces ("stabilons") on lower rear fuselage, large T-tail with tailets on underside of stabilizer.

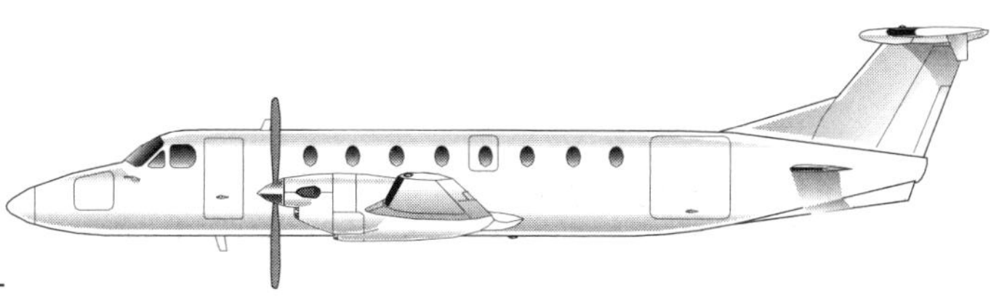

32. Beechcraft 1900D

Powerplants: two Pratt & Whitney PT6 turboprops.

Seats: up to 19 passengers.

Cruise speed: 255 knots.

Distinctive features: similar to the 1900C but with vertically extended "stand-up" cabin, stabilons on lower rear fuselage, large T-tail with tailets on underside of horizontal stabilizer.

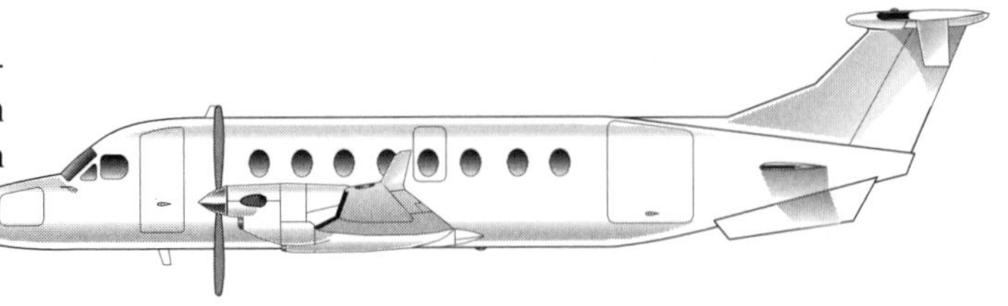

33. Embraer Brasilia EMB-120

Powerplants: two Pratt & Whitney PW115 turboprops.

Seats: up to 30 passengers.

Cruise speed: 300 knots.

Distinctive features: long sloping nose, large wrap-around cockpit windows, T-tail.

34. Fairchild Metroliner

Powerplants: two Garrett TPE 331 turboprops.

Seats: up to 19 passengers.

Cruise speed: 290 knots.

Distinctive features: long tubular fuselage, long nose.

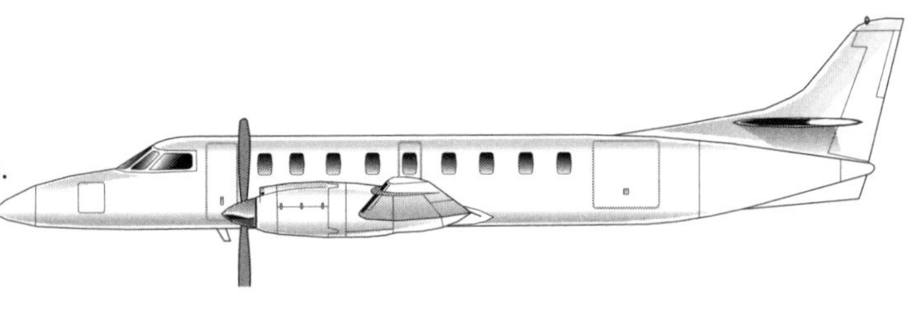

35. British Aerospace Jetstream 31/Super 31

Powerplants: two Garrett TPE 331 turboprops.

Seats: up to 19 passengers.

Cruise speed: 240 knots.

Distinctive features: long nose, short fuselage, baggage pod on belly of aircraft.

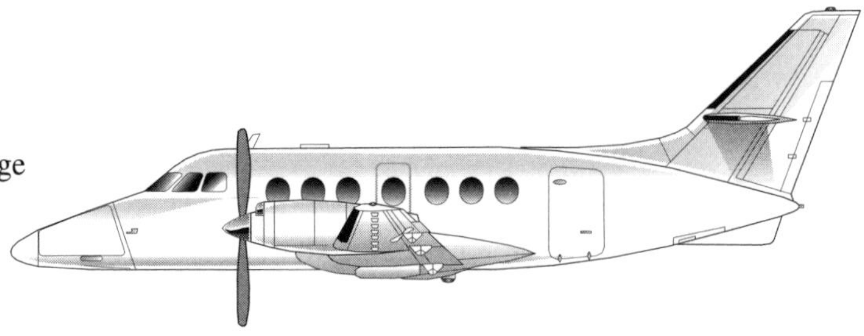

36. British Aerospace Jetstream 41

Powerplants: two Garrett TPE 331 turboprops.

Seats: up to 30 passengers.

Cruise speed: 250 knots.

Distinctive features: two-piece windshield, similar to Jetstream 31 with stretched fuselage.

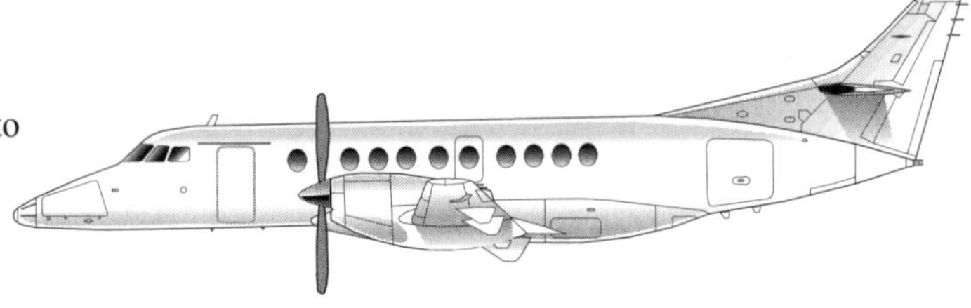

37. Saab 340

Powerplants: two General Electric CT7 turboprops.
Seats: up to 34 passengers.
Cruise speed: 250 knots.
Distinctive features: nose shape, pronounced dihedral of horizontal stabilizer.

38. British Aerospace ATP

Powerplants: two Pratt & Whitney PW126 turboprops.
Seats: up to 72 passengers.
Cruise speed: 220 knots.
Distinctive features: massive size for regional airliner, low horizontal stabilizer with little or no dihedral.

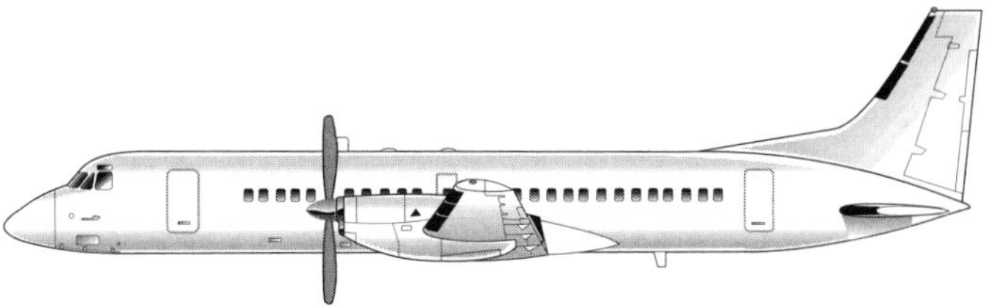

High-Wing Turboprops

39. Dornier 228

Powerplants: two Garrett TPE 331 turboprops.
Seats: up to 19 passengers.
Cruise speed: 230 knots.
Distinctive features: high wings, unique sloped nose, hump on belly to house main landing gear.

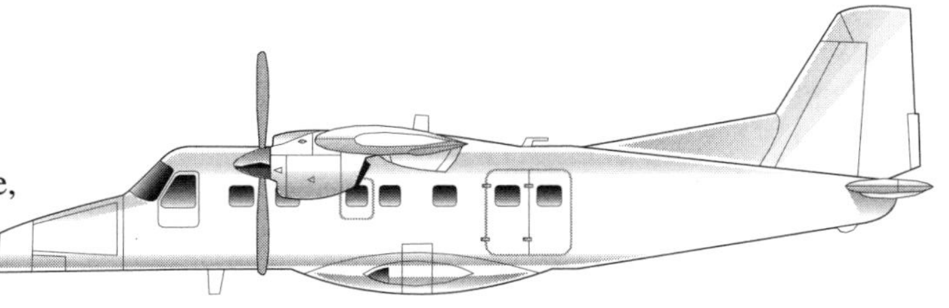

40. Shorts 360

Powerplants: two Pratt & Whitney PT6 turboprops.
Seats: up to 36 passengers.
Cruise speed: 215 knots.
Distinctive features: high wings with struts, boxlike fuselage, pointed nose (earlier 330 models have twin tail).

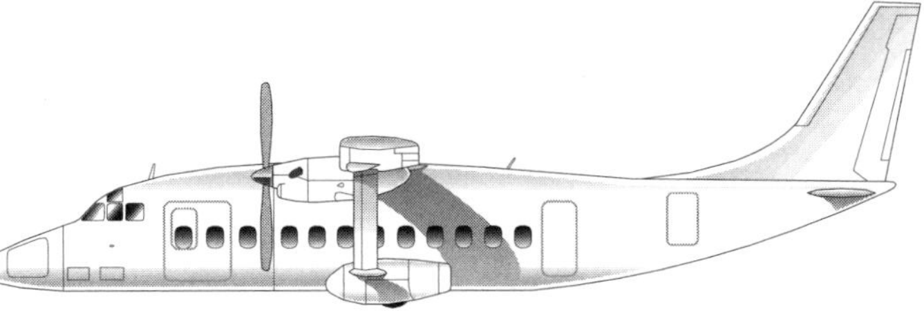

41. de Havilland Dash-7

Powerplants: four Pratt & Whitney PT6 turboprops.
Seats: up to 50.
Cruise speed: 210 knots.
Distinctive features: four engines, high wings, large T-tail.

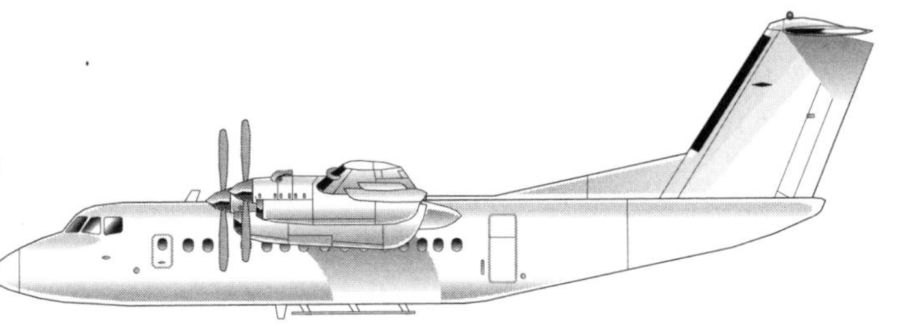

42. Fokker F27/F50

Powerplants: two Rolls-Royce Dart or Pratt & Whitney PW125 turboprops.
Seats: up to 50 passengers.
Cruise speed: 270 knots.
Distinctive features: high wings, long pointed nose.

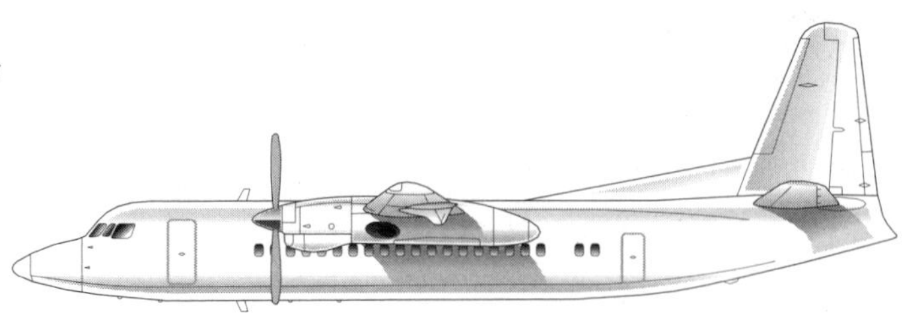

43. Fairchild/Dornier 328

Powerplants: two Pratt & Whitney PW119.
Seats: up to 32 passengers.
Cruise speed: 330 knots.
Distinctive features: supercritical wing, hump on belly that houses main landing gear, unusual slope of nose, distinctive overwing fairing.

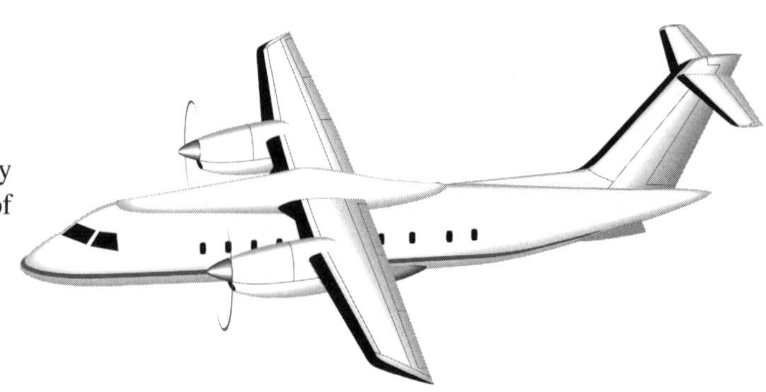

44. ATR 42/72

Powerplants: two Pratt & Whitney PW120 (ATR42), PW124 and PW127 (ATR72).
Seats: up to 50 passengers (ATR42), up to 74 passengers (ATR72).
Cruise speed: 270 knots.
Distinctive features: rounded nose, large vertical tail surface, small wing for an aircraft of this size.

45. de Havilland Dash-8-100/200/300/400

Powerplants: two Pratt & Whitney PW120 or PW123 turboprops.
Seats: up to 56 passengers.
Cruise speed: 260 knots.
Distinctive features: slope of nose, large T-tail, wraparound cockpit windows.

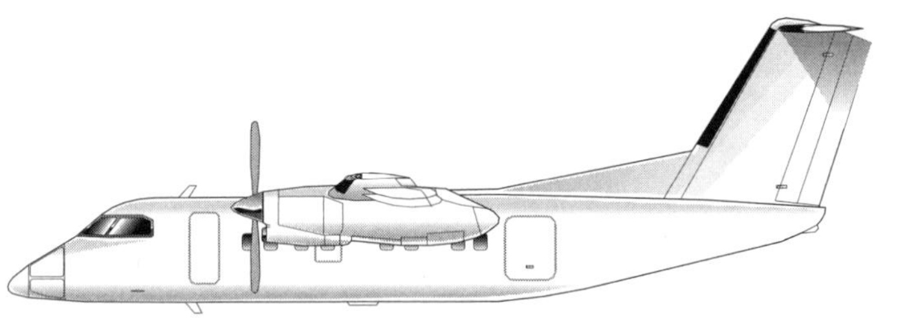

Jet Airliners

Regional Airline Jet Aircraft

46. Canadair Regional Jet CRJ 200/700/900

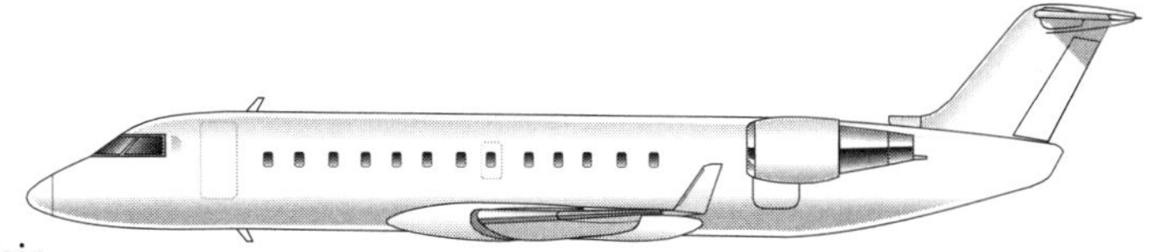

Powerplants: two General Electric CF34 turbofans.
Seats: up to 86 passengers.
Cruise speed: 450 knots.
Distinctive features: stretched version of Canadair Challenger, large wraparound cockpit windows, large turbofan engines, winglets.

47. British Aerospace BAe 146/RJ70/RJ85

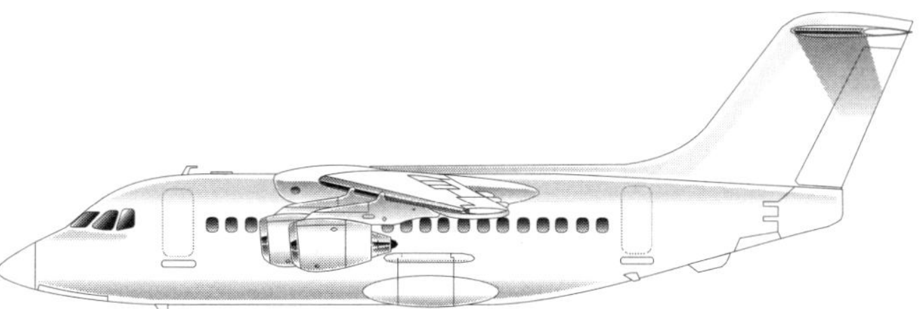

Powerplants: four Lycoming ALF 502 or LF507 turbofans.
Seats: 70–109 passengers.
Cruise speed: 380 knots.
Distinctive features: four turbofan engines, high wings, T-tail.

48. Fairchild/Dornier 328/428/728 Jet

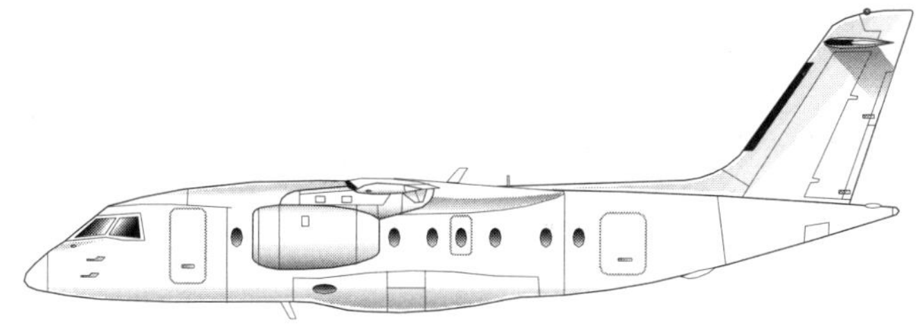

Powerplant: two General Electric CF34 turbofans.
Seats: up to 70 passengers.
Cruise speed: 435 knots.
Distinctive features: similar in appearance to the Fairchild/Dornier 328 turboprop, except it has turbofan engines.

49. Embraer 135/145/170

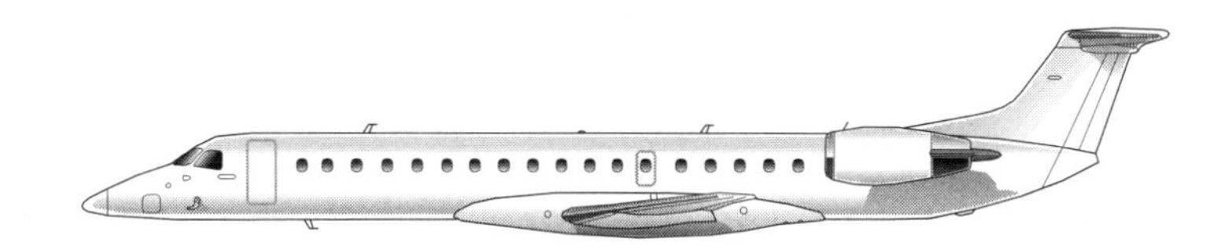

Powerplants: two Allison AE3007 turbofans.
Seats: 37–70 passengers.
Cruise speed: 425 knots.
Distinctive features: long, slender fuselage.

Rear-Engined Jet Airliners

50. McDonnell Douglas DC-9

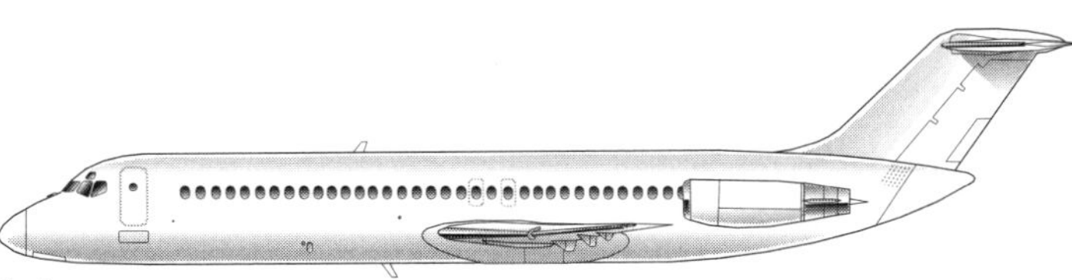

Powerplants: two Pratt & Whitney JT8D engines.
Seats: up to 122 passengers.
Cruise speed: 445 knots.
Distinctive features: tail-mounted engines, rounded nose, cockpit "eyebrow window" over each pilot, T-tail.

51. Boeing 717

Powerplants: two BMW/Rolls-Royce BR 715 turbofans.
Seats: 105.
Cruise speed: 440 knots.
Distinctive features: looks similar to DC-9, large engines, squared-off tail.

52. Fokker F28

Powerplants: two Rolls-Royce Spey turbofans.
Seats: up to 85 passengers.
Cruise speed: 360 knots.
Distinctive features: short fuselage, air brake built into tail cone, T-tail.

53. Fokker F100

Powerplants: two Rolls-Royce Tay turbofans.
Seats: up to 110 passengers.
Cruise speed: 435 knots.
Distinctive features: looks similar to DC-9 but has numerous flap track fairings trailing from the aft portion of each wing, air brake built into tail cone.

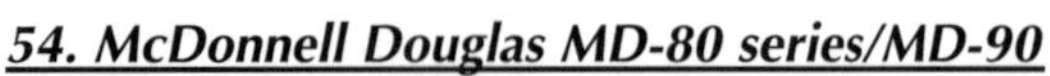

54. McDonnell Douglas MD-80 series/MD-90

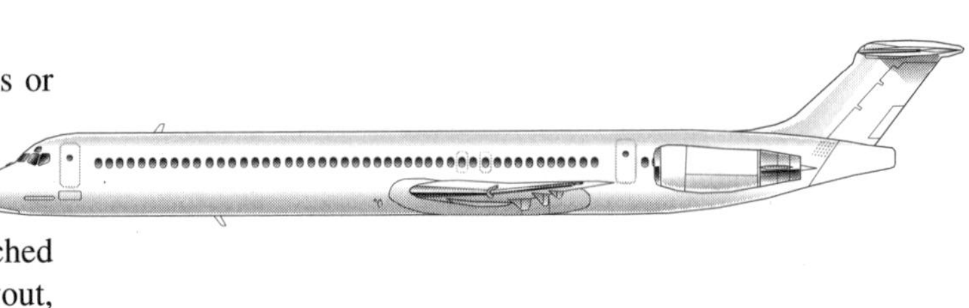

Powerplants: two Pratt & Whitney JT8D turbofans or IAE V2500 turbofans (MD-90).
Seats: up to 155 passengers.
Cruise speed: 470 knots.
Distinctive features: long, narrow fuselage, stretched version of DC-9 with same cockpit window layout, larger engine nacelles than the DC-9.

55. Boeing 727

Powerplants: three Pratt & Whitney JT8D engines.
Seats: up to 190 passengers.
Cruise speed: 490 knots.
Distinctive features: three engines mounted on the tail, the number two engine intake mounted in the vertical stabilizer and utilizing an S-shaped intake duct, unlike number one and number three engines, which have a standard straight axial flow throughout.

Medium Twin Jet Airliners (Underwing Engines)

56. Boeing 737-200

Powerplants: two Pratt & Whitney JT8D engines.
Seats: up to 130 passengers.
Cruise speed: 440 knots.
Distinctive features: long, narrow engine nacelles mounted under wing, pointed nose.

57. Boeing 737-300/400/500 (737 "Classic" Series)

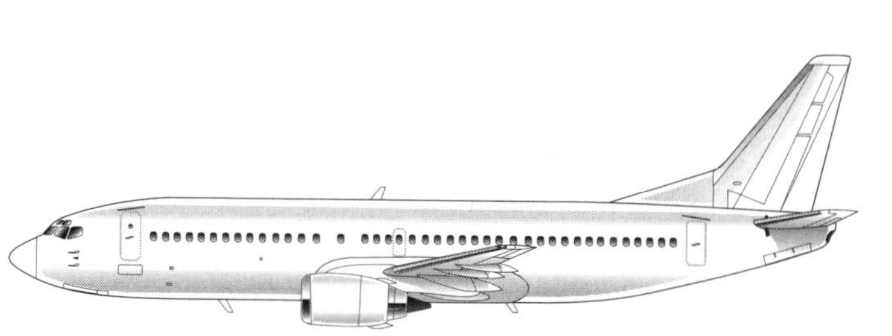

Powerplants: two General Electric/SNECMA CFM56 turbofans.
Seats: up to 170 passengers (737-400).
Cruise speed: 440 knots.
Distinctive features: large engine intakes are flattened at the bottom for ground clearance, not perfectly circular like most turbofan engines.

58. Boeing 737-600/700/800/900 (737 "New Generation")

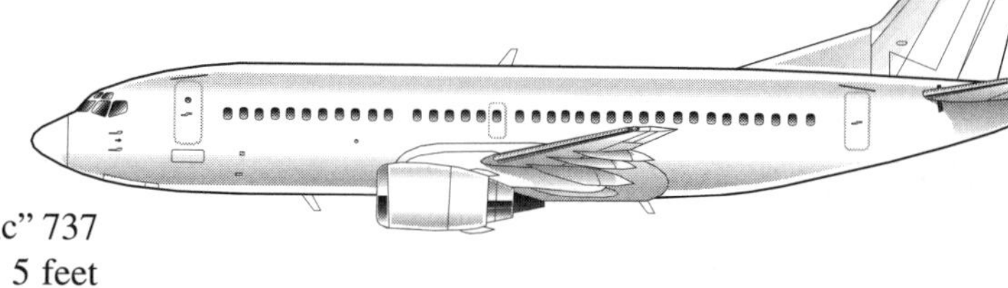

Powerplants: two General Electric/SNECMA CFM 56-7B.
Seats: up to 177 passengers (737-900).
Cruise speed: 440 knots.
Distinctive features: looks very similar to "Classic" 737 series; wingspan is nearly 8 feet longer, tail 5 feet taller, aircraft sits about 18 inches higher on the main gear, so engine intake shape is more rounded than 300/400/500.

59. Airbus A-318/A-319/A-320/A-321 Series

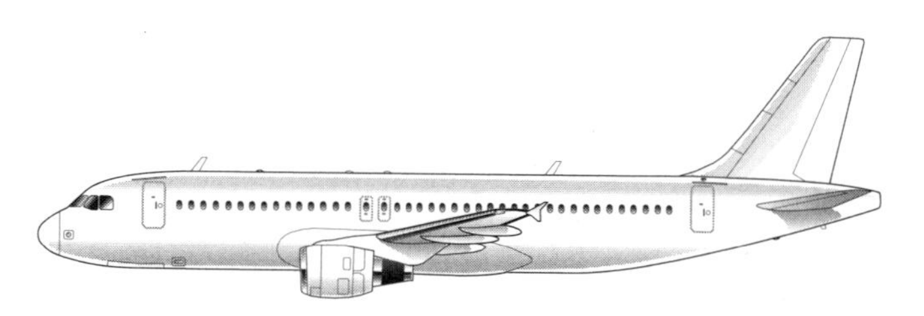

Powerplants: two General Electric/SNECMA CFM56 or IAE V2500 turbofans.
Seats: up to 180 passengers.
Cruise speed: 450 knots.
Distinctive features: extended tail cone, numerous flap extension fairings on trailing edge of each wing, distinctive winglets.

Large To Wide Body Twin-Jet Airliners (Underwing Engines)

60. Boeing 757-200/300

Powerplants: two Rolls-Royce RB211 or Pratt & Whitney PW2037 turbofans.

Seats: up to 225 passengers (757-200), up to 243 passengers (757-300).

Cruise speed: 460 knots.

Distinctive features: long, slender fuselage with very large engine nacelles and tall landing gear, unique nose shape, clean supercritical wing.

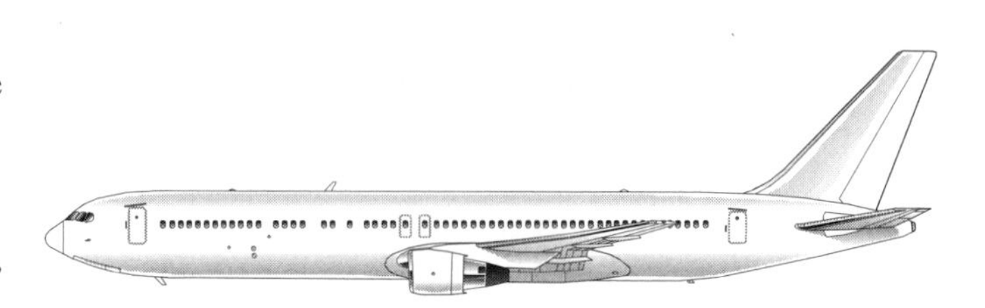

61. Boeing 767-200/300/400

Powerplants: two General Electric CF6 or Rolls-Royce RB211 or Pratt & Whitney JT9D turbofans.

Seats: up to 325 passengers.

Cruise speed: 460 knots.

Distinctive features: wide body relative of Boeing 757, clean supercritical wing design, rounded nose.

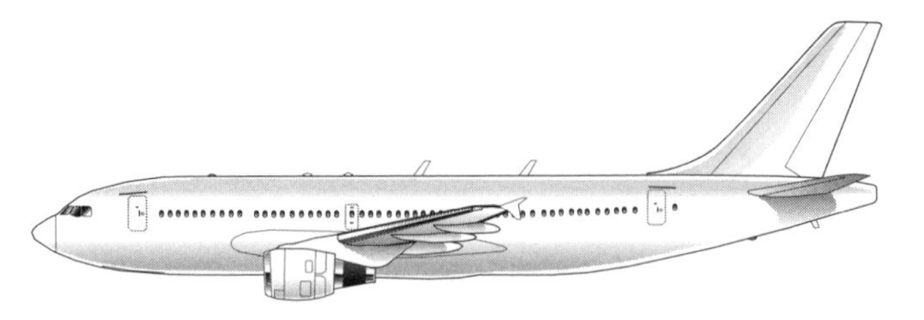

62. Airbus A-310

Powerplants: two General Electric CF6 or Pratt & Whitney JT9D turbofans.

Seats: up to 255.

Cruise speed: 450 knots.

Distinctive features: similar to Boeing 767 but has numerous flap extension fairings on trailing edge of each wing; some have winglets.

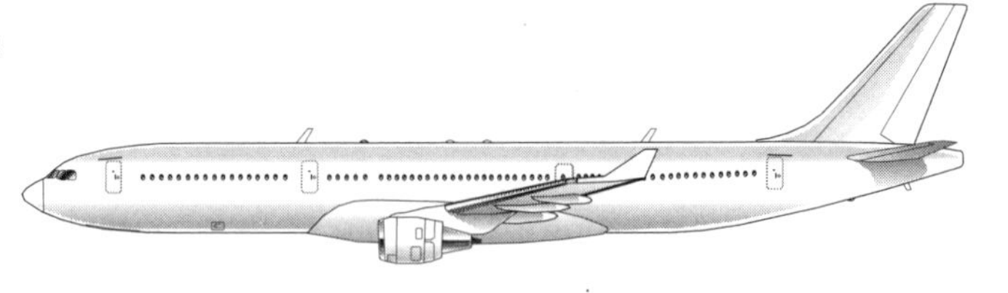

63. Airbus A-330

Powerplants: two General Electric CF6 or Rolls-Royce Trent turbofans.

Seats: up to 330 passengers.

Cruise speed: 490 knots.

Distinctive features: winglets, extended tail cone.

64. Airbus A-300

Powerplants: General Electric CF6 or Pratt & Whitney JT9D.

Seats: up to 320 passengers.

Cruise speed: 490 knots.

Distinctive features: similar to Boeing 767 but has numerous flap extension fairings on the trailing edge of each wing, similar to A-310 but A-300 has longer fuselage.

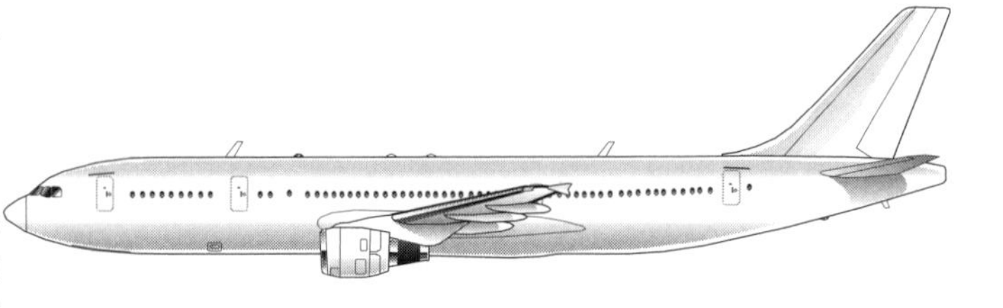

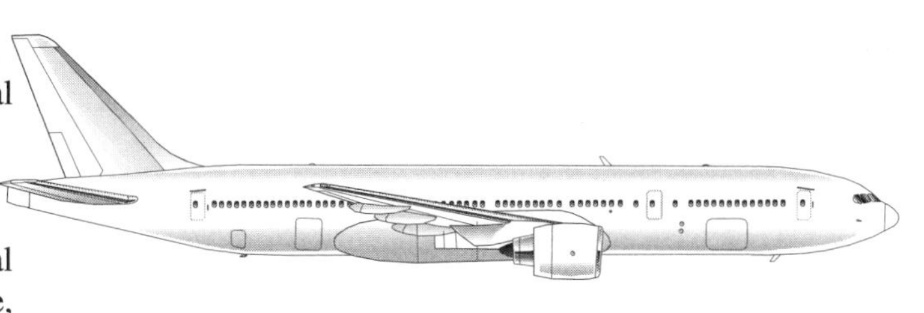

65. Boeing 777

Powerplants: two Pratt & Whitney PW4085, General Electric GE90, or Rolls-Royce Trent turbofans.
Seats: up to 400 passengers.
Cruise speed: 500 knots
Distinctive features: massive size, clean supercritical wing design, tail cone tapers to a flat vertical edge, similar to Boeing 767 or Airbus A-330.

Wide Body Tri-Jets (Two Underwing Engines and One In Tail)

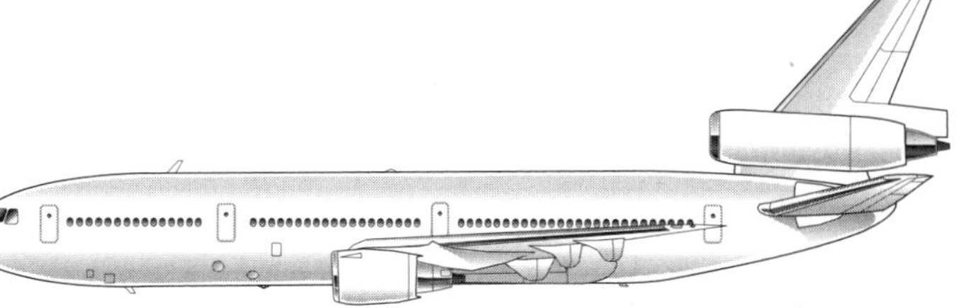

66. McDonnell Douglas DC-10

Powerplants: three General Electric CF6 or Pratt & Whitney JT9D turbofans.
Seats: up to 380 passengers.
Cruise speed: 490 knots.
Distinctive features: number two engine mounted in tail, number two engine nacelle straight from intake to exhaust.

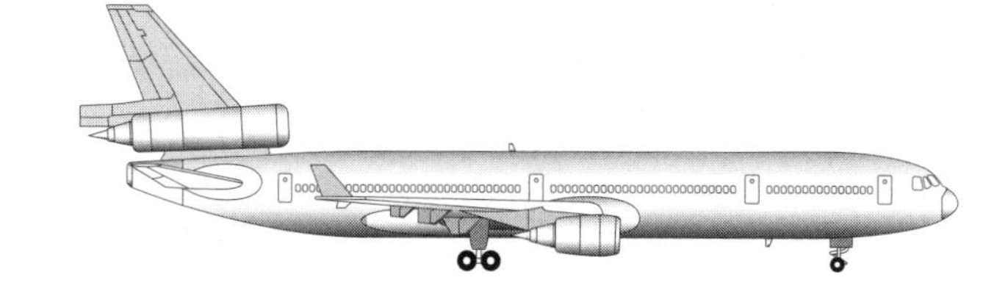

67. McDonnell Douglas MD-11

Powerplants: three General Electric CF6 or Rolls-Royce Trent or Pratt & Whitney PW4360 turbofans.
Seats: up to 405 passengers.
Cruise speed: 490 knots.
Distinctive features: similar to, but larger than, DC-10, large winglets.

68. Lockheed L1011 "TriStar"

Powerplants: three Rolls-Royce RB211 engines.
Seats: up to 400 passengers.
Cruise speed: 490 knots.
Distinctive features: similar in appearance to DC-10, but the number two engine has an S-duct between the intake and the exhaust.

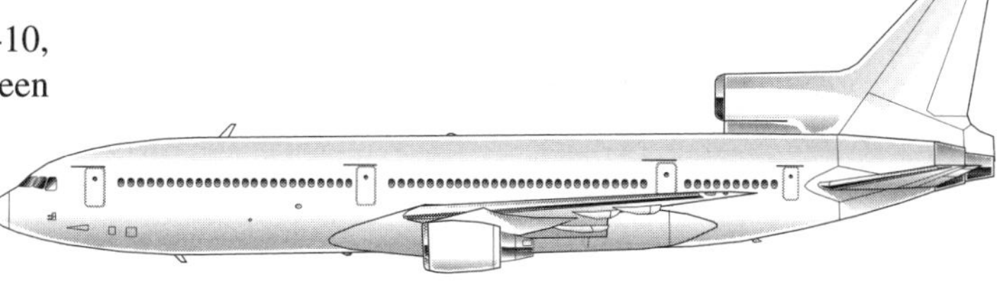

Four-Engined Jet Airliners

69. Boeing 707

Powerplants: four Pratt & Whitney JT3D engines.
Seats: up to 165 passengers.
Cruise speed: 490 knots.
Distinctive features: four engines, long pitot boom at top of tail.

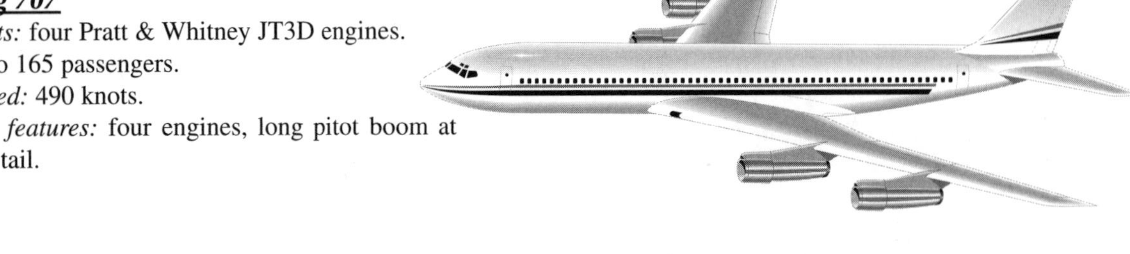

70. McDonnell Douglas DC-8

Powerplants: four Pratt & Whitney JT3D or General Electric/SNECMA CFM56 engines.
Seats: up to 255 passengers.
Cruise speed: 490 knots.
Distinctive features: four engines, long, narrow fuselage, most are currently used to haul freight.

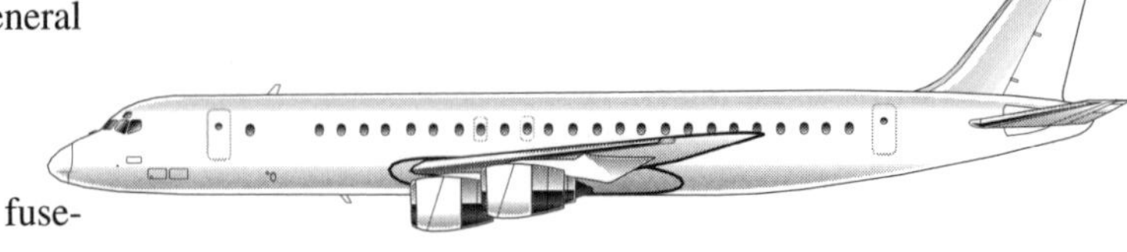

71. Airbus A-340

Powerplants: four General Electric/SNECMA CFM56 turbofans.
Seats: up to 340 passengers.
Cruise speed: 490 knots.
Distinctive features: four engines, winglets.

72. Boeing 747-100/200/300/SP

Powerplants: four Pratt & Whitney JT9D or General Electric CF6 or Rolls-Royce RB211 engines.
Seats: up to 500 passengers.
Cruise speed: 490 knots.
Distinctive features: four engines, bulge on top of fuselage for upper deck seating capacity (747 SP shorter than other models).

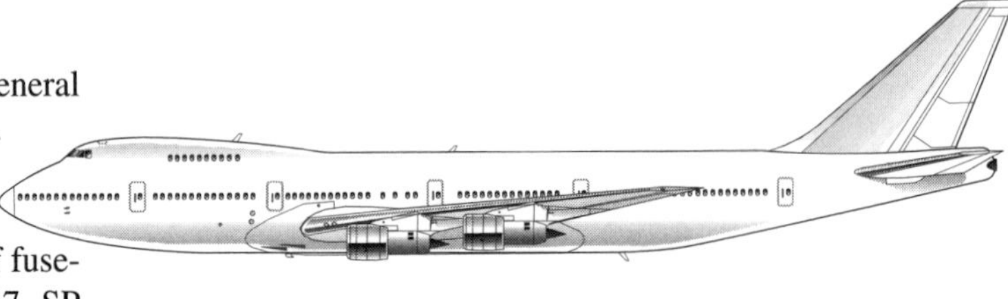

73. Boeing 747-400

Powerplants: four General Electric CF6, Rolls-Royce RB211, or Pratt & Whitney 4256 turbofans.
Seats: up to 600 passengers.
Cruise speed: 485 knots.
Distinctive features: four engines, large winglets, extended upper deck.

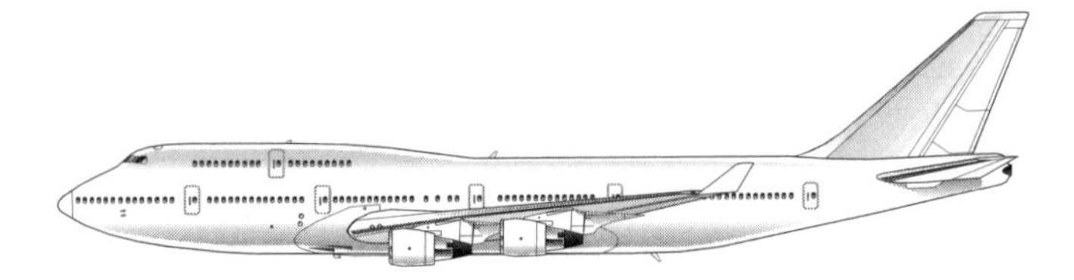

GLOSSARY

Airline and Corporate Aviation Terminology

AS YOU WELL KNOW, aviation is loaded with specialized terminology. Much of it is technical and learned in the course of training, reading, and working in the field. Then there's the matter of nontechnical lingo, which most of us pick up through the school of hard knocks.

Most corporate and Part 135 air-taxi operations use language similar to that found elsewhere in general aviation. If you're coming from the military, corporate and air-taxi terminology will seem comfortable as soon as you're up to speed with civilian operations. The airline industry, on the other hand, has extensive terminology of its own. For that reason, the following glossary covers primarily terms commonly used in normal day-to-day airline operations, though many pertain to other types of operations as well.

ACM: Additional crew member. Normally, this is an abbreviation for a jumpseater but may also include an FAA line check pilot. (Sometimes called XCM, for "extra crew member.")

Alliance Partners: Advanced airline code-sharing agreements between U.S. airlines and their foreign counterparts. Airline alliances join carriers from around the world into a unified route structure. The goal is to be able to provide customers with roundtrip service from virtually anywhere to anywhere on the globe, using single booking through a participating airline.

ALPA: The Airline Pilots Association. The largest union representing pilots in North America. Many airlines have ALPA representation for their pilots, while others have their own in-house unions, and still others are completely nonunion.

AQP: Advanced qualification program. An FAA-approved training curriculum covering indoctrination, qualification, and continuing qualification training.

A-scale: The higher tier of a two-tier pay scale. Pilots on the A-scale are paid more than pilots on the B-scale. Split pay scales resulted from negotiated compromises between pilots and several carriers and effectively lowered the pay of new hires for some period of time, while sustaining the pay of current employees. The A-scale normally applies to pilots who have worked for the same company more than five years.

ATC delay: A delay caused by Air Traffic Control. Normally ATC will delay a flight on the ground before the flight has actually departed. (See also "gate-hold.") Delays may have many different causes (weather, heavy traffic, closed runways, etc.) but most frequently occur due to the need to separate aircraft by a certain distance while enroute or to sequence them into distant terminal areas. Delays may also be encountered while the aircraft is enroute, although ATC, to its credit, tries to avoid this.

Base: Also known as "pilot domicile." This is the airport where a pilot begins and ends all of his or her trips. From the airline's point of view, this is where the pilot lives, so "away" expenses and per diem pay are reckoned from when the pilot leaves the domicile.

Bid awards: The flight schedules or vacation time assigned to pilots by the crew scheduling office.

Bidding, or bids: The method used at airlines and some corporate flight departments for pilots to select their work schedules. Bidding normally occurs monthly and may also include bids for domiciles and vacations. Bids are awarded by order of pilots' seniority. (In other words, pilots with the most seniority are awarded their first choice schedule lines, while the most junior pilots are awarded whatever schedules are left over.)

Block-time: The time between brake release and departure of an aircraft from the departure ramp until it reaches the arrival ramp at the end of flight and brakes are applied.

B-scale: The lower tier of a two-tier pay scale. (See "A-scale.")

Bumped passenger: A passenger that is deplaned from the aircraft or removed from the passenger list. This commonly occurs because there weren't enough seats for the number of passengers booked (that is, the plane was overbooked) or because aircraft performance factors limited the number of passengers that could be carried. Nonrevenue passengers are sometimes bumped when paying passengers show up late for a full flight.

Carry-on bags: These are bags carried onto the airplane by each passenger. Normally passengers are allowed only one or two carry-on bags, depending upon carrier and aircraft type; each bag must fit either in an overhead compartment or under a passenger seat.

CBT: Computer-based training. Certain portions of ground school (e.g., aircraft systems, FMS training, company operations specifications) are now covered by individual computer-based training.

CDL: Configuration deviation list. Similar to an aircraft's minimum equipment list (MEL), an FAA-approved CDL contains allowances and limitations for operation without secondary airframe or engine parts, while still allowing the aircraft to be considered airworthy.

Checked bags: These are bags not carried onto the airplane by passengers. Passengers check bags either at the main ticket counter or with a skycap at curbside. Checked bags are loaded onto aircraft baggage holds by a baggage crew and claimed by passengers at baggage claim after the flight.

Check-in time: Also known as "show time" or "report time," this is when a pilot must be present at the airport for a trip. Typically, check-in time is either forty-five minutes or one hour prior to departure time.

Code-sharing: A marketing agreement between two air carriers that is similar to a partnership. Most code-sharing agreements involve combining the companies' marketing and computer reservations resources. That way, a passenger planning a trip can call either airline or a travel agent and make reservations for a seamless trip, including flight legs on either or both carriers.

COMAT: Company material. This is simply in-house company mail that is carried onboard revenue flights from headquarters to each station and from station to station.

Commuting: Travel to and from a pilot's base or domicile in cases where the pilot's primary residence is not in the same city as his/her domicile. Any pilot who must travel by air from his or her actual home to domicile for trips is known as a "commuter."

CPT: Cockpit procedures training. A CPT trainer is a nonfunctioning mock-up of a particular aircraft's cockpit. New pilots train in the mock-up to gain familiarity with the cockpit layout, prior to starting simulator or flight training.

Crew desk: Also known as "crew scheduling," this office is responsible for daily and long-term scheduling of airline and corporate flight crews.

CVR: Cockpit voice recorder. FAA-required tape recorder that has a microphone in the cockpit. The CVR tapes all communications, both inside the cockpit and on radio, between flight crew members and ATC. CVR tapes are to be used only for accident investigation purposes.

Deadhead(ing): A company-scheduled leg of a trip where a flight crew member rides as a passenger. Deadheading is used by companies to reposition flight crews to meet scheduling needs.

Differences training: Training received by a pilot to qualify on different versions of the same type of aircraft. For example, a pilot qualified to fly a B-737-200 for a given company must receive training on the differences of a B-737-300 before flying that new model.

Dispatch: This office ensures that the company's flight schedule is kept on time. Dispatchers monitor location and status of all company aircraft, as well as fuel, passenger and cargo loads, and enroute and destination weather. They assign specific aircraft for each flight. If a problem develops that could disrupt the company's flight schedule, dispatch decides what action must be taken to best maintain the disrupted schedule. Part 121 operations require the use of FAA-certificated dispatchers in this department. Under Part 121, the dispatcher shares responsibility with the pilot in command for the dispatch of every flight.

DO: Director of flight operations. The DO is a management position in both corporate and airline flight departments. The duties of a DO include policy guidance for all flight crews on company procedures. The DO is responsible for the safe and efficient operation of all aircraft operated by a company.

Duty rig: This term refers to company pay policies that guarantee minimum pay for a given period of work. For example, a pilot may be guaranteed a minimum of four hours of flight pay for any day worked, even though his or her actual flight time was only two hours. Duty rigs are commonly structured by day, by trip, or by week.

Duty time: The time between a flight crew member's check-in time and the time he or she is released from duty after completing a trip.

ETOPS: Extended-range twin-engine operations. ETOPS certification allows a twin-engine aircraft to operate over routes devoid of any suitable alternate/diversionary airports within 120 to 180 minutes flying time.

FE: Flight engineer on a three-pilot crew.

Flight attendant: The flight crew members who attend to the safety and satisfaction of aircraft passengers. (Sometimes called cabin attendants. The term "stewardess" is outdated, is offensive to many, and should never be used.)

Flight control: Most airlines and many corporate operators have flight control offices, which are responsible for operational control of all aircraft, including monitoring the status and activity of each individual airplane. In some companies, flight control is accomplished by the dispatch office.

Flight records (Flight and Rest Records): Pilots are required to keep records of their monthly flight and rest time to ensure that they do not exceed maximum flight and duty time limitations set by FAA regulations. Flight and rest records are normally kept available by both the company and each pilot.

Flight time: The time from when the aircraft begins its takeoff roll to completion of its landing roll.

Flow-release time: An ATC-delayed departure time. Normally, ATC delays flights on the ground before departure to avoid rerouting or holding airborne aircraft. In these cases, ATC will not allow an aircraft to depart until the predetermined flow-release time. (See also "gate-hold" and "ATC delay.")

FTD: Fixed training device. Cockpit mock-up similar to a simulator but with no motion capability. Used for flight crew procedures training prior to actually training in the simulator or aircraft.

Furlough: Aviation lingo for "job layoff." When an airline has too many pilots for its flight schedule, the company may elect to furlough pilots. Normally, this begins at the bottom of the seniority list and moves up. At most airlines, pilots retain their seniority numbers while on furlough. A furlough may last weeks, months, or years, or it may be permanent. All a furloughed pilot can do is get on with life and wait to be called back to work.

Gate-hold: An ATC or company delay requiring the aircraft to hold at the departure gate for a specified amount of time. (Refer to "ATC delay.")

Guarantee: This is the minimum number of flight hours for which a pilot is guaranteed to be paid per month. Different companies have different guarantees. Typically, a guarantee ranges between sixty and eighty-five hours. Pilots are paid for guarantee time or actual flight time, whichever is higher.

HAZMAT: Hazardous materials. There are many state and federal regulations and company policies addressing acceptance, handling, and documentation of hazardous materials. Most larger carriers devote at least one full day of ground school to the topic.

Hub: A city where an airline concentrates a large number of arrivals and departures. Schedules are designed to coordinate arrival and departure times to allow for passenger connections to and from other destinations. In a "hub and spoke" system most or all of a given airline's flights pass through a few hub cities.

Indoctrination: This is a training program included in a pilot's initial ground school. "Indoc" is where the pilot learns about company policies, procedures, and paperwork.

Initial training: Training received as a new-hire pilot. This normally includes extended ground school, simulator training, flight training, and a checkride. Depending on the company, initial training may last anywhere from four to twelve weeks.

IOE: "Initial operating experience," sometimes known as "line training." A new first officer or a recently upgraded captain is required by the FAA, under Part 121 and some 135 operations, and by most companies to fly on line with a designated senior "IOE captain" to learn the ropes. IOE normally lasts for about twenty hours and ends with the IOE captain signing off each pilot as competent.

Jumpseat: An additional crew member seat in the cockpit. An FAA line check pilot or company check pilot may sit in the jumpseat to evaluate a flight crew while on line. When not in use for this purpose, the jumpseat may be used by another airline pilot on personal travel, such as those commuting to or from work. Most major and regional airlines have reciprocating jumpseat agreements that allow pilots from other airlines to travel free of charge. The jumpseat is issued on a space-available basis subject to approval of the captain of the airplane and is considered a professional privilege. A certain amount of "jumpseater" etiquette must be followed. Be sure to research this before trying it for the first time!

Layover: The time between a pilot's arrival at a given destination and departure. Most commonly used to describe an overnight at some destination.

Line: The listing of a crew member's monthly flight schedule. It tells the pilot what trips will be flown and what days will be off.

Line check: An FAA or company check of pilot operations during a routine flight. Line checks are used as much for procedures standardization as for flight crew evaluation.

Load manifest: A detailed list of an aircraft's passenger, cargo, and fuel loads for a given flight.

Maxing out: Sometimes known as "timing out." If a pilot exceeds maximum allowable monthly or yearly flight hours, based on FARs or company policy, that pilot may not accept any more commercial flight time until the end of the applicable period. Affected pilots often get the rest of the month or year off, or they might be assigned additional nonflight-related duties (like revising company navigational charts!).

NBAA: National Business Aircraft Association.

Nonrev passenger: Nonrevenue passenger. Typically this is a passenger traveling on a reduced-rate or no-charge company pass. Nonrev passengers are usually employees or employee relatives traveling on vacation, employees of another airline and their families traveling on "interline passes," or occasionally, company employees traveling on nonurgent business. Nonrev passengers are normally assigned seats on a space-available basis.

OMC: Observer member of the crew. Could be a company pilot, check pilot, FAA observer, or jumpseater.

On and in times: Flight crews for scheduled carriers must record times of touchdown and arrival into the gate. This is used to keep track of airline schedules and may be reported automatically by ACARS (aircraft communications addressing and reporting system), if installed. Otherwise, the pilots report "on's and in's" to their ground crew upon arrival.

OPS: Operations office. This is where pilots go to check in and gather preflight information for each flight. In some locations the flight control office may be located at OPS, as may be pilot mailboxes and a crew lounge.

Out and off times: Flight crews for scheduled carriers must record departure time out of the gate and time off the ground. This is used to keep track of airline schedules and may be reported automatically by ACARS (aircraft communications addressing and reporting system), if installed. Otherwise, the pilots report "out's and off's" to their company by radio after departure.

Pax count: The total number of passengers onboard an aircraft at departure time.

PED: Portable electronic device. Devices such as AM/FM radios, CD players, cellular phones, and laptop computers may interfere with an aircraft's avionics systems. For this reason, FAA regulations now prohibit airline passengers from using PEDs except during cruise flight or when the flight crew determines it is safe to do so.

Per diem pay: Money allotted to a pilot to cover daily expenses incurred on the road. Per diem allowances are often calculated at a flat hourly rate for hours away from domicile, then added to pilot paychecks.

PFE: Professional flight engineer. A career flight engineer, often an A&P mechanic.

Quick turn: This refers to a short turnaround time (usually less than a half hour) between a flight's arrival at the gate and the next departure for that aircraft.

RA: Resolution advisory. A TCAS-commanded vertical maneuver to achieve separation from an aircraft on a projected collision course. (TCAS stands for Traffic collision avoidance system.)

Recurrent training: Regularly scheduled pilot refresher training that occurs either annually or semiannually. For Parts 121 and 135 operations, all crew members must attend recurrent ground school and take company checkrides on a regular basis. Many corporate operators and aircraft insurers require similar recurrent training for Part 91 operations, although it's not federally mandated.

Relief line: The crew scheduling office builds relief lines out of open or extra flight time and unassigned trips. Relief lines also cover pilots on vacation. Normally these are biddable just like regular flight lines.

Reserve: An airline normally maintains extra pilots not assigned normal monthly lines, for backup purposes. Instead they are put on reserve, meaning on-call status. Reserves are the company's insurance that it will maintain schedule. If a pilot becomes ill and cannot fly a trip, a reserve is called in to substitute. Typically, only the most junior pilots are assigned reserve lines, although senior pilots sometimes bid them. (Reserve lines tend to have uncertain schedules due to on-call status but often result in less flying than normal lines.)

Rest records (Flight and Rest Records): Pilots must keep a record of how much rest was received during each twenty-four-hour period, for purposes of meeting FAA minimum-rest regulations. Rest is calculated from the time a pilot is released from duty by dispatch to the time of next check-in for duty.

Revisions: A set of revised pages that must be inserted into a pilot's company or navigational chart manuals; revisions replace outdated pages.

RON: Remain overnight. Refers to a trip where the crew stays away from base overnight, usually with its aircraft. When the aircraft is RON, pilots ensure that the aircraft is "put to bed," meaning gust locks, gear pins, and intake and pitot tube covers are installed to secure the airplane for the night.

Second officer: Flight engineer on a three-pilot crew.

Seniority: A method of ranking pilots by order of date of hire, where the longest employed pilot is most senior (number one in seniority). Historically, seniority number can be the most influential factor of an aviation career. It determines a pilot's domicile, monthly schedule, vacation time, type of aircraft flown, and opportunities for upgrade to captain.

Space-A: Space-available passenger. Space-A passengers may board a flight only if there are seats available after all full fare passengers are onboard. (Refer to "standby passenger.")

Space positive: This (usually company) passenger is guaranteed a seat, even at the expense of paying passengers. Space positive status is used for important company travel, as when crew scheduling must transfer a deadheading pilot to pick up another flight or when mechanics must be sent to an outstation to repair an aircraft.

Standby passenger: A passenger traveling on some type of space-available ticket, often at reduced fare. An airline loads its full fare passengers first and then boards standby passengers to fill any remaining seats. There are different classes of standby tickets, the highest having first priority in boarding.

Stand-up overnight: A trip schedule that brings a flight crew into a destination late at night and has the same crew depart very early the next morning. The flight crew remains officially on duty for the entire night but usually gets some limited amount of sleep. Stand-ups are used when the same crew flies the last late night flight into a destination and then the early bird flight out the next morning.

Stewardess: An outdated term for a flight attendant, considered sexist by many. Do not use it. (See "flight attendant.")

Surplus: Eliminated pilot "seat" or "slot" at a given domicile. Pilots in this situation often have bumping rights to another domicile or seat, based on seniority.

TAFB: Time away from base. TAFB is recorded on a pilot's monthly schedule and flight records for pay purposes. It indicates how many hours the pilot is away from his or her domicile.

TOLD Cards: Takeoff and landing data cards. These quick reference cards include takeoff and landing weights and speeds for a given aircraft, computed from performance charts.

Transition training: Pilot training for a move to a different type of aircraft in the same company (say, transition from a B-737 to an Airbus).

Trip trade: One pilot trades scheduled trips with another. Approval is usually required by crew scheduling. Most often pilots trade trips because one of them wants one or more specific days off.

Turn: Short for "turnaround." Often "turn" refers to a flight out to a destination with a same-day return back to the original departure airport (e.g., "I'm doing a Monterey turn today.")

Unions: As in other industries, many aviation employee groups are represented by unions. Depending on the company, pilots may be represented by a large multicompany union, by in-house unions, or no union at all.

Upgrade training: Pilot training for upgrade from first officer to captain.

BIBLIOGRAPHY

IN ADDITION to the following, pilot training materials were consulted for many different types of aircraft, including Boeing 727, 737, and 767, McDonnell Douglas MD-80 and DC-10, Beechcraft King Air C90, BE-200, and BE-1900 models, de Havilland Dash-8, and British Aerospace Jetstream 31.

"Aerospace Source Book 2000." 2000. *Aviation Week and Space Technology* 152(3):68–74, 96–99.

American Airlines Flight Department. 1971. *Boeing 727 Operating Manual.* Fort Worth, Tex.: American Airlines Flight Academy.

American Airlines Flight Training Department. 1986. *Crew Resource Management Study Guide.* Fort Worth, Tex.: American Airlines Flight Academy.

Caracena, Fernando, Ronald L. Holle, and Charles A. Doswell III. 1990. *Microbursts: A Handbook for Visual Identification.* 2nd ed. Washington, D.C.: U.S. Government Printing Office.

Collins Division Technical Services. 1987. *Collins Weather Radar System.* Revision 8. Cedar Rapids, Iowa: Collins Air Transport Division.

Collins, Richard L. 1983. "Understanding the Jet Stream." *Flying,* November, 53–55.

Crane, Dale. 1991. *Dictionary of Aeronautical Terms.* 2nd ed. Renton, Wash.: Aviation Supplies and Academics.

Degani, Asaf. 1990. *Human Factors of Flight-Deck Checklists: The Normal Checklist.* 1st ed. NASA Report. Coral Gables, Fla.: University of Miami.

Delta Air Lines Training Department. 1996. *Boeing 737-300G Pilot's Reference Manual.* [Delta in-house training manual] Atlanta, Ga.

Delta Air Lines Training Department. 1988. *Lockheed L1011 Pilot's Reference Manual.* [Delta in-house training manual] Atlanta, Ga.

Department of the Air Force. 1986. *Aerospace Physiology/Human Factors (T-37/T-38).* P-V4A-A-AS-SW. Headquarters Air Training Command. Randolph Air Force Base, Texas.

Department of Transportation. 1972. *Aircraft Wake Turbulence.* Advisory circular AC 90-23D. Washington, D.C.: U.S. Government Printing Office.

_____. 1975. *Aviation Weather.* Advisory circular AC 00-6A. Washington, D.C.: U.S. Government Printing Office.

_____. 1979. Aviation Weather Services. Advisory circular AC 00-45C. Washington, D.C.: U.S. Government Printing Office.

_____. 1994. *Airman's Information Manual.* Renton, Wash.: Aviation Supplies and Academics.

Garrett Technical Training Center. 1981. *TPE331 Study Guide 103.* 1st ed. Phoenix, Ariz.: Garrett General Aviation Services Company.

_____. 1989. *TFE731 Study Guide 142.* 1st ed. Phoenix, Ariz.: Garrett General Aviation Services Company.

Grayson, David. 1988. *Terror in the Skies.* Secaucus, N.J.: Citadel Press.

Gunston, Bill. 1992. *Chronicle of Aviation.* 1st ed. London: Chronicle Communications.

Horne, Thomas. 1992. "Counting to X." *AOPA Pilot* 35(9):19–21.

Hurt, Jr., H. H. 1965. *Aerodynamics for Naval Aviators.* 2nd ed. Office of the Chief of Naval Operations Aviation Training Division.

McCarthy, John, and Dr. T. Theodore Fujita. 1987. *Wind Shear Microburst Home Study Guide.* 1st ed. Dallas-Fort Worth, Tex.: American Airlines Training Academy.

McLaren, Grant. 1994. "Turboprop Sales Strong but Shakeout Expected." *Professional Pilot* 28(1):62–66.

Mark, Robert. 1992. "Regional Airline Aircraft." *Airline Pilot* 61(6):12–17.

Nader, Ralph J., and Wesley J. Smith. 1993. *Collision Course.* Blue Ridge Summit, Pa.: TAB Books.

National Aeronautics and Space Administration. 2000. *NASA's Aviation Safety Reporting System: Callback Newsletter.* Issue 250, April, 2000.

National Transportation Safety Board. 1990. *Aviation Accident Report Transcripts: American Airlines Crew Resource Management Handout.* American Airlines Flight Academy.

"1994 Planning and Purchasing Handbook." 1994. *Business and Commercial Aviation* 74(5).

1998 Civil Aviation/Federal Clip Art. One Mile Up, Inc., Annandale, Va.

Pratt & Whitney Technical Services Department. 1988. *The Aircraft Gas Turbine Engine and Its Operation.* 6th ed. United Technologies Aircraft Corporation.

Richardson, John E. 1982. "Wind Shear." *Federal Aviation Administration Accident Prevention Newsletter.* FAA-P-8740-40.

Rogers, Ron. 1994. "Airbus Flight Report A-330." *Airline Pilot* 63(2):16–18.

Wiley, John. 1992. "Beech Super King Air 350." *Professional Pilot* 26(12):82–85.

_____. 1993. "The Beech 1900D." *Professional Pilot* 27(10):50–53.

Wilkinson, Stephan. 1990. "Assaulting the Barrier: Mach One." *Air and Space,* December, 58–72.

Williams, Jack. 1992. *The Weather Book.* 1st ed., edited by Carol Knopes, USA Today. New York: Random House.

Wood, Dereck. 1992. *Jane's World Aircraft Recognition Handbook.* 5th ed. Coulsdon, U.K.: Jane's Information Group.

INDEX

Refer to figure on pages with italicized numbers.

A-scale, 249
Abnormal procedures. *See* Flight procedures (emergency/abnormal)
ACARS. *See* Aircraft communications addressing and reporting system (ACARS)
ACM. *See* Additional crew member (ACM)
ACMs. *See* Air cycle machines (ACMs)
Additional crew member (ACM), 249
ADF. *See* Automatic direction finder (ADF)
Advanced qualification program (AQP), 249
Advisory panels. *See* Annunciator/warning systems
Aerodynamics
  fixed surfaces of, 205, *207,* 209
  sound barrier and, 201, 203
  swept wing, 203, 205
Aeronautical Information Manual (AIM), 103
AFM. *See Aircraft Flight Manual (AFM)*
Ailerons, 78
Airbus
  A-300, 246
  A-310, 246
  A-318/A-319/A-320/A-321 Series, 245
  A-330, 246
  A-340, 248
Airbus A-320, 160, 172
Aircraft communications addressing and reporting system (ACARS), 163, *164*
Aircraft Flight Manual (AFM), 54, 152
Aircraft Maintenance and Flight Records, 152
Aircraft Performance Data, 152
Aircraft Registration Certificate, 152
Aircraft systems
  annunciator/warning systems, 113, *114–115,* 116
  environmental systems, 91–92, 96
  fire protection systems, 116–119, *117*
  flight controls, 75, *76–77,* 78, *79–80,* 81–82, *82–84*
  fuel systems, 96, *97–98,* 99, *100,* 101–102, *102*
  ice/rain protection, 103–110
  landing gear systems, 110–113, *112*
  oxygen systems, 90–91, *91*
  pressurization, 82–83, 85, 87
Air cycle machines (ACMs), 69, 92, *94–95*
Airline Pilots Association (ALPA), 249
Airport Analysis, 142, 152
Air rage, 216
Airspeeds, 120, *121–122,* 123, 132
Air Traffic Control (ATC)
  delay, 249
  flight spoilers/speed brakes and, 78
Air turbine motors (ATM), 65
Airworthiness Certificate, 152
Alcohol/drug use, 9
Alliance partners, 249
ALPA. *See* Airline Pilots Association (ALPA)
Altimetry, 209–210
Altitude level-off, 227–228
Annunciator/warning systems, 113, *114–115,* 116
Antennas, 187, *190*
Anti-icing systems. *See* Ice/rain protection
Approaches, 210
APU. *See* Auxiliary power unit (APU)
AQP. *See* Advanced qualification program (AQP)
Area navigation (RNAV), 163, *167–168,* 167–169, *169–171,* 171, *196*
Asymmetric propeller thrust. *See* Propellers
ATC. *See* Air traffic control (ATC)
ATM. *See* Air turbine motors (ATM)
ATR 42/72, 242
Automatic direction finder (ADF), 196–197, 228–229
Autopilots, 158, *160*
Auxiliary power unit (APU), 26, 65, 69, *72–73,* 74, 118–119. *See also* Electrical systems
Axial-flow compressor, *5,* 13
Azimuth, 187, *188*

B-scale, 250
Balance considerations. *See* Limitations
Balloon engines, 10, *11*
Base, 249
Batteries, 49–50, *50, 58*
Beechcraft
  1900, 113
  1900C, 239
  1900D, 240
  400, 238
  King Air 200/300/350, 235
  King Air Series 90, 234
  Starship, 235
Beta range, 31, 33, *34–35*
Bid awards/bidding, 249
Bleed air, 69–70, *70,* 83, 92, 96
Block-time, 249
Boeing
  707, 248
  717, 81, 244
  727, 140, 244
  737, 78
  737-200, 245
  737/300/400/500 ("Classic Series"), 245
  737/600/700/800/900 ("New Generation"), 245
  747-100/200/300/SP, 248
  747-400, 248
  757-200/300, 246
  767-200/300/400, 246
  777, 160, 247
Boundary-layer wing fences, 205, *208*
Brake fade, 111
Brakes, 78, 81, 111–112, *112. See also* Landing gear systems
British Aerospace
  ATP, 241
  BAe 146/RJ70/RJ85, 243
  HS-125, 237
  Jetstream 31/Super 31, 240
  Jetstream 41, 240
Bumped passenger, 250
Bus bar systems, *53,* 53–54, *59, 61*

Cabin pressurization. *See* Pressurization
Canadair
  Challenger 600/601/603, 239
  Global Express, 239
  Regional Jet CRJ 200/700/900, 243
Carry-on bags, 250
Cascade-type reversers, 29, *30*
CAT. *See* Clear air turbulence (CAT)
Category I/II/III approaches, 210
CBT. *See* Computer-based training (CBT)
CDI. *See* Course deviation indicator (CDI)
CDL. *See* Configuration deviation list (CDL)
CDU. *See* Control display unit (CDU)
Centrifugal-flow compressors, 12–13, *14,* 16
Cessna
  Caravan, 110, 234
  Citation I/II/V/Bravo/Encore/Excel/Sovereign, 237
  Citation III/VI/VII/X, 238
  Citation Jet/Jet2, 238
  Citation S-II, 105
  Conquest I/II, 234

CFIT. *See* Controlled flight into terrain (CFIT)
CG (center of gravity). *See* Limitations
Checked bags, 250
Check-in time, 250
Checklists/callouts, 127–128, *128–129,* 130–132, *135–136,* 152, *153–154,* 155
Check valves, 72
Circuit breakers (CBs), *48,* 54. *See also* Electrical systems
Circuits. *See* Electrical systems
Civilians and transitioning military aviators, 4–5
Clamshell reversers, 29, *30*
Clear air turbulence (CAT), 224, *225,* 226
Cockpit oxygen breathing systems, 90, *91*
Cockpit procedures trainer (CPT), 8, 250. *See also* Flight procedures (normal)
Cockpit voice recorder (CVR), 250
Code-sharing, 250
Coffin corner, 203, *204,* 224
Collector bays, 96
Collision avoidance systems. *See* Traffic alert and collision avoidance system (TCAS)
COMAT. *See* Company material (COMAT)
Commander 840/980/900/1000, 235
Commuting, 250
Company material (COMAT), 250
Compass Deviation Cards, 152
Compressibility, 201–202
Compressors
  multispool engines and, 16, *17*
  multistage, 16
  of reciprocating engines, 10, 12–13, *12–15,* 16
Computer-based training (CBT), 7, 250
Condensers. *See* Vapor cycle machines (VCMs)
Condition levers, 24
Configuration deviation list (CDL), 152, 155, 250
Control devices (electrical), 50–51. *See also* Flight control
Control display unit (CDU), *175*
Controlled flight into terrain (CFIT), 200
Core turbine engines, 16, *18*
Course deviation indicator (CDI), 156, 167, 228–229
CPT. *See* Cockpit procedures trainer (CPT)
Crew desk, 250
Crew resource management (CRM)
  as contemporary issue, 9
  normal flight procedures and, 124–127
  and transitioning military aviators, 4
Critical engines, 38, 40, *41*
CRM. *See* Crew resource management (CRM)
Current. *See* Electrical systems
CVR. *See* Cockpit voice recorder (CVR)

Dassault-Breguet
  Falcon 20, 236
  Falcon 2000, 236
  Falcon 50/900, 238
DC-8-61, 126
DC aircraft. *See* McDonnell Douglas
de Havilland
  Dash-7, 241
  Dash-8-100/200/300/400, 241
Deadhead, 250
Deicing systems. *See* Ice/rain protection
DHC-6 Twin Otter, 110
Differences training, 250
Director of flight operations (DO), 250
Discrimination, 9
Dispatch, 151, 250
Distance measuring equipment (DME), 156, 210
DME. *See* Distance measuring equipment (DME)
DO. *See* Director of flight operations (DO)
Documents, 152
Doppler radar, 194, *195. See also* Radar
Dornier
  Dornier 228, 241
  Fairchild/Dornier 328, 242
  Fairchild/Dornier 328/428/728 Jet, 243
Dorsal fins, 209
Double-cue systems, 158
Drift-down procedures, 140, *142*
Dump valves, 87, *88,* 145
Dutch roll, 205, *206*
Duty rig, 250
Duty time, 250

Echo height/shapes, 192–193, *194*
EGPWS. *See* Enhanced ground proximity warning systems (EGPWS)
Electrical discharge detectors, 196–198, *197*
Electrical systems. *See also* Flight procedures (emergency/abnormal)
  basic circuitry, *47*
  basic components, *48*
  batteries and, 49–50
  checklists/callouts and, 130
  circuit protection for, 51, *53,* 53–54
  control devices for, 50–51
  faults, *52*
  fire protection and, 117–118
  generators as power source, 46, 49
  reading diagram for, 54, *55–59*
  troubleshooting, 60, *61*
Electronic flight instrumentation systems (EFIS), 158, 160, *161–162,* 163, *176*
Elevation, 187
Embraer
  135/145/170, 243
  Brasilia EMB-120, 240
Emergency descents, 89
Emergency procedures. *See* Flight procedures (emergency/abnormal)
Engine pressure ratio (EPR), 22, *24*
Engines, 120. *See also* Balloon engines; Gas turbine engines
Enhanced ground proximity warning systems (EGPWS), 163, 200
Environmental conditions/systems, 91–92, 96, 123, 230–231
EPR. *See* Engine pressure ratio (EPR)
Equivalent shaft horsepower (eshp), 22
Ethylvinylacetate, 70
ETOPS. *See* Extended-range twin-engine operations (ETOPS)
EVA. *See* Ethylvinylacetate
Evaporators. *See* Vapor cycle machines (VCMs)
Extended-range twin-engine operations (ETOPS), 211, 250

FADECs. *See* Full-authority digital engine controls (FADECs)
FAF. *See* Final approach fix (FAF)
Fairchild
  Merlin, 234
  Metroliner, 240
Fairchild/Dornier
  328, 242
  328/428/728 Jet, 243
FCU. *See* Fuel control unit (FCU)
FE. *See* Flight engineer (FE)
Feathering, 37
Final approach fix (FAF), 131
Fins, *208,* 209
Fire protection
  bleed hazards/protections and, 72
  detection/extinguishing systems and, 116–119, *117*
First officer (FO), 124, 152, 216
Fixed training device (FTD), 251
Flaps, 75, *77, 122*
Flash cards, 7
Flight attendant, 250
Flight control, 75, *76–77,* 78, *79–80,* 81–82, *82–84,* 250–251
Flight director, 158, *159–160*
Flight dispatch, 151
Flight engineer (FE), 250
Flight idle, 28
Flight levels, 209
Flight management computer (FMC), 172, *173–174, 177*
Flight management system (FMS), 171–172, *173–179,* 180, *181–183,* 184
Flight procedures (emergency/abnormal)
  abnormal procedures, 134, 136, *136*
  definition, 133
  documents and, 152
  emergency procedures, 89, 133–134, *135*
Flight procedures (normal)
  checklists/callouts for, 127–128, *128–129,* 130–132
  crew coordination and, 124, *125*
  crew resource management (CRM) and, 124–127
  documents and, 152
Flight profiles, 8
Flight records, 251
Flight spoilers, 78, *80,* 81
Flight time, 251
Flow-release time, 251
Fluidic coupling, 21
Fly-by-wire control systems, 81–82, *83–84*
Flying pilot (FP), 124, *125,* 133–134, 136, 216
FMC. *See* Flight management computer (FMC)

FMS. *See* Flight management system (FMS)
FO. *See* First officer (FO)
Fokker
  F100, 244
  F27/F50, 242
  F28, 244
FP. *See* Flying pilot (FP)
Freon units, 92
FTD. *See* Fixed training device (FTD)
Fuel
  filling procedure for, 151–152
  heaters, 101
  pumps, 99
  quantity measurement, 101, *102*
  systems of, 96, *97–98,* 99, *100,* 101–102, *102,* 227
  valves, 99, 101
  vents, 101–102
Fuel control unit (FCU), 99
Full-authority digital engine controls (FADECs), 99
Furlough, 251

Gas generators, 10, 16, *18*
Gas turbine engines
  characteristics in flight of, 27–28
  cockpit controls of, 22, 24
  core or gas, 16, *18*
  igniters for, 26, *28*
  operating principles of, 10, *11, 13*
  starting of, 24–26
Gate-hold, 251
GCUs. *See* Generator control units (GCUs)
Generator control units (GCUs), 51, 54
Generators, 10, 16, *18,* 46, *47–48,* 49, 90–91
Global positioning system (GPS), 156, *168,* 168–169
Governors, 31, *32–33*
GPS. *See* Global positioning system (GPS)
GPU. *See* Ground power unit (GPU)
GPWS. *See* Ground proximity warning systems (GPWS)
Ground power unit (GPU), 49
Ground proximity warning systems (GPWS), 200
Ground school preparation, 7–8
Ground spoilers, 78, *80,* 111
Guarantee, 251
Gulfstream II/III/IV/V, 238

Harassment, 9
Hawker 800/1000, 237
Hazard avoidance systems. *See also* Radar
  EGPWS, 200
  GPWS, 200
  lightning detectors, 196–198, *197*
  TCAS, 163, 198, *199,* 252
Hazardous conditions, 70, 72. *See also* Ice/rain protection; Weather conditions
Hazardous materials (HAZMAT), 251
HAZMAT. *See* Hazardous materials (HAZMAT)
Head-up displays (HUD), 163, *165–166*
Header tanks, 96
Heat exchangers, 92, *93*
HGS. *See* Head-up displays (HUD)
High-pressure compressors/turbines, 16
High-speed buffet, 202–203, *204*
Holding speeds, 210
Horizontal situation indicator (HSI), 156, *157*
Hot start, 25–26
HSI. *See* Horizontal situation indicator (HSI)
Hub, 251
HUD. *See* Head-up display (HUD)
Hydraulic power systems
  backup pumps for, 65, *68,* 69
  benefits of, 60, 62
  components, 62, *64, 67*
  power transmission properties, *62–63*
  reservoirs/accumulators for, 65, *66*

Ice/rain protection
  brake antiskid/anti-ice systems and, 111–112
  deicing fluid characteristics, 109
  engine icing and, 106–107, 109
  fuel system icing, 109
  ground icing and, 109–110
  high altitudes and, 221
  in-flight structural icing, 103–106
  pilot and, 109
  rain and, 110
  system limitations and, 123
  thermal leading edge anti-ice systems and, *105*
Igniters, 26, *28*
ILS. *See* Instrument landing system (ILS)
Indoctrination training, 5–6, 251
Inertial navigation system (INS), 156, 158, 169
Initial operating experience (IOE), 7, 251
Initial training, 251
INS. *See* Inertial navigation system (INS)
Instrument Flight Rules (IFR), 209–210
Instrument landing system (ILS), 156
International flight operations, 211
IOE. *See* Initial operating experience (IOE)
Isolation valves, 72
Israel Aircraft Industries
  Astra Jet, 237
  Westwind, 237

Jet routes, 209
Jetstream, 222, *223,* 224
Jumpseat, 251

Laminar-flow wings, 203
Landing gear systems, 110–113, *112*
Landing performance. *See* Performance
Latitude/longitude, 169, *170–171,* 171
Layover, 251
Lead-acid batteries. *See* Batteries
Leading edge devices (LEDs), 75, *77*
Lear
  23/24/25/35/36, 239
  28/31/45/55/60, 239
LEDs. *See* Leading edge devices (LEDs)
Lightning detectors, 196–198, *197*
Limitations
  airspeeds, 120, *121–122,* 123
  balance considerations, 147–148, *148–149,* 150
  of engines, 120
  load manifest, *149*
  passenger weights and, 150
  random loading programs and, *149,* 150
  training and, 5
  weight categories, 145, 147
Line, 251
Line check, 251
Line oriented flight training (LOFT), 127
LLWAS. *See* Low-level wind shear alerting systems (LLWAS)
Load manifest, *149,* 152, 251
Lockheed
  Jetstar, 236
  L1011 "TriStar," 78, 126, 247
LOFT. *See* Line oriented flight training (LOFT)
Long-range area navigation (LORAN), 167–168
LORAN. *See* Long-range area navigation (LORAN)
Low-level wind shear alerting systems (LLWAS), 218, 221

Mach numbers/tuck, 201, 203, 228
Manual reversion, 81
Manuals, 54, 152
Maximum differential (max diff), 83
Maximum landing weight (MLW), 145
Maximum takeoff weight (MTOW), 99, 101, 145, *149*
Maximum zero-fuel weight (MZFW), 145, *146*
Maxing out, 251
McDonnell Douglas, 142
  DC-10, 127, 247
  DC-8, 248
  DC-8-61, 126
  DC-9, 81, 243
  MD-11, 160, 247
  MD-80 series/MD-90, 81, 244
Mean aerodynamic chord (MAC), *147,* 147–148
MEAs. *See* Minimum enroute altitudes (MEAs)
MEL. *See* Minimum equipment list (MEL)
MFD. *See* Multifunction display (MFD)
Microbursts, 217–218, *219–220*
Military aviators, 4–5
Minimum enroute altitudes (MEAs), 140
Minimum equipment list (MEL), 152, 155, 250
Mitsubishi MU-2, 78, 235
MLW. *See* Maximum landing weight (MLW)
Motive flow. *See* Fuel
Mountain waves, 224, *225*
MTOW. *See* Maximum takeoff weight (MTOW)
Multifunction display (MFD), 163
Multispool engines, 16, *17–18*
Multistage compressors, 16
MZFW. *See* Maximum zero-fuel weight (MZFW)

National Business Aircraft Association (NBAA), 251

National Transportation Safety Board (NTSB), 126
Navigation control systems
  ACARS, 163, *164*
  area navigation (RNAV), 163
  autopilots, 158, *160*
  flight director, 158, *159–160*
  FMS, 171–172, *173–179,* 180, 184
  GPS, 156, *168,* 168–169
  HSI, 156, *157*
  HUD, 163, *165–166*
  LORAN, 167–168
  RNAV, 163, *167,* 167–169, *169–171,* 171
  VOR/DME-based RNAV, 167
NBAA. *See* National Business Aircraft Association (NBAA)
Negative valves, 87, *88*
NFP. *See* Nonflying pilot (NFP)
Ni-cads (nickel-cadmium) batteries. *See* Batteries
Nonflying pilot (NFP), 124, *125,* 133–134, 136
Nonrev passenger, 251
Normal, Abnormal, and Emergency Checklists, 152
Nozzles, 10
NTSB. *See* National Transportation Safety Board (NTSB)

Observer member of the crew (OMC), 251
OMC. *See* Observer member of the crew (OMC)
On and in times, 252
Operations office (OPS), 252
OPS. *See* Operations office (OPS)
Oxygen systems, 90–91, *91,* 118

P-factor. *See* Propellers
PACK (aircraft environmental system), 96, *98*
Packs, 69
Panel lighting. *See* Annunciator/warning systems
Passengers
  air rage and, 216
  bumped, 250
  nonrev, 251
  oxygen systems for, 90–91
  weight limitations and, 150
Pax count, 252
PED. *See* Portable electronic device (PED)
Per diem pay, 252
Performance
  cruise, 142–144
  FMS and, 180
  landing/braking, *139,* 140
  routine planning for, 140, 142–144
  takeoff/climb, 138, *139,* 140
Petal door fan reversers, 29
PF. *See* Nonflying pilot (NFP)
PFD. *See* Primary flight display (PFD)
PFE. *See* Professional flight engineer (PFE)
Piaggio Avanti, 236
Pilatus Aircraft LTD. PC-12, 233
Pilot domicile, 249
Pilots, 124, *125,* 133–134, 136, 152, 216
Pilot's Operating Handbook (POH), 54
Pilot weather reports (PIREPs), 103
Piper Cheyenne, 235
PIREPs. *See* Pilot weather reports (PIREPs)
Pitch. *See* Propellers
Pneumatic leading edge deice boots, 103–104, *104*
Pneumatic power systems, 69–70, *70–74*
PNF. *See* Nonflying pilot (NFP)
POH. See *Pilot's Operating Handbook (POH)*
Portable electronic device (PED), 252
Positive pressure relief valves, 87, *88*
Power systems
  depiction of, in pilot training, 42
  electrical, *47–48,* 49–51, *50,* 52–*53,* 53–54, *55–59,* 60
  hydraulic, 60, *61–64,* 62, 65, *66–68,* 69
  pneumatic, 69–70, *70–74*
  waterwheels, 43–44, *43–45,* 46
Precipitation gradient, 187, *192*
Preflight procedures. *See* Checklists/callouts
Pressure systems. *See* Pneumatic power systems
Pressurization, 82–83, 85, *85–86,* 87, *88,* 89–91, *91*
Primary flight display (PFD), 160, *162*
Procedures. *See* Flight procedures
Professional flight engineer (PFE), 252
Profile descent, 209
Prop beat, *36*
Propellers
  asymmetric thrust of, 37
  auto-feather systems of, 33, 36
  beta range and, 31, 33
  governors of, 31, *32–33*
  p-factor of, 37, *39*
  pitch of, 37, *38*
  slipstream effects of, 38, *40*
  synchronizer/synchrophasers of, 36, *36*
  tractor *v.* pusher, 40
  transitioning military aviators and, 5
Pumps. *See* Fuel

Quick turn, 252

RA. *See* Resolution advisory (RA)
Radar, 185, *186,* 187, *188–197,* 198
Radio magnetic bearing pointer (RMI), 156, 228–229
Radio Station License, 152
Rain protection. *See* Ice/rain protection
Ram air turbines (RAT), 65, 69
Rapid decompression (RD), 89
RAT. *See* Ram air turbines (RAT)
Raytheon Hawker jet series, 105
RD. *See* Rapid decompression (RD)
Reciprocating engine, 10, *12*
Recurrent training, 252
Redundancy, 81–82
Relief line, 252
Remain overnight (RON), 252
Report time. *See* Check-in time
Reserve, 252
Resolution advisory (RA), 252
Rest records, 252
Reversers. *See* Thrust reversers
Revisions, 252
Rheostats, 50
RMI. *See* Radio magnetic bearing pointer (RMI)
RNAV. *See* Area navigation (RNAV)
Rockwell Sabreliner, 236
Roll spoilers, 78, *79*
RON. *See* Remain overnight (RON)
Runaway battery, 50

Saab 340, 241
Safety valves, 87
SAT. *See* Standard air temperature (SAT)
Second officer, 252
Seniority, 252
Shaft horsepower (shp), 22
Shock wave, 202, *202*
Shorts 360, 241
Show time. *See* Check-in time
Sidestick controllers, *84*
Simulator/flight training, 6, 8
Single-cue systems, 158
Slats, 75, 78
Socata-Groupe Aerospatiale Matra TBM700, 234
Space-A, 252
Space positive, 252
Speeds. *See also* Airspeeds
  descent rate calculations and, 229–230
  estimation of ground, 228
  wind adjustments and, 229
Spoilers, 78, *79–80,* 81
Squat switches, 87, *88*
SR-71, 26
Stabilons, *208,* 209
Stalls, 209
Stand-up overnight, 252
Standard air temperature (SAT), 230–231
Standby passenger, 252
Sterile cockpit, 132
Stewardess, 252
Storm conditions. *See* Weather conditions
Strakes, 209
Super-critical wings, 203
Surge tanks, 96
Surplus, 253
System jamming, 81

T-handles. *See* Fire protection
TAFB. *See* Time away from base (TAFB)
Tailets, 205, *208*
Takeoff and landing data (TOLD) cards, 142, *143,* 152, 253
Takeoff/climb performance. *See* Performance
Takeoff configuration warning system (TOCWS), 116
Taxiing, 130–131
TCAS. *See* Traffic collision avoidance system (TCAS)
Temperature mixing valves, 96

Temperatures. *See* Environmental conditions/systems
Thermal leading edge anti-ice systems, 104–105, *105*
Thermal runaway, 50
Thrust, 22, *26,* 28
Thrust reversers, 29, *30,* 31
Time away from base (TAFB), 253
Timing out. *See* Maxing out
TKS liquid ice protection systems, 105, *106*
TOCWS. *See* Takeoff configuration warning system (TOCWS)
TOLD cards. *See* Takeoff and landing data (TOLD) cards
Torque effect, 38
Torque levers, 24
Traffic alert warnings. *See* Traffic collision avoidance system (TCAS)
Traffic collision avoidance system (TCAS), 163, 198, *199,* 252
Training
  indoctrination training, 5–6, 251
  initial training, 251
  limitations, systems, and procedures, 5
  new-hire, 7–9
  recurrent, 252
  simulator/flight, 7
  transition, 37–38, *38–41,* 40, 253
  upgrade, 253
Trip trade, 253
Turbofan engines, *19,* 20
Turbojet engines, 18, *19*
Turboprop engines, *19, 21, 23, 25, 27*
  corporate aircraft and, 233–236
  operating principles of, 20–22
  pneumatic leading edge deice boots and, 103–104, *104*
  thrust reverse on, 29, *30,* 31
Turbulance, 211, 213. *See also* Radar
Turn (turnaround), 253

Union affiliations, 9, 253
Upgrade training, 253

V-speeds, 120, *121,* 132, 134, *136,* 137–142, *139, 143*
Vapor cycle machines (VCMs), 92, *94,* 96, *97*
VCMs. *See* Vapor cycle machines (VCMs)
VDP. *See* Visual descent point (VDP)
Ventral fins, *208,* 209
Vertical navigation (VNAV), 156, 158
Vertical scan, *189*
Visual descent point (VDP), 229
VNAV. *See* Vertical navigation (VNAV)
VOR. *See* Course deviation indicator (CDI)
Vortex generators, 205, *208*
Vortices, 211, *212,* 213, *215*
Vortilons, 205, *208,* 209

Wake turbulence, 213
Warning systems. *See* Annunciator/warning systems
Waterwheel power systems, 43–44, *43–45,* 46
Weather avoidance systems
  lightning detectors, 196–198, *197*
  radar, *186,* 187, *188–197,* 198
Weather conditions
  bleed hazards/protections and, 70
  clear air turbulence (CAT), 224, *225,* 226
  high-altitudes and, 221–222, *223,* 224, *225,* 226
  ice/rain protection and, 103–110
  jetstream and, 222, *223,* 224
  microbursts, 217–218, *219–220*
  radar, 185
  thunderstorm avoidance, 228
  wind shear, 217–218
Weeping wing, 105
Weight. *See* Limitations
Wet wings, 96
Wind conditions. *See* Weather conditions
Windmilling propeller, 37
Wind shear, 217–218, 221
Winglets, 205, *207*
Wing tip vortices, 211, *212,* 213, *215*

Yaw damper, 205